Intermetallic Compound (Volume II)

Intermetallic Compound (Volume II)

Editor

Jacek Ćwik

Basel • Beijing • Wuhan • Barcelona • Belgrade • Novi Sad • Cluj • Manchester

Editor
Jacek Ćwik
Institute of Low Temperature
and Structure Research
Polish Academy of Sciences
Wroclaw, Poland

Editorial Office
MDPI
St. Alban-Anlage 66
4052 Basel, Switzerland

This is a reprint of articles from the Special Issue published online in the open access journal *Crystals* (ISSN 2073-4352) (available at: https://www.mdpi.com/journal/crystals/special_issues/HN35J585H5).

For citation purposes, cite each article independently as indicated on the article page online and as indicated below:

Lastname, A.A.; Lastname, B.B. Article Title. *Journal Name* **Year**, *Volume Number*, Page Range.

ISBN 978-3-0365-9546-7 (Hbk)
ISBN 978-3-0365-9547-4 (PDF)
doi.org/10.3390/books978-3-0365-9547-4

© 2023 by the authors. Articles in this book are Open Access and distributed under the Creative Commons Attribution (CC BY) license. The book as a whole is distributed by MDPI under the terms and conditions of the Creative Commons Attribution-NonCommercial-NoDerivs (CC BY-NC-ND) license.

Contents

Preface . . . . **vii**

Santosh Sampath, Vignesh Pandian Ravi and Srivatsan Sundararajan
An Overview on Synthesis, Processing and Applications of Nickel Aluminides: From Fundamentals to Current Prospects
Reprinted from: *Crystals* **2023**, *13*, 435, doi:10.3390/cryst13030435 . . . . **1**

Junqing Han, Wentao Yuan, Yihan Wen, Zuoshan Wei, Tong Gao, Yuying Wu and Xiangfa Liu
Morphology and Growth Mechanism of β-Rhombohedral Boron and Pentagonal Twins in Cu Alloy
Reprinted from: *Crystals* **2022**, *12*, 1516, doi:10.3390/cryst12111516 . . . . **23**

Sara J. Yahya, Mohammed S. Abu-Jafar, Said Al Azar, Ahmad A. Mousa, Rabah Khenata, Doha Abu-Baker and Mahmoud Farout
The Structural, Electronic, Magnetic and Elastic Properties of Full-Heusler Co_2CrAl and Cr_2MnSb: An Ab Initio Study
Reprinted from: *Crystals* **2022**, *12*, 1580, doi:10.3390/cryst12111580 . . . . **31**

N. V. Kostyuchenko, I. S. Tereshina, A. I. Bykov, S. V. Galanova, R. V. Kozabaranov, A. S. Korshunov, et al.
Field-Induced Transition in $(Nd,Dy)_2Fe_{14}B$ in Ultrahigh Magnetic Fields
Reprinted from: *Crystals* **2022**, *12*, 1615, doi:10.3390/cryst12111615 . . . . **53**

André Götze, Siobhan Christina Stevenson, Thomas Christian Hansen and Holger Kohlmann
Hydrogen-Induced Order–Disorder Effects in $FePd_3$
Reprinted from: *Crystals* **2022**, *12*, 1704, doi:10.3390/cryst12121704 . . . . **61**

Galina Politova, Irina Tereshina, Ioulia Ovchenkova, Abdu-Rahman Aleroev, Yurii Koshkid'ko, Jacek Ćwik and Henryk Drulis
Investigation of Magnetocaloric Properties in the $TbCo_2$-H System
Reprinted from: *Crystals* **2022**, *12*, 1783, doi:10.3390/cryst12121783 . . . . **89**

Christopher Dickens, Adam O. J. Kinsella, Matt Watkins and Matthew Booth
The Presence of Charge Transfer Defect Complexes in Intermediate Band $CuAl_{1-p}Fe_pS_2$
Reprinted from: *Crystals* **2022**, *12*, 1823, doi:10.3390/cryst12121823 . . . . **99**

Milad Takhsha Ghahfarokhi, Federica Celegato, Gabriele Barrera, Francesca Casoli, Paola Tiberto and Franca Albertini
Dewetting Process in Ni-Mn-Ga Shape-Memory Heusler: Effects on Morphology, Stoichiometry and Magnetic Properties
Reprinted from: *Crystals* **2022**, *12*, 1826, doi:10.3390/cryst12121826 . . . . **119**

Jonathan Gjerde and Radi A. Jishi
Hyperbolic Behavior and Antiferromagnetic Order in Rare-Earth Tellurides
Reprinted from: *Crystals* **2022**, *12*, 1839, doi:10.3390/cryst12121839 . . . . **129**

Chenxiao Ye, Jiantao Che and Hai Huang
Boundary Effect and Critical Temperature of Two-Band Superconducting FeSe Films
Reprinted from: *Crystals* **2023**, *13*, 18, doi:10.3390/cryst13010018 . . . . **141**

Yun Wei, Ben Niu, Qijun Liu, Zhengtang Liu and Chenglu Jiang
First-Principles Calculations of Structural and Mechanical Properties of Cu–Ni Alloys
Reprinted from: *Crystals* **2023**, *13*, 43, doi:10.3390/cryst13010043 . . . . **151**

Jingjing Wang, Hongji Meng, Jian Yang and Zhi Xie
GPU-Based Cellular Automata Model for Multi-Orient Dendrite Growth and the Application on Binary Alloy
Reprinted from: *Crystals* **2023**, *13*, 105, doi:10.3390/cryst13010105 . . . . **159**

Lei Cao, Desheng Chen, Xiaomeng Sang, Hongxin Zhao, Yulan Zhen, Lina Wang, et al.
Direct Evidence for Phase Transition Process of VC Precipitation from $(Fe,V)_3C$ in Low-Temperature V-Bearing Molten Iron
Reprinted from: *Crystals* **2023**, *13*, 175, doi:10.3390/cryst13020175 . . . . **173**

Tianyi Han, Jiantao Che, Chenxiao Ye and Hai Huang
Ginzburg–Landau Analysis on the Physical Properties of the Kagome Superconductor CsV_3Sb_5
Reprinted from: *Crystals* **2023**, *13*, 321, doi:10.3390/cryst13020321 . . . . **185**

Hui Wang, Fuyong Su and Zhi Wen
The Effect of Cr Additive on the Mechanical Properties of Ti-Al Intermetallics by First-Principles Calculations
Reprinted from: *Crystals* **2023**, *13*, 488, doi:10.3390/cryst13030488 . . . . **195**

Dheyaa F. Kadhim, Manindra V. Koricherla and Thomas W. Scharf
Room and Elevated Temperature Sliding Friction and Wear Behavior of $Al_{0.3}CoFeCrNi$ and $Al_{0.3}CuFeCrNi_2$ High Entropy Alloys
Reprinted from: *Crystals* **2023**, *13*, 609, doi:10.3390/cryst13040609 . . . . **207**

Halit Sübütay and İlyas Şavklıyıldız
Effect of High-Energy Ball Milling in Ternary Material System of (Mg-Sn-Na)
Reprinted from: *Crystals* **2023**, *13*, 1230, doi:10.3390/cryst13081230 . . . . **219**

Preface

This Special Issue entitled "Intermetallic Compound (Volume II)" contains seventeen articles of which one article is a review article. The articles presented are a continuation of the first part and focus on the broadly understood physical properties of intermetallic compounds.

In the first article, Halit Sübütay et al. present experimental studies on the nature of the ball-milling mechanism in a ternary materials system (Mg-Sn-Na) for proper mechanical alloying. In 12 milling conditions, the homogenous secondary phase distribution is achieved, which eventually supplies the highest relative density (95%), modulus of elasticity (34.5 GPa), and hardness (89 HV) values in this ternary material system. In Dheyaa F. Kadhim et al.'s research, processing–structure–property relations were systematically investigated at room and elevated temperatures for two FCC $Al_{0.3}CoFeCrNi$ and $Al_{0.3}CuFeCrNi_2$ high-entropy alloys. The structure, elastic properties and electronic structure of Ti-Al intermetallics including Ti_3Al (space group P63/mmc), TiAl (space group I4/mmm) and TiAl3 (space group P4/mmm) are systematically studied by first-principles calculations by Fuyong Su et al. Tianyi Han et al.'s theoretical work presents Ginzburg–Landau analysis on the physical properties of the Kagome superconductor CsV_3Sb_5. In Lei Cao et al.'s research, the morphology and structure of V-rich carbides in V-bearing pig iron were studied, and the precipitation characteristics of V-rich carbides in molten iron were discussed. The next paper by Jingjing Wang et al. aims to develop a high-performance CA model incorporating the decentered square capture rule to simulate dendrite growth with different orientations more efficiently. Chenglu Jiang et al. present first-principles calculations of structural and mechanical properties of Cu–Ni alloys. Based on two-band Bogoliubov–de Gennes theory, Hai Huang et al. study the boundary effect of an interface between a two-gap superconductor FeSe and insulator. In Jonathan Gjerde and Radi A. Jishi's worl, the optical and magnetic properties of several rare-earth tellurides have been examined. In Milad Takhsha Ghahfarokhi et al.'s research, dewetting process has been investigated in shape-memory Heuslers. Christopher Dickens et al. use density functional theory to investigate Fe substitution in $CuAlS_2$. The magnetocaloric effect in the TbCo2 -H system in the region of the Curie temperature was studied both by direct and indirect methods in external magnetic fields up to 1.4 and 14 T by Galina Politova et al. Hydrogen-induced order–disorder effects in $FePd_3$ was presented by Holger Kohlmann et al. In N. V. Kostyuchenko's research, the peculiarities of the magnetization process in the ferrimagnetic intermetallic compound $(Nd_{0.5}Dy_{0.5})_2Fe_{14}B$ have been studied theoretically and experimentally using ultrahigh magnetic fields. Next theoretical paper by Sara J. Yahya et al. presents structural, elastic, magnetic and electronic properties of inverse and conventional Heusler (Co_2CrAl, Cr_2MnSb) compounds. In Junqing Han et al.'s work, boron particles with β-rhombohedral structure were prepared in Cu-4B alloy and the morphology and growth mechanism of β-B and pentagonal twins were analyzed. In the last article Santosh Sampath et al. present an overview on synthesis, processing and applications of Nickel aluminides.

I hope that the presented set both theoretical and experimental articles will arouse genuine interest among readers and, perhaps, push them to their own successful research in the field of intermetallic compounds.

Thanks to all contributing authors of this Special Issue and the Editorial staff of Crystals.

Jacek Ćwik
Editor

Review

An Overview on Synthesis, Processing and Applications of Nickel Aluminides: From Fundamentals to Current Prospects

Santosh Sampath *, Vignesh Pandian Ravi and Srivatsan Sundararajan

Department of Mechanical Engineering, Sri Sivasubramaniya Nadar (SSN) College of Engineering, Old Mahabalipuram Road, Kalavakkam 603110, Tamilnadu, India
* Correspondence: santoshs@ssn.edu.in

Abstract: Nickel aluminides have desirable properties for use in high-temperature applications. Nickel aluminides have certain desirable qualities, but for almost a decade in the 1990s, those benefits were overshadowed by the challenges of processing and machining at room temperature. Manufacturing improvements, increased knowledge of aluminide microstructure and deformation processes, and developments in micro-alloying have all contributed to the development of nickel aluminides. Key developments in nickel aluminides, such as their microstructure, alloy addition and alloy development, are given and discussed at length. Methods of production from the past, such as ingot metallurgy and investment casting and melting are addressed, and developments in powder metallurgy-based production methods are introduced. Finally, the difficulties of producing nickel aluminides and possible solutions are examined. This paper gives an overview of the fundamentals, preparation, processing, applications and current trends in nickel aluminides.

Keywords: nickel aluminides; intermetallics; processing; applications

1. Introduction

Intermetallic compounds are a class of metallic materials, which are now the subject of extensive research by scientists and engineers working in the field of materials science. Intermetallic compounds are used in higher temperature engineering applications, as they have properties intermediate of metals and ceramics. These materials are now necessary in a wide variety of applications and have the potential to provide additional advances in performance in a variety of domains such as magnetic materials, hydrogen storage materials, and high-temperature structural materials (>1200 °C), etc. [1–3]. Most intermetallic compounds have high melting temperatures and are brittle at normal temperature. Because of the low number of separate slip systems necessary for plastic deformation, intermetallics typically fracture in a cleavage or intergranular manner. However, some intermetallics, such as Nb-15Al-40Ti, exhibit ductile fracture modes. Alloying with additional elements can increase grain boundary cohesion, resulting in increased ductility in other intermetallics [4,5]. Figure 1 shows few compounds of current interest by comparing melting temperature and density. Examples of some intermetallics based on their properties are shown in Figure 2.

Sometimes, intermetallics are categorized based on crystal formation, and have highly complicated atomic arrangements with common structures adhering to the simple stoichiometric formulae AB, AB$_2$ and C [5,6]. The immense promise of intermetallics, particularly aluminides, arises from their numerous desirable features, including excellent oxidation resistance, corrosion resistance, comparatively low densities, stiffness at increased temperatures and the ability to preserve strength [7–9]. Despite its usefulness, poor ductility—especially at low and intermediate temperatures—is a major drawback of intermetallics. Different compounds have different reasons for lacking ductility, which is presented in Figure 3 [8–11].

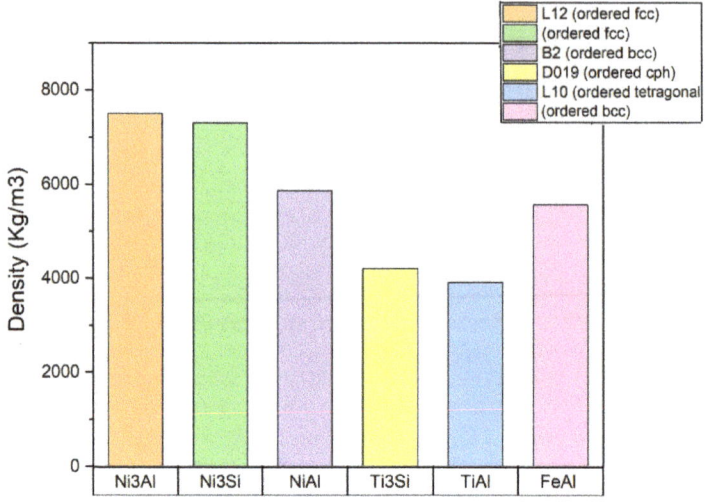

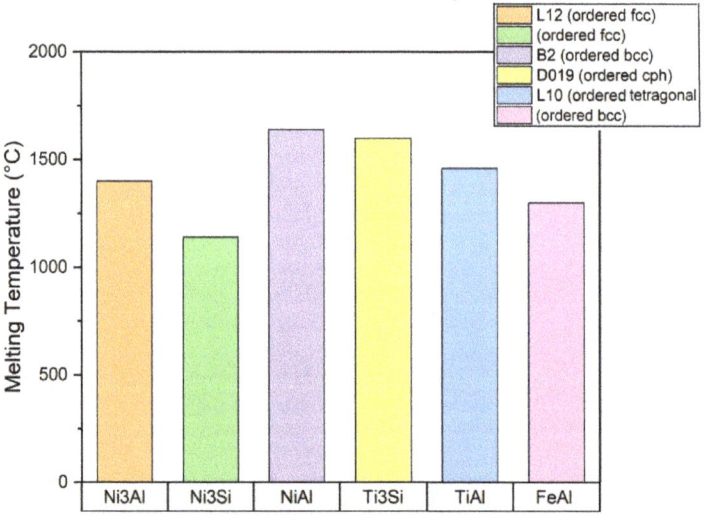

Figure 1. Few intermetallic compounds of current interest by comparing melting temperature and density.

Small amounts of alloying additives, however, have shown to improve ductility several intermetallics: Boron in Ni_3Al, Manganese in TiAl, and Niobium in TiAl [8]. Titanium aluminides and nickel aluminides systems have been the focus of the majority of research in the field of intermetallics [7]. Ni_3Al and NiAl are the two important aluminides that are found in the nickel–aluminum system. As a possible structural alloy, Ni_3Al has garnered a significant amount of attention recently. The majority of superalloys contain Ni_3Al, which functions as a strengthening phase [7,9].

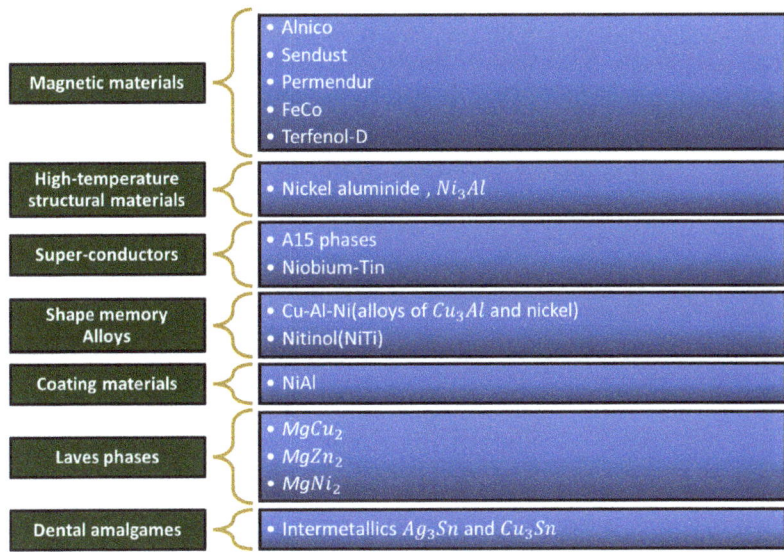

Figure 2. Classification of intermetallics based on properties.

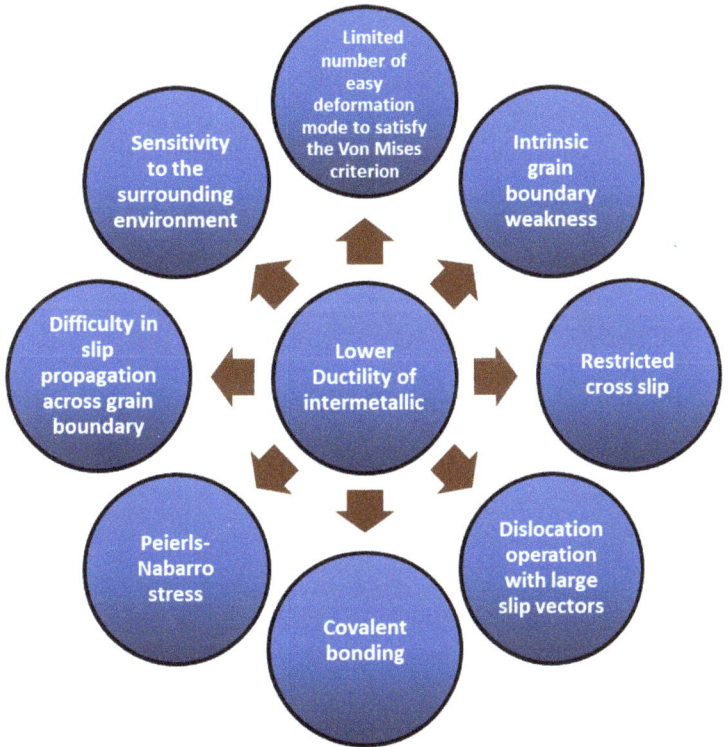

Figure 3. Factors that cause low ductility of intermetallics.

When exposed to oxygen-rich environments, the aluminide of a transition metal will produce a continuous, totally adhering alumina coating on its surface. Aluminides typically include aluminum concentrations between 10 and 30 wt%, which is much greater than the aluminum content of standard superalloy and alloy. Alumina layer that forms upon nickel surface and iron aluminides is what allows these materials to retain their superior oxidizing and carburizing resistivity at temperatures of 1000 °C or higher [10]. Therefore, aluminides do not always need chromium for producing layer of oxide upon material surface to counter higher temperature oxidizing and rust, in contrast to typical steels and superalloys based on Fe, Co and Ni [11]. The characteristics of alumnides are shown in Figure 4.

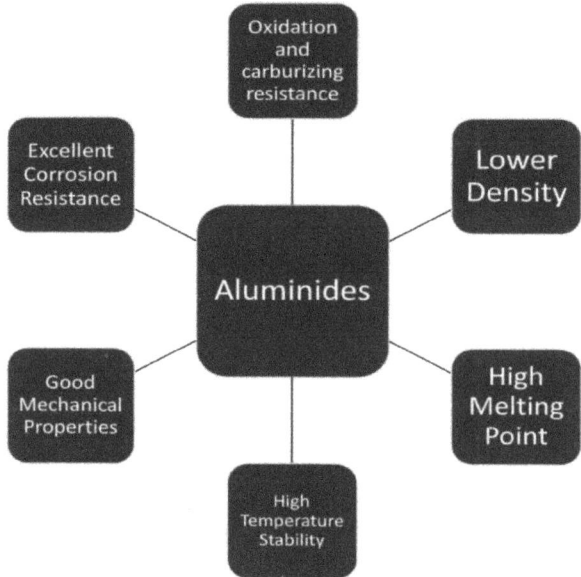

Figure 4. Characteristics features of Aluminides [12–15].

There are often many intermetallic equilibrium aluminide phases present in metal-aluminum binary systems. In thin-film bilayers, it is typically found that only a single phase is growing at any one moment [16,17]. This is in contrast to bulk diffusion couples, in which, after adequate indurating, all of the equilibrium stages normally occur. Located somewhere in the middle of these two patterns of behavior are lateral diffusion couples.

2. Importance of Nickel Aluminides

Nickel aluminides have a low density and great resistance to oxidation. They also keep their strength well at higher temperatures. Because of these characteristics, they are a good choice for high-temperature structural applications. One of Ni_3Al's most notable characteristics is the fact that its yield stress rises as its temperature rises to a maximum temperature of 600 °C, as shown in Figure 5 [8,18]. Table 1 indicates the weight percentage and melting point of aluminum based intermetallics. This behavior has been noticed in other L_{12} intermetallics as well. This effect is caused by the cross slip of screw dislocations, which are thermally triggered, moving from the planes labelled (1 1 1) to the planes labeled (1 0 0), which is where the antiphase boundary (apb) energy lies. Observations of apb energies using electron microscopy that are given in Table 2 illustrate that the apb energy on {1 0 0} declines with increasing amounts of aluminum content. This affects the composition dependency of the strength, which is shown in Figure 5. A significant work hardening rate is also caused by the cross-slipping of screw displacements by {1$\bar{1}\bar{1}$} planes with cube planes.

Table 1. Intermetallics weight percentages of aluminum, the temperatures at which they form, and their melting points [19].

Intermetallics	Weight Percent (wt%) of Aluminum	Heat of Formation <298 (kcal/mol)	Melting Point (°C)
Ni_3Al	13.28	-66.6 ± 1.2	1395
NiAl	31.49	28.3 ± 1.2	1639
Ni_2Al_3	40.81	67.5 ± 4.0	1133
$degNiAl_3$	57.96	36.0 ± 2.0	854

Table 2. Anti-Phase Boundary Energies in Ni_3Al [8].

Alloy	$\gamma 111$ (mJ/m^2)	$\gamma 111/\gamma 100$	$\gamma 100$ (mJ/m^2)
Ni-23.5Al + 0.25B	170 ± 13	1.37	124 ± 8
Ni-26.5Al	175 ± 12	1.51	113 ± 10
Ni-25.5Al	175 ± 13	1.31	134 ± 8
Ni-24.5Al	179 ± 15	1.25	143 ± 7
Ni-23.5Al	183 ± 12	1.17	157 ± 8

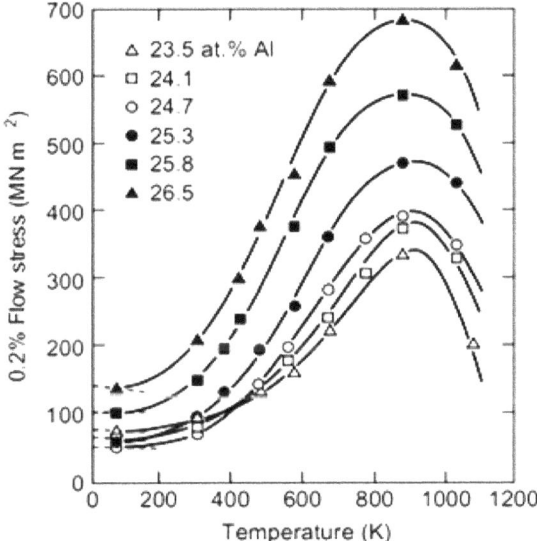

Figure 5. The influence of aluminum content on the temperature dependence of flow stress in Ni_3Al [8]. Reproduced with permission from Elsevier.

3. Challenges Involved in Nickel Aluminides

Nickel aluminide is a long range ordered intermetallic. Consequently, due to longer-ranged orders, there is a significant problem of lower ductility and inelastic intergranular fissure at room temperatures [19,20]. A limited number of simple slip systems, restricted cross-slip, large slip vectors and adversity of transferring slip through grain boundary are some of factors that may be the reason for the brittle failure of intermetallic alloys [8]. In spite of this, several metallurgical processes such as processing control, grain refining, micro and macro alloying, and quick solidification [6,8,20], have culminated noticeable improvements into ductility and toughness of material. For instance, it has been found that adding a minuscule amount of boron to Ni$_3$Al increases the grain boundary adhesion level, which in turn reduces the tendency for the polycrystalline material to crack along its brittle intergranular boundaries.

There are at least three different ways that may be improved upon for increasing ductility of NiAl as shown in Figure 6 [21–27]. The structural properties of NiAl and Ni$_3$Al are shown in Table 3.

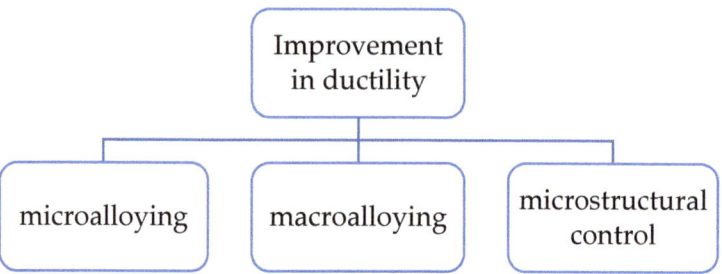

Figure 6. Different methods for improving ductility of NiAl.

Table 3. Structural properties of NiAl and Ni$_3$Al [28–38].

Particulars	NiAl	Ni$_3$Al
Lattice stelecructure	Ordered body-centered cubic	Ordered face-centered cubic
Phase formation	composition range of ~45–60 at% Ni below ~400 °C	23–28 at% Al
Strukturbericht-superstructure	B2, (ordered crystal structure of simple CsCl prototype)	L1$_2$, (systematized crystal structure of simple AuCu$_3$ prototype)
Space group	pm-3m (221)	cubic pm-3m (221)
Lattice parameter	2.887 A	0.356 nm (No ternary addition)—Bradley and Taylor 0.357 nm—by Mishima et al. and y Guard and Westbrook
sublattices (alpha and beta)	Ni in corners (0,0,0) Al atoms into center body positioning (1/2,1/2,1/2)	Al atoms into (0,0,0) lattice locations are coexisting with nickel atoms into (0,1/2,1/2, 1/2,0,1/2, and 1/2,1/2,0) lattice positions. Linear dependency of lattice constraints upon LRO constraints
Ordering behavior	nonlinear second-order transition behavior [34]	Order–order relaxation had been observed for one of the very first times in the Ni$_3$Al phase of an intermetallic compound [33,34]
Density	5.85 g/cm^3	7.50 g/cm^3
Youngs Modulus (GPa)	294	179

4. Phase Diagram of Nickel Aluminides

The phase diagram of NiAl is shown in Figure 7. Ni has a poor solubility in Al, making it very hard to obtain compounds with a higher availability of Al; nevertheless, Al becomes significantly soluble in Ni, being accountable for the formation of Ni-rich complexes upon its adding. When it comes to the primary phase regions, Al-Ni phase diagram is a very precise structure. It is possible to come across Ni_3Al with a percentage of Al between 73% and 76%. As per the binary phase diagram of Al-Ni, the compounds of Al_3Ni, Al_3Ni_2, AlNi, $AlNi_5$, and $AlNi_3$ are produced progressively with increasing Ni content. There are two eutectic processes and three peritectic zones in the Al-Ni phase diagram, with Al_3Ni, AlNi, and Ni_3Al as intermetallic compounds and $NiAl_3$ as an intermetallic compound with constant composition.

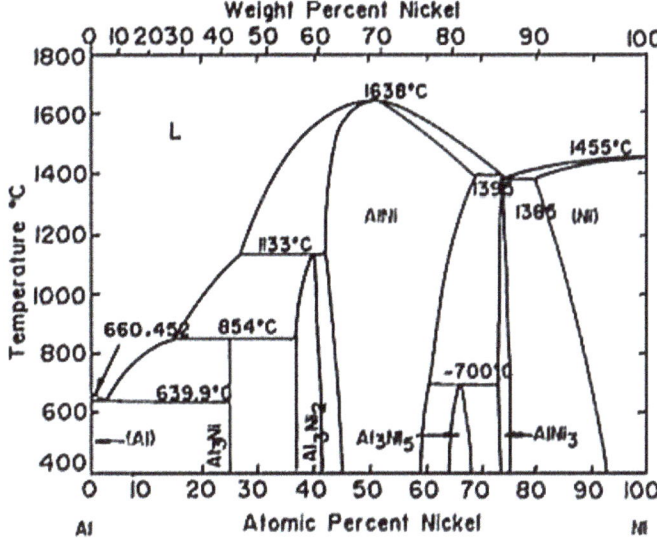

Figure 7. The Phase Relationships in the Al-Ni System: A Plot of the Al-Ni Binary Phase Diagram [7]. Reproduced with permission from Elsevier.

There has been a long period of development for the Al-Ni phase diagram, during which it has been tweaked and improved upon on numerous occasions by a number of experts. Evolution and development of the Ni-Al phase diagram is shown in Figure 8 [5,39].

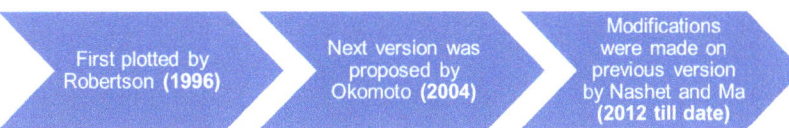

Figure 8. Evolution and development of Ni–Al phase diagram [5].

5. Properties of Nickel Aluminides

Comparisons of the mechanical and physicochemical characteristics of Ni_3Al intermetallic alloys comparing with traditional metallic materials have been subject of a significant amount of research in the scientific literature. The Ni_3Al alloys are, for the most part, exceptional when compared with commercialized alloys, particularly in the category of higher-temperature application, in conditions that are both oxidizing and carburizing.

5.1. Hardness

Unalloyed nickel aluminides show composition-dependent hardness, with stoichiometric NiAl having lower hardness than off-stoichiometric compositions [39,40]. The existence of triple defects was responsible for the shift in hardness at the stoichiometric ratio of 2.4 to 3.2 GPa. Triple defects, consisting of two vacancies with one sublattice and antisite upon other, are specific to intermetallic complexes. To no one's surprise, high hardness attributes observed for mildly Al- or Ni-rich stoichiometric complexes of $Ni_{52}Al$ and $Ni_{48}Al$ could be explained by the existence of thermally flustered vacancies upon the Al-rich side of stoichiometric NiAl and antisite kinds of deformities for the Ni-rich side of stoichiometric NiAl alloy [41,42].

Hardness deliberations of both stoichiometric and non-stoichiometric composites were also reported by Guard and Westbrook.

- Hardness was found to be lower for stoichiometric compositions than for Al-rich compounds with non-stoichiometric compositions, which had greater hardness values.
- Guard and Westbrook also looked at how hardness changed with temperature for materials of the same composition, finding that measures of hardness were lowest at low temperatures, and highest with a Ni:Al ratio of 3 [43,44].

5.2. Magnetic Properties

Ni_3Al is either highly paramagnetic or weakly itinerant ferromagnetic, with Tc (curie temperature) varying as a function of Al content [45,46]. Due to the presence of a larger number of nonmagnetic Al atoms, NiAl, like Ni_3Al, is a weekly ferromagnetic material whose magnetic moment diminishes by an upsurge in Al concentration. Whether alloying atoms are located into the Ni site or the Al site has no effect on whether adding Mn and Fe improves the total magnetic moment [47,48].

5.3. Electrical Properties

NiAl's electrical conductivity at normal temperature is composition dependent. 13×10^6 S/m at stoichiometric composition, but 6×10^6 S/m for Ni and Al-rich near stoichiometric configurations [49,50].

Despite having the same conduction, as-cast specimens were 50% less conductive than homogenized NiAl-Ag alloys. Due to higher Ag solubility into NiAl lattice, alloying with Ag reduces electric conduction at ambient temperatures or above 5 at%. Precipitation and coarsening increase conductivity in homogenized alloys [49–54].

5.4. Grain-Boundary Embrittlement

It is noteworthy that single crystals of Ni_3Al have a ductile structure, but pure polycrystalline Ni_3Al has a brittle structure at an ambient temperature due to intergranular fracture. In traditional materials, brittle intergranular fracture is typically followed by isolation of impurity elements like sulphur, phosphorus, and oxygen, which results in embrittlement at the grain boundaries. However, in sufficiently pure polycrystalline Ni_3Al, no evidence of such segregation has been detected. This leads one to believe that grain boundary is intrinsically friable. It is noticed that grain border fragility is linked to both a lack of grain-boundary cohesiveness and environmental fragility. Grain-boundary cohesion absence is connected to differences in the energy ordering, electronegativity, vacancies and size of atoms that exist amongst atomic components that make up the intermetallics. The formation of atomic hydrogen as a result of the interaction of Ni_3Al with moisture is what causes grain-boundary embrittlement [55–57].

5.5. Creep Behaviour

According to the findings of a few investigations, both single- and polycrystalline Ni_3Al exhibits the characteristic "inverse creep" behavior into average temperatures. In this case, creep curves have a transitory primary phase lasting until the 1% strain, and is distinguished with a drop-in strain rate having a rising strain. This stage is trailed with an "inverse" tertiary phase that exhibits increased creep that ultimately leads to failure. These creep curves do not display the steady phase creep stage anywhere in their progression. In the case of samples consisting of a singular crystal, creeping failure is not by the formation of voids but by necking.

In addition, creep studies conducted on single crystals of Ni$_3$Al with 1% Ta content revealed the existence of a steady-state creep stage for all orientations tested. It has been discovered, which is quite fascinating, that the steady-state creep rate of single crystal specimens orientated in a variety of directions scales, having resolved the shear stress of cube cross-slip planes. TEM experiments did show evidence of slip upon octahedral planes while in the primary phase of creep, and upon cube cross-slip planes while in the secondary creep phase. This finding is consistent with what was anticipated [58].

In Mo, Fe, and Co additions increase the proportion of metallic bonds into intermetallic framework of NiAl, hence shifting the electron concentration at the Fermi level. Peierls energy U_p and interrelated Peierls hindrance of plastic deforming R_p drop as covalent component of interatomic bonds decreases:

$$R_p = \frac{2pU_p}{ba}$$

here a: lattice constraint; b: Burgers vector.

Reduction of R_p enhances alloy plasticity and diminishes its strength. Such impact is supported with strong link amongst the microhardness and electronic structural properties of NiAl-based alloys as depicted in Figure 9 [4].

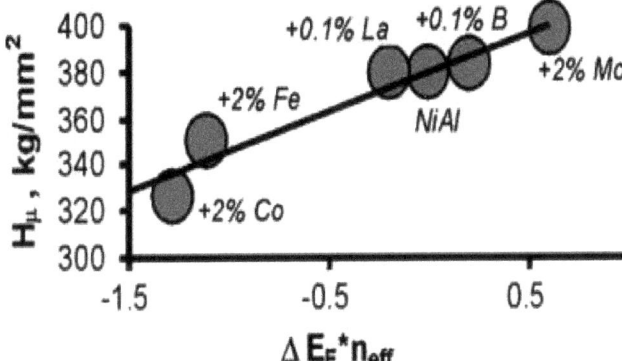

Figure 9. Correlation between micro-hardness and electronic structure characteristics of alloys on base of NiAl [4]. Reproduced with permission from Elsevier.

Examining the mechanical characteristics of nickel aluminide alloy in strain, compression, and impact toughness yielded the findings depicted in Figure 10. A cold fragility threshold for nickel aluminide compound is, as expected, in the range of a 0.43–0.45 of melting point—Tm. Every sample failed brittle as in tensile tests with temperatures under 500 °C, and high elongation values, following failures, were observed around 450–650 °C. The effect of alloying on NiAl's brittle/ductile transition temperature is most pronounced in tensile trials [4].

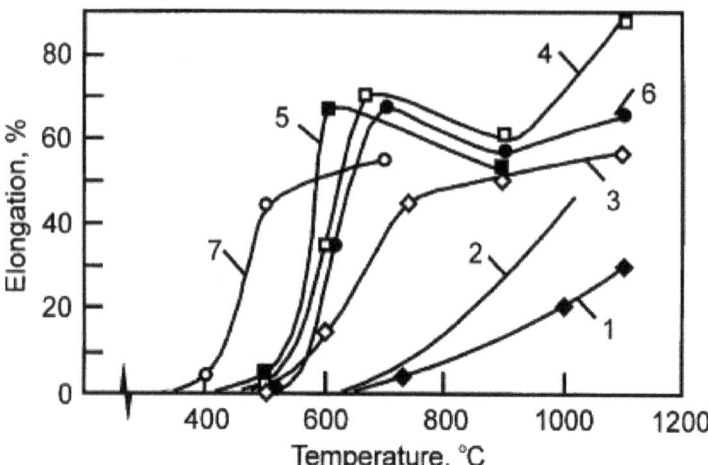

Figure 10. Elongation of NiAl alloys tested with 400–1200 °C. (1) casting of NiAl; (2) extruded NiAl; (3) NiAl(Mo); (4) NiAl(W); (5) NiAl(Fe); (6) NiAl(Cr); (7) NiAl (Co, B, La) [4]. Reproduced with permission from Elsevier.

6. Impact of Alloying upon Strength and Ductility

When Ni_3Al is alloyed with ternary, quaternary, and quinary elements, the oxidation resistance, scale adhesion, capacity to create an Al_2O_3 scale, and oxidation processes are dramatically altered. The effect of alloying elements is presented in detail in Table 4.

Table 4. Importance of Alloying elements on Nickel Aluminides [50–70].

Elements	Description	Reference
Boron	<ul><li>Improved ductility.</li><li>Due to the substantial creation of geometric voids upon the substrate surface, adding B to Ni_3Al did not increase oxidizing resistivity or the oxide scale adhesion, but it did improve the aqueous corrosion resistance [70–72].</li><li>Improve strength with addition of elements like Hf, Ti etc.</li></ul>	[51–58]
Chromium	<ul><li>The general oxidation behavior of Ni3Al is marginally improved by Cr's presence, however this improvement is negated at temperatures above 1300 °C due to the creation of blisters caused by the transition of Cr_2O_3 into volatile CrO_3.</li><li>Nevertheless, oxidation rates may decrease at low temperatures when 8 at.% Cr is present. This is due to the fact that Cr enhances the capacity to produce a healing layer of Al_2O_3, which prevents further damage.</li><li>According to one traditional theory, Cr might play the role of the secondary getter of oxygen, therefore lowering flow of oxygen in alloy in event that the primary getter (Al) is destroyed because of corrosion.</li></ul>	[59–64]
Titanium	<ul><li>When combined with B, Ti enhances the scale's adhesion, but this often leads to worse oxidation behavior due to large weight increases from the formation of Ti-containing oxides and the disruption of the Al_2O_3 scale.</li><li>although the addition of titanium alone to Ni_3Al at a weight percentage of 2.99% has a tendency to lower the cyclic oxidation resistance</li></ul>	[65]

Table 4. *Cont.*

Elements	Description	Reference
Lithium	• The higher temperature oxidizing resistivity of Ni_3Al alloys may be significantly improved by the addition of lithium. • According to the rule of Hauffe, replacing Ni with Li can bring about a reduction in the concentration of cation vacancy in p-type NiO. This, in turn, can bring about a slowdown in the rate of oxidation. In addition, adding Li altered the morphology of oxide scales, reduced the size of oxide grains, made the oxide more homogenous, densified oxide scales, and increased the mechanical characteristics of oxide scales.	[66]
Molybdenum	• The overall oxidation behavior is reduced due to the limited solubility of the molybdenum, and as a result, oxidation of the molybdenum-rich phases leads to the formation of volatile species. This is true even though the addition of 3 weight percent of molybdenum does reduce the overall oxide weight gain.	[66]
Reactive elements	• Improves strength at low and high temperatures [6]. • Adherence of oxide scales can be enhanced with the existence of reactive elements including Hf, Y, and Zr. Incorporating Y as an oxide dispersion keeps the favorable advantages of adding a reactive element. Individually and in combination with B enhancements, Hf and Zr seem to give the greatest overall behavior. Ni_3Al's isothermal oxidation behavior was investigated by Kuenzly and Douglass from 900 to 1200 °C in air with and without Y addition (0.5 wt.%). They found scaling behavior in alloys following strict parabolic rule, because adding Y had no effect on the steady-state scale ratio of Ni_3Al.	[71–74]

Table 5 displays the effect of alloying on properties like ductility and strength for nickel aluminide. Compression testing at room temperature represents metal's soft stress condition. As a result, all of the samples into compression testing demonstrated adequately higher/lower temperature ductility [75–78].

Table 5. Effect of alloying on the strength and ductility characteristics of nickel aluminide (compression testing at room temperature) [4,11,19].

Alloy	$\sigma_{0.2}$ (MPa)	ε (%)	ψ (%)
NiAl	292	12.0	0
NiAl (B)	400	25.6	0
NiAl (La)	311	29.5	70.0
NiAl (Fe)	396	28.0	65.0
NiAl (Co)	384	30.8	69.0
NiAl (Cr)	421	24.8	60.8
NiAl (Mo)	340	26.0	17.0

When tested in air at room temperature, it was found that adding B to polycrystalline Ni_3Al with 25% Al increased its tensile ductility by a lot, so much so that the way it broke changed from intergranular to transgranular [79–82]. Ni_3Al microalloyed with 0.1 wt% B broke with a tensile strain of more than 50% in air [79]. Table 6 shows some of the results of tensile tests that were done on Ni_3Al, with and without B in different environments. [79,81–83].

Two ideas have been put forward to explain how adding B makes a material more ductile: (i) rise into cohesive strength at the grain boundary due to the addition of B [84–87] and (ii) slip transferring transversely with grain boundary [88–90].

Table 6. Tensile properties of Ni$_3$Al and Ni$_3$Al-B Alloys under different environments [8,10,79–81].

Alloy Composition	Strain Rate (s^{-1}) and Environment	Yield Strength (MPa)	Ultimate Tensile Strength (MPa)	Elongation to Failure (%)
Tests at room temperature				
Ni-24 Al	3.3×10^{-3}, air	280	333	2.6
Ni-24 Al	3.3×10^{-3}, oxygen	279	439	7.2
Ni-24 Al-500 ppm B	3.3×10^{-3}, air	290	1261	41.2
Ni-24 Al-500 ppm B	3.3×10^{-3}, oxygen	289	1316	39.4
Ni-24.8 Al-500 ppm B	3.3×10^{-3}, air	290	671	18.1
Ni-24.8 Al-500 ppm B	3.3×10^{-3}, oxygen	306	801	25.4
Ni-25.2 Al-500 ppm B	3.3×10^{-3}, air	221	300	8.4
Ni-25.2 Al-1000 ppm B	3.3×10^{-3}, air	344	552	10.2
Tests at -196 °C (77 K)				
Ni-23.4 Al		254–269	672–762	31.3–31.8

7. Processing of Nickel Aluminides

When deciding on a method of processing, it is important to consider the characteristics of the final product. A coating, for instance, would use a thin-film processing technology, while near-net-shaped bulk materials could benefit more from ingot metallurgy processing, which includes melting and casting, or appropriate powder metallurgy processing. For intermetallics, in order to prevent the oxidation of constituent elements and contamination of the products during processing, which could result in inadequate densification and the production of unwanted or deleterious phases as impurities, vacuum or inert gas conditions are typically recommended. Composites and alloys based on Ni aluminides can be treated with the ingot and powder metallurgical methods. Many commercially viable uses of these alloys have been developed because of their undeniable benefits over "conventional" materials [11,91]. Any material's microstructure and characteristics rely on how it is processed. As certain intermetallic alloys have complicated crystal structures, their characteristics are affected by stoichiometry, impurities, and defects. The intermetallic compounds have an ordered structure and distinctive chemistry. These alloys must have a certain atomic ratio of elements for a specified crystal structure and mechanical properties. In alloys with a variety of stoichiometry, microstructures and characteristics rely on atomic ratios.

7.1. Melting and Casting

Ni$_3$Al and NiAl have differing melting points, hence special consideration must be given to melting and casting each material. For instance, NiAl has a greater melting point than either Al or Ni individually. Because of the reaction of alloying elements with H and the absence of grain-boundary cohesion, fabrication of Nickel Aluminides by casting was not easily obtained. It was also attempted to employ fluxes for casting. However, this could lead to the creation of brittle compounds that weaken the grain boundary [92] due to reactivity amongst flux components and alloys. Maxwell and Grala [93] were effective in melting and casting in the 31.5–33 wt% Al range, but not the 31.5–34 wt% Al range. For Al contents above 35 wt%, castings failed and alloys shattered completely. Ordnance Research and Development Laboratory (ORNL) developed Exo-Melt casting for such alloys, which makes the advantage of heat reactivity in the efficacious cast of Ni$_3$Al alloy [93,94].

In 1996, ORNL came up with the "Exo-melt" process to lessen these effects [11]. In the process known as ExoMeltTM, the melt stock is divided into numerous sections and then loaded into the furnace in such a way that an extremely exothermal reactivity having higher adiabatic combusting temperature is preferred at the beginning. This results in the production of a molten product (Figure 11). For Ni$_3$Al, forming NiAl is an extremely exothermal process and the melting point of NiAl corresponds to the temperature at which it may be burned in an adiabatic reaction [12–15,94].

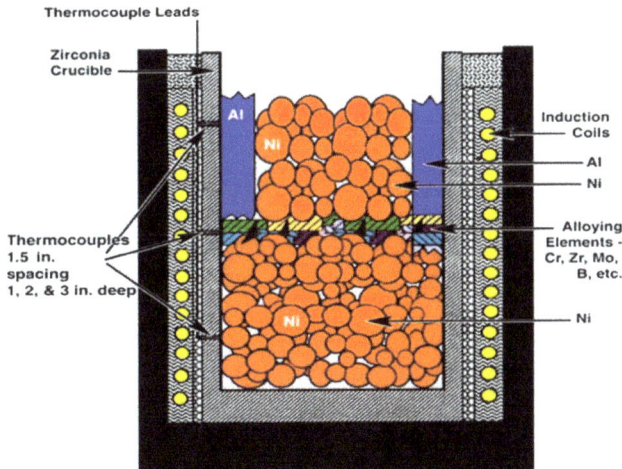

Figure 11. Exo-melt process [77]. Reused from MDPI under Creative Commons Attribution license.

Furthermore, the Exo-Melt procedure is even helpful into cutting production expenses. It saves about 50% both for energy and timing.

Casting processes like sand, investment, centrifugal casting, respectively, and directional solidifying can all be utilized in the production of aluminides. Other casting methods include directional solidification. Cast aluminides are then subjected to a subsequent processing step.

Utilizing a variety of metal formation procedures like hot extrusion, swaging, forging, flat and bar rolling, cold flat and bar rolling, and cold drawing in tube, rod and wire, all of which contribute to the microstructural refining and augmentation of mechanical characteristics of the metal. For instance, temperature ranging from 1050–1150 °C is used for the hot forging process when alloys of Ni3Al comprising less than 0.3 at.% Zr. It is possible to duce the reactive cast of NiAl-based intermetallic alloys with a mix of Ni and Al or NiCo and Al in liquid in air form, and then allowing the mixture to solidify to form a compound that is either NiAl or NiAl-Co, depending on which compound is desired. This approach was also used to cast the Fe-containing NiAl, and neither the failure of the casting due to cracking nor cracks presence had been recorded.

The hot fabricating of Ni_3Al intermetallics is negatively impacted with excess inclusion of Hf and Zr at levels greater than 103, which results in the development of surface fissures and early failure. Both ductileness and strength of nickel aluminides were demonstrated to be improved with adding alloy elements B, Cr, Co, C, and Ce, as well as by the strengthening element TiB2. The majority of nickel aluminide alloys used in the production of products comes from the commercial sector [91,95]. These alloys are used to make bars, wires, sheets, and strips.

A further noteworthy accomplishment was the invention of the casting process utilizing the software known as ProCast (Figure 12a). Because of its lower fluidic nature and shrinking of the material after it has been cast, casting alloys based on Ni3Al can be quite challenging. This fact should be brought to your attention. On the other hand, it was stated that a particular version of the ProCast software makes it possible to cast components that are free of flaws while having a complex shape (Figure 12b) [19,91,92].

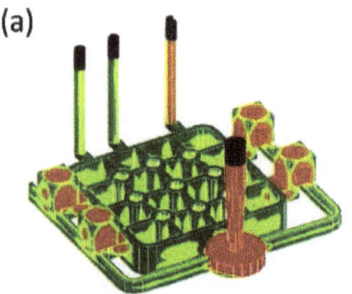 

Figure 12. (**a**) Modeling in ProCast software and (**b**) actual casting [77]. Reused from MDPI under Creative Commons Attribution license.

7.2. Powder Metallurgy

Processing powder metallurgy can be done through spark plasma sintering, pressureless sinter, uni/multi-axial hot press, liquid phase-assist or reactive sintering by application or non-application of pressure, or uniaxial or multiaxial hot pressing. Atomization carried out in an environment devoid of oxygen results in the production of aluminide powders such as Ni_3Al.

After that, the powders are packed into cans and hot extruded at temperatures ranging from 1100 °C to 1200 °C using a reduction ratio of between 8 and 1. The products that are consequently created from these powder metallurgical procedures often has a tiny grain size as a result of dynamic recrystallization, and as a result, they are able to be molded using superplastic techniques in order to obtain near-net shapes.

7.3. Solid State Sintering

This is a common powder metallurgy process for producing Nickel Aluminide. Longer sintering makes compacts denser and grains increase. If hardness increases during sintering, the Kirkendall effect may make it tougher to obtain full density and good mechanical characteristics as it increases intermetallic phase volume percentage [5,21,24]. Powder metallurgy (P/M) was used to make B-alloyed Ni_3Al, and the effect of alloying was studied by adding Fe, Cr, Zr, and Mo while keeping the Al content at 23 at% [24]. The main problems with P/M-processed Ni_3Al alloys are their sensitivity to strain rates below 10^4 s^{-1} and their microstructures (FCC solid solutions) [5,21–23,95].

7.4. Mechanical Alloying

Mechanical alloying also was employed to effectively create nickel aluminides, but it is time-consuming and costly, and unalloyed aluminides are vulnerable to impurity contamination and oxide development. An Ni-containing Al-supersaturated solid solution containing unreacted Ni and Al is the first kind of intermetallic to emerge during mechanical alloying of Ni–Al mixtures. Milling parameters, such as milling duration and power, largely determine the final product's chemical makeup. To reach intermetallic NiAl, this phase must first be milled into Al_3Ni, where it may coexist with Ni_3Al and $AlNi_2$ [25–30].

7.5. Reaction Synthesis

In this method, the heat from an exothermal reaction amongst Ni and Al is used to make intermetallic. High temperatures, between 500 °C and 750 °C, are applied to contents of a container, while the container is kept under a vacuum [31,95]. During the reaction synthesis process, the reaction is often not complete. The unreacted parts may also make the final product stronger, since intermetallic powders are usually fragile and need more pressure to pack them together. Metallic powder size is an important factor in this process [96].

8. Applications of Nickel Aluminides

Many commercially viable uses of these alloys have been developed because of their undeniable benefits over "conventional" materials. Applications for Ni_3Al-based alloys are diverse. This is shown in Table 7. Commercialization of Ni_3Al alloys for specified applications is expected to happen very soon, since the degree of research into this material has been significantly higher than that of other aluminides.

Table 7. Overall Applications of nickel aluminides [80–97].

Application	Description	Example
Automotive: • dyes for hot press Fe-B-Nd magnetic powders • Turbocharger rotors Diesel trucks • Car bodies	1. able to withstand high temperatures without weakening, oxidation resistance, chemical compatibility low cost and improved fatigue life 2. Compared to common automotive materials, this one is more corrosion-resistant and can withstand high temperatures without deforming. It is also lighter and five times stronger than stainless steel. Since this is the case, Ni_3Al alloys may be utilized for a wide variety of purposes, including those requiring high strength or the absorption of energy, such as in the construction of car bodies.	IC-221M
Hydroturbine rotors	• Good cavitation • erosion resistance • Because of their great corrosion resistance, these materials may also be employed as working elements in a seawater environment.	Alloy IC-50
Glass processing	• Oxidizing resistivity, • higher temperature strength • higher weariness life	IC-221M
Chemical processing	• strength • corrosion resistance	IC-218LZr
Metal processing • As a Die material for isothermal forging. • The rollers for steel slab heating furnaces	• higher temperature oxidizing resistivity (1100 °C or below) • high strength	IC-221 IC-221 M
Binder for ceramics	suitable for use in tungsten carbide systems in place of cobalt. Nickel aluminide-bonded tungsten carbide provides better higher- and lower-temperature strength and cutting capabilities	IC-50 IC218LZr
Roller bearings	• From room temperature to 650 °C, nickel aluminide's wear resistance improves by over a factor of 1000.	IC218LZr
Steel Industry • Transfer rolls in furnace for hydrogenation, carburization and used as a roll continuous casting process • Sometimes used as a replacement for currently used stainless steel	• considerable energy cost reductions by eliminating the need for water cooling • prolonging the operating life four to six times over presently utilized materials	IC-221M

Table 7. *Cont.*

Application	Description	Example
Compressor and Turbine Blades in Aircraft Engines	• high temperature structural materials • Ni$_3$Al base alloy, commercialized under the name IC6 to utilize into higher-performance jet engine turbine vanes and blades working at temperatures between 1050 °C and 1100 °C, was created. Second stage gas turbine vanes are being made from this material with a NiCrAlYSi coating.	IC6 IC10 VKNA's

8.1. Nickel Aluminide Coating

Coatings made of nickel-aluminide have received a lot of interest recently because of the fact that they offer a variety of potential applications in technology and science. NiAl has a long history of use as a protective coating for machinery and buildings. Its main purpose is to improve coating adherence, and its secondary purpose is to reduce thermo-mechanical stress at the substrate-coating interface. NiAl's lengthy history of usage may be attributed to the material's low density, high melting point, outstanding thermal conductivity, and great oxidizing resilience. [79,80,91].

NiAl coatings' high-temperature oxidation behavior in moving air is seen around 750 °C and 850 °C, according to previous studies [81,82]. The aerospace industry, along with other high-performance applications, has increased demand for nickel-aluminum alloys and its derivatives. This is because, for certain alloys, an increase in temperature also results in an increase in yield strength. The second major nickel aluminide, Ni$_3$Al, has also been receiving considerable notice of late. Ni$_3$Al is an essential component of NiAl that serves as a stiffening agent, and the two elements together are extensively used for higher temperature structural material for aircraft engines and aerospace applications. [79,80,93]. Table 8 shows the applications and properties of nickel aluminide coatings.

Table 8. Application and properties of nickel aluminide coating [80–85].

Applications	Properties
Furnace rollers for heating steel slabs	• High temperature strength • good oxidation • corrosion resistance
Hydro turbine rotors	• Excellent vibration • cavitation resistance in water
Jet engines turbine blades vanes	• Superior strength • Creep resistance
Cutting tools	• High and low temperature cutting tool strength

8.2. Ni$_3$Al Thin Foils

Ni$_3$Al intermetallics like thin foils and tapes are anticipated in contributing the production of highly advanced tools of MEMS and MECS. This is because Ni$_3$Al possess unique physical and chemical properties in addition to a relatively low weight. A comparison is shown in Figure 13.

Ni$_3$Al alloys do, however, have a few drawbacks, the majority of which are related to the fact that they have a lower vulnerability in plastic deforming and higher propensity in getting brittle crack. Because of these disadvantages, the manufacturing sector is unlikely to ever be able to mass-produce components, having a thickness of lesser than 400 μm [53,91]. However, two processing methods have matured to the point that they might be employed in a laboratory setting:

- directional solidifying and cold rolling
- directional crystallization: deliberated upon meticulous deforming of traditional cast of alloys.

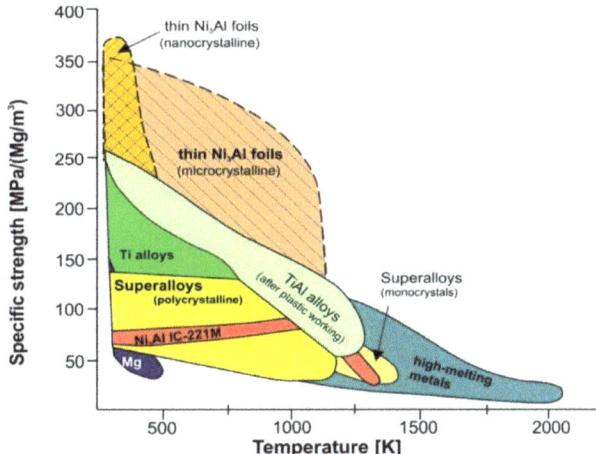

Figure 13. Temperature vs. specific strength for comparing Ni thin foils with other metal alloys [77]. Reused from MDPI under Creative Commons Attribution license.

Mechanical and electrical components (such as an actuator, a sensor, and a microprocessor) that can withstand their environments are integrated in MEMS and MECS systems, allowing for the fabrication of a device with both controlling and specialized capabilities [91–93].

It has been noticed that there is an increase in people's curiosity into Ni_3Al intermetallics with thin foils because these intermetallics have excellent explicit strength, higher environment resistivity, and higher catalyst activities. Additionally, the creation of composite materials has been reported by Ni_3Al-based alloys serving as the matrix and being toughened by elements such as TiC, ZrO_2, WC, SiC, and graphene [94–98].

Uses of foils and strips made of Ni_3Al-based alloys that are extremely promising include those known as MEMS or MECS devices. A comparison in mass gain and hydrogen production is shown in Figure 14 for Ni and Ni_3Al foils. It would appear that the creation of microsensors/systems of chemical separators, heat exchanger and micropumps would benefit enormously from the particular qualities that they possess [99–101].

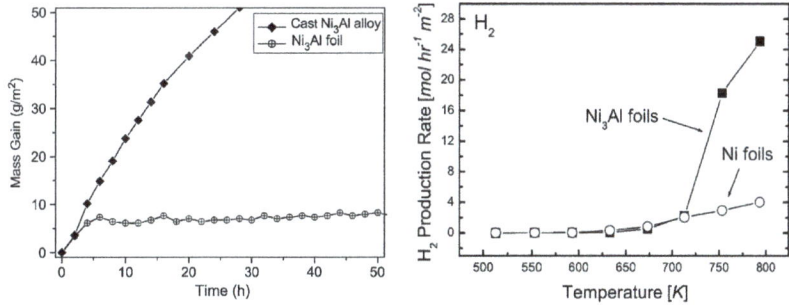

Figure 14. Comparison of production rates of H_2 in methanol decomposition of Ni_3Al foils and Ni foils [77]. Reused from MDPI under Creative Commons Attribution license.

9. Conclusions

This review paper seeks to improve understanding of the nickel aluminide structure, properties, and applications, as well as their scope, characteristics, advantages, and disadvantages. In addition, current alloy applications were summarized. The effect of alloying elements on phase transformation, mechanical properties, and corrosion was investigated. Furthermore, the most significant barriers to the widespread use of nickel aluminide were considered. To overcome the difficulties faced by alloys, different metal processing method were discussed. Properties and application of thin foil of nickel aluminides were discussed. Finally, characteristics of nickel coating were studied.

Author Contributions: Conceptualization, S.S. (Santosh Sampath) and V.P.R.; methodology, S.S. (Santosh Sampath), S.S. (Srivatsan Sundararajan) and V.P.R.; formal analysis, S.S. (Santosh Sampath); investigation, S.S. (Santosh Sampath), S.S. (Srivatsan Sundararajan) and V.P.R.; resources, S.S. (Santosh Sampath); data curation, V.P.R.; writing—S.S. (Santosh Sampath) and V.P.R.; writing—review and editing, S.S. (Santosh Sampath) and S.S. (Srivatsan Sundararajan); visualization, V.P.R.; supervision, S.S. (Santosh Sampath); project administration, S.S. (Santosh Sampath). All authors have read and agreed to the published version of the manuscript.

Funding: This research received no external funding.

Data Availability Statement: No data was used in this article.

Conflicts of Interest: The authors declare no conflict of interest.

References

1. Ward-Close, C.M.; Minor, R.; Doorbar, P.J. Intermetallic-matrix composites—A review. *Intermetallics* **1996**, *4*, 229. [CrossRef]
2. Stoloff, N.S.; Liu, C.T.; Deevi, C.S. Emerging applications of intermetallics. *Intermetallics* **2000**, *8*, 1313–1320. [CrossRef]
3. Sikka, V.K.; Deevi, S.C.; Viswanathan, S.; Swindeman, R.W.; Santella, M.L. Advances in processing of Ni$_3$Al-based intermetallics and applications. *Intermetallics* **2000**, *8*, 1329–1337. [CrossRef]
4. Kovalev, A.I.; Barskaya, R.A.; Wainstein, D.L. Effect of alloying on electronic structure, strength and ductility characteristics of nickel aluminide. *Surf. Sci.* **2003**, *532–535*, 35–40. [CrossRef]
5. Talaş, Ş. Nickel aluminides. In *Intermetallic Matrix Composites*; Woodhead Publishing: Cambridge, UK, 2018; pp. 37–69. [CrossRef]
6. Deevi, S.C.; Sikka, V.K.; Liu, C.T. Processing, properties, and applications of nickel and iron aluminides. *Prog. Mater. Sci.* **1997**, *42*, 177–192. [CrossRef]
7. Biswas, A.; Roy, S.K.; Gurumurthy, K.R.; Prabhu, N.; Banerjee, S. A study of self-propagating high-temperature synthesis of NiAl in thermal explosion mode. *Acta Mater.* **2002**, *50*, 757–773. [CrossRef]
8. Smallman, R.E.; Ngan, A.H. *Selected Alloys. Modern Physical Metallurgy*; Elsevier: Amsterdam, The Netherlands, 2014; pp. 529–569. [CrossRef]
9. Barrett, C.; Massalski, T.B. Chapter 10—The Structure of Metals and Alloys. In *Structure of Metals*, 3rd ed.; Pergamon: New York, NY, USA, 1980; pp. 223–269.
10. Mitra, R. *Structural Intermetallics and Intermetallic Matrix Composites*, 1st ed.; CRC Press: Boca Raton, FL, USA, 2015. [CrossRef]
11. Sikka, V.K.; Mavity, J.T.; Anderson, K. Processing of nickel aluminides and their industrial applications. In *High Temperature Aluminides and Intermetallics*; Elsevier: Amsterdam, The Netherlands, 1992; pp. 712–721. [CrossRef]
12. Nieh, T.G.; Stephens, J.J.; Wadsworth, J.; Liu, C.T. Chemical compatibility between silicon carbide and a nickel aluminide. In Proceedings of the International Conference on Composite Interfaces, Cleveland, OH, USA, 13–17 June 2018. No. CONF-880671-1.
13. Liu, C.T.; Sikka, V.K.; Horton, J.A.; Lee, H. *Alloy Development and Mechanical Properties of Nickel Aluminide Ni 3Al Alloys, ORNL-6483*; Martin Marietta Energy Systems, Inc.; Oak Ridge National Laboratory: Oak Ridge, TN, USA, 1988.
14. Sikka, V.K. Nickel Aluminides-New Advanced Alloys. *Mater. Manufact. Process.* **1989**, *4*, 1–24. [CrossRef]
15. Sikka, V.K.; Loria, E.A. *Nickel Metallurgy, Vol. II, Industrial Applications of Nickel*; Canadian Institute of Mining and Metallurgy: Montreal, QC, Canada, 1986; p. 293.
16. Gosele, U.; Tu, K. Growth kinetics of planar binary diffusion couples: Thin-film case versus bulk cases. *J. Appl. Phys.* **1982**, *53*, 3252. [CrossRef]
17. Tu, K.-N. Interdiffusion in thin films. *Annu. Rev. Mater. Sci.* **1985**, *15*, 147–176. [CrossRef]
18. Yang, J.-M.; Kao, W.H.; Liu, C.T. Development of nickel aluminide matrix composites. *Mater. Sci. Eng. A* **1989**, *107*, 81–91. [CrossRef]
19. Deevi, S.C.; Sikka, V.K. Nickel and iron aluminides: An overview on properties, processing, and applications. *Intermetallics* **1996**, *4*, 357–375. [CrossRef]
20. Sikka, V.K.; Santella, M.L.; Orth, J.E. Processing and operating experience of Ni$_3$Al-based intermetallic alloy IC-221M. *Mater. Sci. Eng. A* **1997**, *239–240*, 564–569. [CrossRef]
21. Miracle, D.B. Overview No. 104 The physical and mechanical properties of NiAl. *Acta Metall. Mater.* **1993**, *41*, 649–684. [CrossRef]
22. Godlewska, E.; Mitoraj, M.; Leszczynska, K. Hot corrosion of Ti–46Al–8Ta (at.%) intermetallic alloy. *Corros. Sci.* **2014**, *78*, 63–70. [CrossRef]
23. Wu, Y.T.; Li, C.; Li, Y.F.; Wu, J.; Xia, X.C.; Liu, Y.C. Effects of heat treatment on the microstructure and mechanical properties of Ni3Al-based superalloys: A review. *Int. J. Miner. Metall. Mater.* **2021**, *28*, 553–566. [CrossRef]

24. Tiwari, R.; Tewari, S.N.; Asthana, R.; Garg, A. Mechanical properties of extruded dual-phase NiAl alloys. *J. Mater. Sci.* **1995**, *30*, 4861–4870. [CrossRef]
25. Choudry, M.S.; Dollar, M.; Eastman, J.A. Nanocrystalline NiAl-processing, characterization and mechanical properties. *Mater. Sci. Eng. A* **1998**, *256*, 25–33. [CrossRef]
26. Cammarota, G.P.; Casagrande, A. Effect of ternary additions of iron on microstructure and microhardness of the intermetallic NiAl in reactive sintering. *J. Alloys Compd.* **2004**, *381*, 208–214. [CrossRef]
27. Guo, J.T.; Du, X.H.; Zhou, L.Z. Preparation of nanocrystalline NiAl compounds and composites by mechanical alloying. In *Materials Science Forum*; Trans Tech Publications Ltd.: Wollerau, Switzerland, 2005; Volume 475, pp. 749–754.
28. Pabi, S.K.; Murty, B.S. Mechanism of mechanical alloying in NiAl and CuZn systems. *Mater. Sci. Eng. A* **1996**, *214*, 146–152. [CrossRef]
29. Kubaski, E.T.; Cintho, O.M.; Antoniassi, J.L.; Kahn, H.; Capocchi, J.D.T. Obtaining NiAl intermetallic compound using different milling devices. *Adv. Powder. Technol.* **2012**, *23*, 667–672. [CrossRef]
30. Ying, D.Y.; Zhang, D.L. Effect of high energy ball milling on solid state reactions in Al–25 at.-%Ni powders. *Mater. Sci. Technol.* **2001**, *17*, 815–822. [CrossRef]
31. German, R.M.; Bose, A.; Sims, D. Production of reaction sintered nickel aluminide material. U.S. Patent 4,762,558, 9 August 1988.
32. McCoy, K.P.; Shaw, K.G.; Trogolo, J.A. Analysis of residual phases in nickel aluminide powders produced by reaction synthesis. In *MRS Fall Meeting-Symposium L—High- Temperature Ordered Intermetallic Alloys V*; Materials Research Society: Warrendale, PA, USA, 1992; Volume 288, pp. 909–914.
33. Liu, H.C.; Mitchell, T.E. Irradiation induced order-disorder in Ni_3Al and NiAl. *Acta Metall.* **1983**, *31*, 863–872. [CrossRef]
34. Chen, G.; Ni, X.; Nsongo, T. Lattice parameter dependence on long-range ordered degree during order–disorder transformation. *Intermetallics* **2004**, *12*, 733–739. [CrossRef]
35. Bradley, A.J.; Taylor, A. Electric and magnetic properties of B2 structure compounds: NiAl, CoAl. *Proc. R. Soc.* **1937**, *159A*, 56–72.
36. Foiles, S.M.; Daw, M.S. Application of the embedded atom method to Ni_3Al. *J. Mater. Res.* **1987**, *2*, 5–15. [CrossRef]
37. Pike, L.M.; Chang, Y.A.; Liu, C.T. Point defect concentrations and hardening in binary B2 intermetallics. *Acta Mater.* **1997**, *45*, 3709–3719. [CrossRef]
38. Fleischer, R.L. High-strength, high-temperature intermetallic compounds. *J. Mater. Sci.* **1987**, *22*, 2281–2288. [CrossRef]
39. Nagpal, P.; Baker, I. Effect of cooling rate on hardness of FeAl and NiAl. *Metal Trans. A* **1990**, *21*, 2281–2282. [CrossRef]
40. Westbrook, J. Temperature dependence of hardness of the equi-atomic iron group alu- minides. *J. Electrochem. Soc.* **1956**, *103*, 54–63. [CrossRef]
41. Kogachi, M.; Minamigawa, S.; Nakahigashi, K. Determination of long range order and vacancy content in the NiAl β'-phase alloys by x-ray diffractometry. *Acta Metall. Mater.* **1992**, *40*, 1113–1120. [CrossRef]
42. Cahn, R.W. Lattice parameter changes on disordering intermetallics. *Intermetallics* **1999**, *7*, 1089–1094. [CrossRef]
43. Talaş, Ş.; Göksel, O. Characterization of TiC and TiB2 reinforced Nickel Aluminide (NiAl) based metal matrix composites cast by in situ vacuum suction arc melting. *Vacuum* **2020**, *172*, 109066. [CrossRef]
44. Guard, R.W.; Westbrook, J.H. The alloying behavior of Ni_3Al. *Trans. Metall. Soc. AIME* **1959**, *215*, 807–814.
45. De Boer, F.R.; Schinkel, C.J.; Biesterbos, J.; Proost, S.; Ning, B.; Weaver, M.L. Exchange-enhanced paramagnetism and weak ferromagnetism in the Ni_3Al and Ni_3Ga phases: Giant moment inducement in Fe-doped Ni_3Ga. *J. Appl. Phys.* **1969**, *40*, 1049–1055. [CrossRef]
46. Sasakura, H.; Suzuki, K.; Masuda, Y. Curie temperature in itinerant electron ferromagnetic Ni_3Al system. *J. Phys. Soc. Jpn.* **1984**, *53*, 754–759. [CrossRef]
47. Manga, V.R.; Saal, J.E.; Wang, Y.; Crespi, V.H.; Liu, Z.-K. Magnetic perturbation and associated energies of the antiphase boundaries in ordered Ni_3Al. *J. Appl. Phys.* **2010**, *108*, 103509. [CrossRef]
48. Lazar, P.; Podloucky, R. Ductility and magnetism: An ab-initio study of NiAl–Fe and NiAl–Mn alloys. *Intermetallics* **2009**, *17*, 675–679. [CrossRef]
49. Zhou, J.; Guo, J.T. Effect of Ag alloying on microstructure, mechanical and electrical properties of NiAl intermetallic compound. *Mater. Sci. Eng. A* **2003**, *339*, 166–174. [CrossRef]
50. Terada, Y.; Ohkubo, K.; Mohri, T.; Suzuki, T. Thermal conductivity of intermetallic com- pounds with metallic bonding. *Mater. Trans.* **2002**, *43*, 3167–3176. [CrossRef]
51. Liu, C.; White, C.; Horton, J. Effect of boron on grain-boundaries in Ni3Al†. *Acta Metall.* **1985**, *33*, 213–229. [CrossRef]
52. Doychak, J.; Nesbitt, J.A.; Noebe, R.D.; Bowman, R.R. Oxidation of Al_2O_3 continuous fiber-reinforced/NiAl composites. *Oxid. Met.* **1992**, *38*, 45–72. [CrossRef]
53. Shigeji, T.; Shibata, T. Oxidation behavior of Ni2Al-0.1 B containing 2Cr. *Oxid. Met.* **1987**, *28*, 155–163.
54. Pan, Y.C.; Chuang, T.H.; Yao, Y.D. Long-term oxidation behaviour of Ni_3Al alloys with and without chromium additions. *J. Mater. Sci.* **1991**, *26*, 6097–6103. [CrossRef]
55. Liu, C.T.; Stiegler, J.O. Ductile ordered intermetallic alloys. *Science* **1984**, *226*, 636–642. [CrossRef]
56. Shigeji, T.; Shibata, T. Cyclic oxidation behavior of Ni3Al-0.1 B base alloys containing a Ti, Zr, or Hf addition. *Oxid. Met.* **1986**, *25*, 201–216.
57. Guo, J.; Sun, C.; Li, H.; Guan, H. Correlation between oxidation behaviour and boron content in Ni_3Al. *Chin Shu Hsueh Pao* **1989**, 25.

58. Yuan, Z.; Song, S.; Faulkner, R.G.; Yu, Z. Combined effects of cerium and boron on the mechanical properties and oxidation behaviour of Ni3Al alloy. *J. Mater. Sci.* **1988**, *33*, 463–469. [CrossRef]
59. Wood, G.C.; Stott, F.H. Oxidation of alloys. *Mater. Sci. Technol.* **1987**, *3*, 519–530. [CrossRef]
60. Dongyun, L. Effects of solution heat treatment on the microstructure, oxidation, and mechanical properties of a cast Ni3Al-based intermetallic alloy. *Met. Mater. Int.* **2006**, *12*, 153–159.
61. Hippsley, C.A.; Strangwood, M.; DeVan, J.H. Effects of chromium on crack growth and oxidation in nickel aluminide. *Acta Metall. Et Mater.* **1990**, *38*, 2393–2410. [CrossRef]
62. Wagner, C. Passivity and inhibition during the oxidation of metals at elevated temperatures. *Corros. Sci.* **1965**, *5*, 751–764. [CrossRef]
63. Zhai, W.; Shi, X.; Yao, J.; Ibrahim, A.M.M.; Xu, Z.; Zhu, Q.; Xiao, Y.; Chen, L.; Zhang, Q. Investigation of mechanical and tribological behaviors of multilayer graphene reinforced Ni$_3$Al matrix composites. *Compos. Part B Eng.* **2015**, *70*, 149–155. [CrossRef]
64. Kear, B.H.; Pettit, F.S.; Fornwalt, D.E.; Lemaire, L.P. On the transient oxidation of a Ni-15Cr-6Al alloy. *Oxid. Met.* **1971**, *3*, 557–569. [CrossRef]
65. Choi, S.C.; Cho, H.J.; Lee, D.B. Effect of Cr, Co, and Ti additions on the high-temperature oxidation behavior of Ni$_3$Al. *Oxid. Met.* **1996**, *46*, 109–127. [CrossRef]
66. Kainuma, R.; Ohtani, H.; Ishida, K. Effect of alloying elements on martensitic transformation in the binary NiAl (β) phase alloys. *Metall. Mater. Trans. A* **1996**, *27*, 2445–2453. [CrossRef]
67. Matsuura, K.; Kitamura, T.; Kudoh, M.; Itoh, Y. Changes in Microstructure and Mechanical Properties during Solid Sintering of Ni–Al Mixed Powder Compact. *Mater. Trans. JIM* **1996**, *37*, 1067–1072. [CrossRef]
68. Black, R.; Carolan, R.; Li, C.-Y.; Sikka, V.K.; Liu, C.T. Load relaxation studies of grain boundary effects in two Ni$_3$Al alloys at elevated temperatures. *Scr. Metall.* **1987**, *21*, 1675–1680. [CrossRef]
69. Wright, R.N.; Sikka, V.K. Elevated temperature tensile properties of powder metallurgy Ni$_3$Al alloyed with chromium and zirconium. *J. Mater. Sci.* **1988**, *23*, 4315–4318. [CrossRef]
70. Ko, H.; Hong, K.T.; Kaufmann, M.J.; Lee, K.S. The effect of long range order on the ac- tivation energy for atomic migration in NiAl alloys: Resistivity study. *J. Mater. Sci.* **2002**, *37*, 1915–1920. [CrossRef]
71. Cardellini, F.; Mazzone, G.; Montone, A.; Antisari, M.V. Solid state reactions between Ni and Al powders induced by plastic deformation. *Acta Met. Mater.* **1994**, *42*, 2445–2451. [CrossRef]
72. Ivanov, E.; Grigorieva, T.; Golubkova, G.; Boldyrev, V.; Fasman, A.B.; Mikhailenko, S.D.; Kalinina, O.T. Synthesis of nickel aluminides by mechanical alloying. *Mater. Lett.* **1988**, *7*, 51–54. [CrossRef]
73. Coreño Alonso, O.; Cabañas-Moreno, J.G.; Cruz-Rivera, J.J.; Florez-Diaz, G.; De Ita, A.; Quintana-Molina, S.; Falcony, C. Al-Ni intermetallics produced by spontaneous reaction during milling. *J. Metastab. Nanocryst. Mater.* **2000**, *343–346*, 290–295.
74. Kuenzly, J.D.; Douglass, D.L. The oxidation mechanism of Ni$_3$Al containing yttrium. *Oxid. Met.* **1974**, *8*, 139–178. [CrossRef]
75. Orth, J.E.; Sikka, V.K. Commercial casting of nickel aluminide alloys. *Adv. Mater. Processes* **1995**, *148*.
76. Sikka, V.K.; Wilkening, D.; Liebetrau, J.; Mackey, B. Melting and casting of FeAl-based cast alloy. *Mater. Sci. Eng. A* **1998**, *258*, 229–235. [CrossRef]
77. Jozwik, P.; Polkowski, W.; Bojar, Z. Applications of Ni$_3$Al Based Intermetallic Alloys—Current Stage and Potential Perceptivities. *Materials* **2015**, *8*, 2537–2568. [CrossRef]
78. Hirano, T.; Demura, M.; Kishida, K.; Hong, H.U.; Suga, Y. Mechanical properties of cold-rolled thin foils of Ni$_3$Al. In Proceedings of the 3rd International Symposium on Structural Intermetallics, Jackson Hole, WY, USA, 23–27 September 2001; pp. 765–774.
79. Schafrik, R.E. A perspective on intermetallic commercialization for aero-turbine applications. In Proceedings of the 3rd International Symposium on Structural Intermetallics, Jackson Hole, WY, USA, 23–27 September 2001; pp. 13–17.
80. Han, Y.F.; Chen, R.Z. R&D of cast superalloys and processing for gas turbine blades in BIAM. *Acta Metall. Sin.* **1996**, *9*, 457–463.
81. Malik, A.U.; Ahmad, R.; Ahmad, S.; Ahmad, S. High temperature oxidation behaviour of nickel aluminide coated mild steel. *Anti-Corros. Methods Mater.* **1991**, *38*, 4–10. [CrossRef]
82. Dey, G.K. Physical metallurgy of nickel aluminides. *Sadhana* **2003**, *28*, 247–262. [CrossRef]
83. Brandl, W.; Marginean, G.; Maghet, D.; Utu, D. Effects of specimen treatment and surface preparation on the isothermal oxidation behaviour of the HVOF-sprayed MCrAlY coatings. *Surf. Coat. Technol.* **2004**, *188–189*, 20–26. [CrossRef]
84. Hsiung, L.; Stoloff, N. Point defect model for fatigue crack initiation in Ni3Al+B single crystals. *Acta Metall. Et Mater.* **1990**, *38*, 1191–1200. [CrossRef]
85. Sglavo, V.M.; Marino, F.; Zhang, B.-R. The preparation and mechanical properties of Al$_2$O$_3$/ Ni$_3$Al composites. *Compos. Sci. Technol.* **1999**, *59*, 1207–1212. [CrossRef]
86. Gao, M.X.; Oliveira, F.J.; Pan, Y.; He, Y.; Jiang, E.B.; Baptista, J.L.; Vieira, J.M. The oxidation behaviour of TiC matrix Ni$_3$Al and Fe$_{40}$Al toughened composites at high temperatures. *Mater Sci. Forum* **2006**, *514–516*, 657–661. [CrossRef]
87. Xu, Y.; Kameoka, S.; Kishida, K.; Demura, M.; Tsai, A.-P.; Hirano, T. Catalytic properties of Ni$_3$Al intermetallics for methanol decomposition. *Mater. Trans.* **2004**, *45*, 3177–3179. [CrossRef]
88. Darolia, R.; Walston, E.S.; Noebe, R.; Garg, A.; Oliver, B.F. Mechanical properties of high purity single crystal NiAl. *Intermetallics* **1999**, *7*, 1195–1202. [CrossRef]
89. Ebrahimi, F.; Hoyle, T.G. Brittle-to-ductile transition in polycrystalline NiAl. *Acta Mater.* **1997**, *45*, 4193–4204. [CrossRef]

90. Gehling, M.G.; Vehoff, H. Computation of the fracture stress in notched NiAl-polycrystals. *Mater. Sci. Eng. A* **2002**, *329–331*, 255–261. [CrossRef]
91. Kinsey, H.V.; Stewart, M.T. Nickel Aluminium-Molybdenum alloys for service at elevated temperatures. *Trans. Am. Soc. Metals* **1951**, *43*, 193–219.
92. Maxwell, W.A.; Grala, P.F. *Investigation of Nickel Aluminium Alloys Containing from 14 to 34 Percent Aluminium*; NASA technical note, 3259; NASA: Washington, DC, USA, 1954.
93. Sikka, V.K.; Deevi, S.C.; Vought, J.D. Exo-Melt: A commercially viable process. *Adv. Mater. Process* **1995**, *147*, 29–31.
94. Deevi, S.; Sikka, V.K. Exo-Melt process for melting and casting of intermetallics. *Intermetallics* **1997**, *5*, 17–27. [CrossRef]
95. Chaithanya, M. Processing and Characterization of Ni-Al Coating on Metal Substrates. Master's Thesis, National Institute of Technology, Rourkela, India, 2007.
96. Xanthopoulou, G.; Marinou, A.; Vekinis, G.; Lekatou, A.; Vardavoulias, M. Ni-Al and NiO-Al Composite Coatings by Combustion-Assisted Flame Spraying. *Coatings* **2014**, *4*, 231–252. [CrossRef]
97. Hirano, T.; Demura, M.; Kishida, K. Method for Manufacturing Ni_3Al Alloy Foil. JP2. Patent No. 003,034,832, 7 February 2003.
98. Demura, M.; Suga, Y.; Umezawa, O.; Kishida, K.; George, E.P.; Hirano, T. Fabrication of Ni_3Al thin foil by cold-rolling. *Intermetallics* **2001**, *9*, 157–167. [CrossRef]
99. Demura, M.; Kishida, K.; Suga, Y.; Takanashi, M.; Hirano, T. Fabrication of thin Ni_3Al foils by cold rolling. *Sci. Mater.* **2002**, *47*, 267–272. [CrossRef]
100. Intermetallic Compound. Encyclopedia Britannica. Available online: http://www.britannica.com/EBchecked/topic/290430/intermetallic-compound (accessed on 2 January 2015).
101. Varin, R.A. Intermetallics: Crystal structures. *Encycl. Mater. Sci. Technol.* **2011**, 4177–4180.

Disclaimer/Publisher's Note: The statements, opinions and data contained in all publications are solely those of the individual author(s) and contributor(s) and not of MDPI and/or the editor(s). MDPI and/or the editor(s) disclaim responsibility for any injury to people or property resulting from any ideas, methods, instructions or products referred to in the content.

Article

Morphology and Growth Mechanism of β-Rhombohedral Boron and Pentagonal Twins in Cu Alloy

Junqing Han [1,†], Wentao Yuan [1,†], Yihan Wen [1], Zuoshan Wei [2], Tong Gao [1], Yuying Wu [1,*] and Xiangfa Liu [1]

1. Key Laboratory of Liquid-Solid Structure Evolution and Processing of Materials, Ministry of Education, Shandong University, Jinan 250061, China
2. Shandong Key Laboratory of Advanced Aluminum Materials and Technology, Binzhou Institute of Technology, Binzhou 256600, China
* Correspondence: wuyuying@sdu.edu.cn
† These authors contributed equally to this work.

Abstract: In this work, boron particles with β-rhombohedral structure were prepared in Cu-4B alloy. The morphology and growth mechanism of β-B and pentagonal twins were analyzed. Results show that boron crystals possessed an approximate octahedral structure which consisted of two planes belonging to {001} facet and a rhombohedron formed by {101} planes. The morphology of the boron crystal was determined by the position and size of {001} planes. During growth, parts of boron crystal formed twins to reduce surface energy. Five particular single crystals can shape a pentagonal twin. The morphological distinction between pentagonal twins mainly came from the difference in morphology of single crystal. When the {001} exposed planes were large and showed a hexagonal shape, the boron crystal often formed parallel groupings and polysynthetic twins to reduce surface energy.

Keywords: β-rhombohedral boron; twins; pentagonal twin; growth mechanism

Citation: Han, J.; Yuan, W.; Wen, Y.; Wei, Z.; Gao, T.; Wu, Y.; Liu, X. Morphology and Growth Mechanism of β-Rhombohedral Boron and Pentagonal Twins in Cu Alloy. *Crystals* 2022, 12, 1516. https://doi.org/10.3390/cryst12111516

Academic Editor: Petros Koutsoukos

Received: 10 September 2022
Accepted: 23 October 2022
Published: 25 October 2022

Copyright: © 2022 by the authors. Licensee MDPI, Basel, Switzerland. This article is an open access article distributed under the terms and conditions of the Creative Commons Attribution (CC BY) license (https:// creativecommons.org/licenses/by/ 4.0/).

1. Introduction

In the aviation industry, boron (B) is considered the most ideal combustion aid for jet fuel, due to having the highest volumetric heating value and extremely high gravimetric heating, which is second only to that of beryllium (Be) [1]. 10B, isotope of boron, possesses a relatively strong neutron absorbing capacity [2], so 10B and its composite materials are widely used for boron neutron capture therapy [3] and thermal neutron detectors [4]. Many different shapes of boron have been successfully prepared by chemical and physical methods, such as boron nanotubes, [5,6] boron nanowires, [7,8] boron nanocones, [9] and boron nanoribbons. [10,11] In addition, in our previous studies, we also spheroidized eutectic boron by varying the cooling rate and adding alloying elements to prepare submicron boron spheres and hollow boron spheres. [12] However, the morphology and structure of boron have a great influence on its application. For instance, Evgeni S. Penev discovered that boron had the potential to transform a superconductor when the boron possessed a two-dimensional structure [13].

Allotropy of boron has been widely reported in recent years, as seen with γ-B [14] and t-B [15], but the growth model and mechanism of boron crystal are still focused on the field of simulation. Wataru Hayami calculated the surface energy of α-B [16] and t-B [17], and gave the crystallographic monomorph of these two allotropies. However, the structures of t-B and γ-B are different from those of rhombohedral boron. Under normal temperature and pressure, boron often exists in the form β-B (a = 10.145 ± 0.015 Å, α= 65°17′ ± 8′) [18] and each lattice point is occupied by an icosahedron (B12) [19,20]. Both α-B and β-B have a rhombohedral structure, the space group of $R\bar{3}m$ (group no. 166). The B12 in α-B is distributed at each lattice point of the rhombohedral unit cell, while the B12 of β-B is not only distributed at the lattice points, but also at the center of the edge of the rhombohedral unit cell. In addition, there are two triple-fused B28 polyhedrons in the center of β-B unit cell, and these two polyhedrons are linked by a gap boron atom [20,21]. In an experiment of using aluminum-doped β-rhombohedral boron, a three-dimensional framework made of B_{12} icosahedra with voids being occupied by the B_{28}–B–B_{28} units was found [22]. Sun [23] et al. studied the growth mechanisms of alpha-boron (α-B) and beta-boron (β-B) in Cu-B alloys in copper melts and observed lamellar growth traces and twin structures of alpha-boron (α-B) and beta-boron (β-B) and produced a model for the growth of β-B, but there is little information about their monotype and exposed surface.

Because boron has a special three-center bond structure [24,25], the dislocations are difficult to form. To reduce the surface energy and release stress, boron crystals often twin during crystallization [23] and ball milling [26]. Some particular pentagonal twin structures have been found in B_4C [27] and gold nanocrystal [28], but there are almost no reports for boron. In this work, the morphology of β-B and pentagonal twins were reported and analyzed.

2. Experimental Section

Pure Cu (>99.7 wt%) and pure B (99.9 wt%) were used to prepare Cu-4B (with the same weight percentage as that reported below, unless otherwise specified) alloy. Pure Cu was melted using a high-frequency stove. After Cu melt, the B enfolded by copper foil was added to the melt. Then, the melt was poured into a cast-iron mold to obtain Cu-4B alloy. Small alloy blocks cut from Cu-4B alloy were re-melted by a high-frequency induction coil to obtain blow-cast alloys and rapidly solidifying alloy strips. During the preparation of rapidly solidifying alloy, the rotating speeds of the rotated copper mold with a perimeter of 690 mm were 1500 r/min and 3000 r/min. The cooling of Cu-4B alloy ingots was 100 K/s, and the cooling rate of alloy ingots, blow-cast alloys and alloy strips increased in turn.

The Cu in Cu-4B alloy ingots, blow-cast alloys and alloy strips was eroded by 50% HNO_3, and the B powders remained. The B powders were repeatedly cleaned with deionized water until the PH reached 7. The micro-morphology of boron was characterized by field emission scanning electron microscope (JSM-7800F SEM, Japan). The chemical composition of boron was analyzed using the JEM-2100F high-resolution transmission electron microscope linked with an energy dispersive X-ray spectroscopy (EDS) attachment and Oxford XMax80 spectrometer (SU-70, Japan). The crystal structure was analyzed using a transmission electron microscope (TEM, JEM-2100F, Japan).

3. Results and Discussion

The morphology of Cu-4B alloy and boron particles extracted from Cu-4B ingots, blow-cast alloys and alloy strips (1500 r/min and 3000 r/min) is shown in Figure 1. The white arrows in Figure 1a indicate some special structures, such as hexagonal boron, twins and pentagonal twins. The particles shown in Figure 1b–f were typical primary boron [23]. As shown in Figure 1b, the diameter of boron particles was about 5–10 μm. The EDS result in Figure 1c confirms that these crystals were indeed boron, and the existence of the concentration of Au was caused by the gold spray treatment before the SEM test. As shown in Figure 1d, compared with Figure 1b, the diameter of boron particles extracted from blow casting alloy was relatively small. The size of boron particles extracted from alloy strips was about 3–5 μm, no matter what rotation rate was used. As for the Cu-B binary phase diagram [29], the primary boron in Cu-4B wt% (about Cu-20B at%) alloy was precipitated at 1050 °C, while the eutectic boron was precipitated at 1013 °C, and the cooling rate was relatively high; therefore, the time needed for crystals to grow was very short. Both in blow-cast alloy and alloy strips, the crystals did not grow enough, so the size of particles extracted from blow casting alloy and alloy strips was similar. Although the size of the boron particles extracted from ingots alloys or strips was slightly different, there was almost no distinction in terms of morphology. This indicates that in this work, the cooling rate did not affect the crystal structure of boron. Therefore, the source of boron crystal was not considered in the following crystal morphology analysis.

Figure 1. (**a**) Backscattered electron (BE) image of Cu-4B alloy and the enlarged images of the boron particles with pentagonal twins. Secondary electron (SE) images of boron particles extracted from (**b**) alloy ingots, (**c**) EDS result of boron crystal extracted from alloy ingots, (**d**) blow-cast alloy, (**e**) alloy strips (1500 r/min) and (**f**) alloy strips (3000 r/min).

Figure 2 shows several typical crystal morphologies of boron, with partial planes and facet families were marked in Figure 2a–f. According to the calculation results of Hayami et al. [16], the {001} and {101} have lower surface energy, and the (001) facets have the lowest surface energy for the trigonal structure β-B. The exposed surfaces of these boron crystals were {001} and {101} planes. These crystals can be considered as a structure formed by a completed rhombohedron composed of {101} planes and cut by (001) plane. The morphological difference of these boron crystals is mainly due to the location of {001} planes. As shown in Figure 2a, when the (001) plane closed to the top of the crystal, the shape of the (001) plane was a triangle and the {101} planes were pentagons. As the (001) plane descended, the size of the (001) plane was gradually increased. When the (001) plane dropped to the position shown in Figure 2b, the size of the triangular (001) plane reached its maximum. At this moment, the morphology of the boron crystal was similar to that of an octahedron [30]. Together with {001} planes, all {101} planes were triangles. When the (001) plane was in the middle of the boron crystal, the shape of {001} and {101} planes transformed into hexagons and trapezoids, respectively, as shown in Figure 2c,f. In order to further confirm the crystal structure of boron crystal, TEM and selected electron diffraction were used. As shown in Figure 2g, the morphology of boron crystals was mainly hexagonal under TEM. A relatively thin hexagonal boron crystal, as shown in Figure 2h, was chosen for selected electron diffraction. The diffraction pattern confirmed that the boron particle possessed β-B structure, and the diffraction spots correspond to the (201), (306) and (105) planes of β-B. The results of selected area electron diffraction are consistent with those of Sun et al. [23], suggesting that the pentagonal B particles also have the structure of β-B.

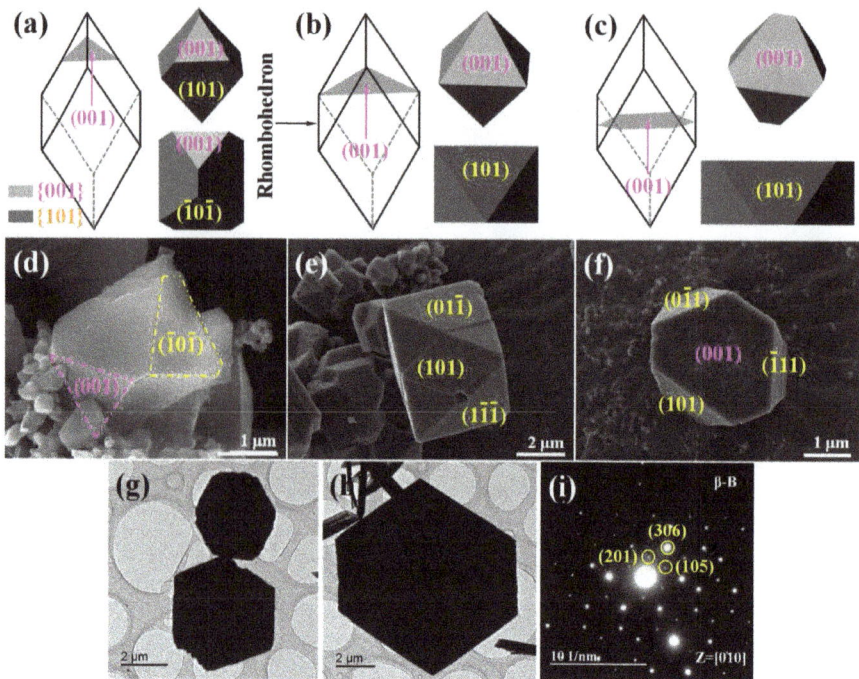

Figure 2. (**a**–**c**) Schematic diagrams of three typical crystal morphologies of boron crystal; (**d**–**f**) SEM images of boron crystals correspond to Figure 2a–c, respectively. (**g**,**h**) TEM images of boron crystals extracted from Cu-4B alloy ingot. (**i**) Diffraction pattern of boron crystal in Figure 2h. All the exposed surfaces of completed rhombohedron in (**a**–**c**) were {101} planes.

During the growth of boron crystal, twins are very common [31]. Figure 3 shows some of the twin structures of boron. The pentagonal twin in Figure 3a was shaped by complete rhombohedral boron crystals. To form a pentagonal structure, angle 1 must be close to 72°. The (101) plane is marked in Figure 3a. Another pentagonal twin, which was formed by the crystals shown in Figure 2a,b, can be seen in Figure 3b. Like the pentagonal twin shown in Figure 3a, the theoretical value of angle 2 was also close to 72°. The pentagonal twin was not perfect; there were grooves and protrusions in the twin, as shown by the white arrows. A side view of the pentagonal twin is shown in Figure 3c, and the crystals that formed the pentagonal twin are marked by yellow and green lines, respectively. The theoretical value of the plane angle between $(01\bar{1})$ and $(\bar{1}11)$ was 72.8°, which was closed to angles 1 and 2. This pentagonal twin crystal has morphological similarity with the penta-twinned gold nanocrystals discovered by Zhang et al. [32] As shown in Figure 3d, when the {001} planes showed the hexagonal shape in Figure 2c, it was difficult for boron crystals to form pentagonal twins. This type of boron crystal is prone to take shape original twins, parallel grouping, as well as the polysynthetic twin that can be observed in the white box in Figure 3c.

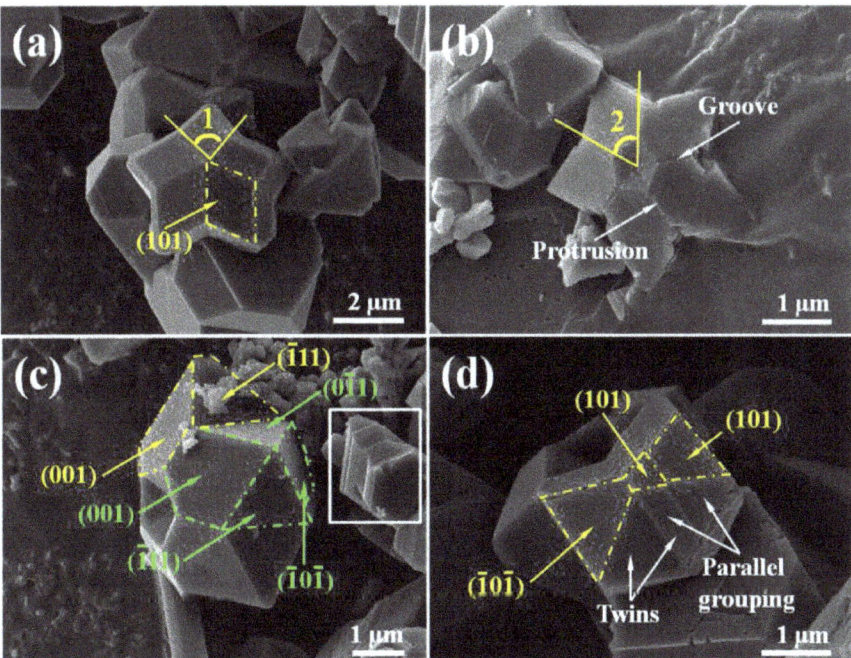

Figure 3. (**a,b**) Pentagonal twins, (**c**) the side view of pentagonal twin, (**d**) twins and parallel grouping. Parts of planes belonging to {101} and {001} are marked in Figure 3.

Figure 4 is the schematic diagram of the growth of several twins. Different morphological crystals can form distinct twins. As shown in Figure 4a, the original rhombohedral boron crystals could form the pentagonal twin I shown in Figure 3a without being cut by {001} planes. First, rhombohedral boron crystal was twinned, and the twin plane was $(01\bar{1})$ exposed planes. Then, the crystal 2, 3 and 5 emerged by twinning, and their twin planes were $(01\bar{1})$ or $(1\bar{1}\bar{1})$ exposed face. Finally, the pentagonal twin was finished. The formation process of pentagonal twin II shown in Figure 4b was similar to that of pentagonal twin I, both adopting $(01\bar{1})$ or $(1\bar{1}\bar{1})$ planes as the twin plane. Most of its monomers are the crystals shown in Figure 2e (simple form 2), that is, octahedral like crystals, and sometimes the crystals shown in Figure 2d also exist. For example, the pentagonal twins in Figure 3b were all composed of octahedral like crystals, while the pentagonal twin II in Figure 3c had the crystals shown in Figure 2d, which are marked with yellow dotted lines. However, the twin plane of pentagonal twin II was located inside crystal 1. As shown in Figure 4b, when the twin plane is inside, the edges of crystal 1 and crystal 2 will intersect. Therefore, after twinning, two new edges were formed on (001) and (101) planes. Edge 1 corresponded to the groove shown in Figure 3b and edge 2 participated to shape angle 2 in Figure 3b. After the formation of rhombohedral boron crystals 3–5, pentagonal twin II with a sunken pentagon in the center formed. The five new edge 2s and the original edges of the crystal together formed the pentagonal funnel shaped defect in Figure 3b.

Boron crystals that have hexagonal {001} planes can take the shape of common twins, parallel groupings, and polysynthetic twins. As shown in Figure 4c, boron crystals like simple form 3 can take the (001) plane as the twin plane to form twin. The twin can continue to grow in this shape, or can turn into a parallel grouping and polysynthetic twin, as shown in the white box in Figure 3c. If single crystal 1 forms a pentagonal twin, the large {001} planes will be exposed, and there will be a hole in the middle of this pentagonal twin. These two phenomena are both harmful in reducing the overall surface energy. On the contrary, the formation of twins as shown in Figure 4c by sharing {001} planes can significantly reduce the surface energy.

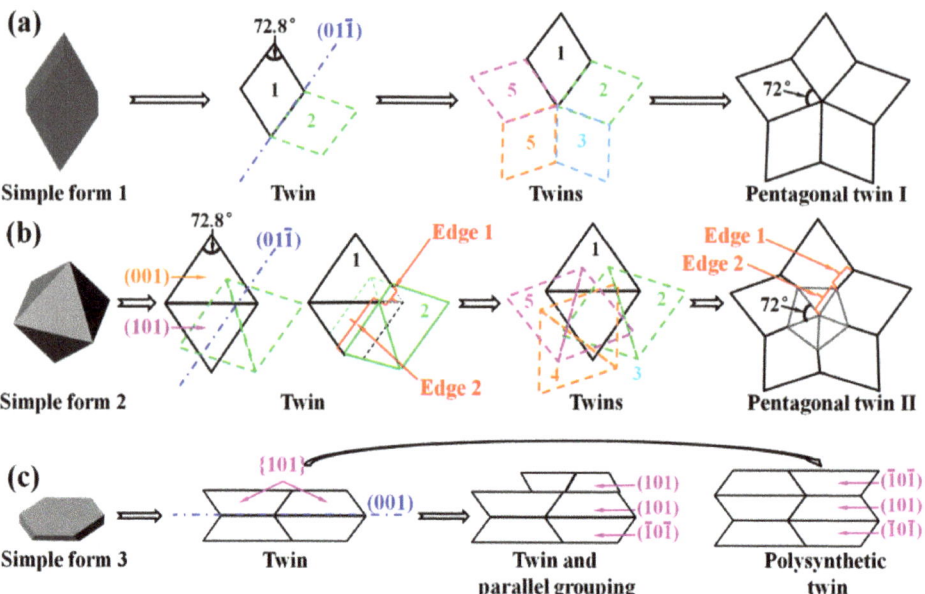

Figure 4. Schematic diagram of twin growth of (**a**) pentagonal twin I, (**b**) pentagonal twin II and (**c**) twins, parallel grouping, and polysynthetic twin.

4. Conclusions

In the Cu-4B alloy, boron particles mainly possessed the β-B structure. Three typical boron crystal morphologies were analyzed and their exposed surfaces were {001} and {101} planes. The morphological difference between them was due to the position of {001}. Boron was prone to twinning during growth, and the complete rhombohedron formed by {101} planes could form relatively complete pentagonal twins. Boron crystals enclosed by triangular {001} planes and {101} planes could take the shape of pentagonal twins with a sunken pentagon in the center. When the {001} planes transformed into hexagons, it was easy for boron crystals to form parallel grouping and polysynthetic twin to reduce surface energy.

Author Contributions: Conceptualization, J.H. and W.Y.; methodology, W.Y.; software, W.Y. and J.H; formal analysis, W.Y. and J.H.; writing—original draft preparation, W.Y., J.H., Y.W. (Yuying Wu), Y.W. (Yihan Wen) and X.L.; writing—review and editing, J.H., W.Y., Y.W. (Yuying Wu), Z.W., T.G., Y.W. (Yihan Wen) and X.L.; visualization, J.H., W.Y. and Z.W.; supervision, Y.W. (Yuying Wu); project administration, Y.W. (Yuying Wu); funding acquisition, Y.W. (Yuying Wu) All authors have read and agreed to the published version of the manuscript.

Funding: The National Key R&D Program of China (2021YFB3400800), the Key Research and Development Program of Shandong Province (Grant No. 2021ZLGX01 and 2021SFGC1001), and the Shandong University Climbing Program Innovation team.

Institutional Review Board Statement: Not applicable.

Informed Consent Statement: Not applicable.

Data Availability Statement: The data presented in this study are available in article.

Acknowledgments: This study was financially supported by the National Key R&D Program of China (2021YFB3400800), the Key Research and Development Program of Shandong Province (Grant No. 2021ZLGX01 and 2021SFGC1001), and the Shandong University Climbing Program Innovation team.

Conflicts of Interest: The authors declare no conflict of interest.

References

1. Ojha, P.K.; Karmakar, S. Boron for liquid fuel Engines-A review on synthesis, dispersion stability in liquid fuel, and combustion aspects. *Prog. Aerosp. Sci.* **2018**, *100*, 18–45. [CrossRef]
2. Rolf, F.B.; Coderre, J.A.; Vicente, M.G.H.; Blue, T.E. Boron neutron capture therapy of cancer: Current status and future prospects. *Clin. Cancer Res.* **2005**, *11*, 3987–4002.
3. Kalot, G.; Godard, A.; Busser, B.; Pliquett, J.; Sancey, L. Aza-BODIPY: A new vector for enhanced theranostic boron neutron capture therapy applications. *Cells* **2020**, *9*, 1953. [CrossRef] [PubMed]
4. Nikolić, R.J.; Conway, A.M.; Reinhardt, C.E.; Graff, R.T.; Wang, T.F.; Deo, N.; Cheung, C.L. 6:1 aspect ratio silicon pillar based thermal neutron detector filled with B10. *Appl. Phys. Lett.* **2008**, *93*, 133502. [CrossRef]
5. Wu, Y.Y.; Li, Y.F.; Chen, H.W.; Sun, Z.X.; Wang, N.; Qin, J.Y.; Li, H.; Bian, X.F.; Liu, X.F. Growth of single crystalline boron nanotubes in a Cu alloy. *CrystEngComm* **2017**, *19*, 4510–4518. [CrossRef]
6. Patel, R.; Chou, T.; Iqbal, Z. Synthesis of Boron Nanowires, Nanotubes, and Nanosheets. *J. Nanomater.* **2015**, *2015*, 243925. [CrossRef]
7. Bai, H.; Zou, H.H.; Chen, G.X.; Yu, J.H.; Nishimura, K.; Dai, W.; Jiang, N. Nucleation and growth of boron nanowires on diamond particles. *Appl. Surf. Sci.* **2014**, *313*, 132–137. [CrossRef]
8. Yang, Q.; Sha, J.; Wang, L.; Su, Z.; Ma, X.; Wang, J.; Yang, D. Morphology and diameter controllable synthesis of boron nanowires. *J. Mater. Sci.* **2006**, *41*, 3547–3552. [CrossRef]
9. Li, C.; Tian, Y.; Hui, C.; Tian, J.; Bao, L.; Shen, C.; Gao, H.J. Field emission properties of patterned boron nanocones. *Nanotechnology* **2010**, *21*, 325705. [CrossRef]
10. Zhang, Z.W.; Xie, Y.; Peng, Q.; Chen, Y.P. Phonon transport in single-layer boron nanoribbons. *Nanotechnology* **2016**, *27*, 445703. [CrossRef]
11. Xu, T.T.; Zheng, J.; Wu, N.; Nicholls, A.W.; Roth, J.; Dikin, D.A.; Ruoff, R.S. Crystalline Boron Nanoribbons: Synthesis and Characterization. *Nano Lett.* **2004**, *4*, 963–968. [CrossRef]
12. Yuan, W.T.; Wu, Y.Y.; Zhang, G.D.; Wu, C.C.; Zhao, S.; Liu, X.F. Study on spheroidization and the growth mechanism of eutectic boron in Cu-B alloys. *CrystEngComm* **2020**, *22*, 6993–7001. [CrossRef]
13. Penev, E.S.; Kutana, A.; Yakobson, B.I. Can two-dimensional boron superconduct? *Nano Lett.* **2016**, *16*, 2522–2526. [CrossRef]
14. Oganov, A.R.; Chen, J.; Gatti, C.; Ma, Y.; Ma, Y.; Glass, C.W.; Liu, Z.; Yu, T.; Kurakevych, O.O.; Solozhenko, V.L. Ionic high-pressure form of elemental boron. *Nature* **2009**, *457*, 863–867. [CrossRef] [PubMed]
15. An, Q.; Reddy, K.M.; Xie, K.Y.; Hemker, K.J.; Goddard, W.A. New ground-state crystal structure of elemental boron. *Phys. Rev. Lett.* **2016**, *117*, 085501. [CrossRef] [PubMed]
16. Hayami, W.; Otani, S. The role of surface energy in the growth of boron crystals. *J. Phys. Chem. C* **2007**, *111*, 688–692. [CrossRef]
17. Hayami, W.; Otani, S. Surface energy and growth mechanism of β-tetragonal boron crystal. *J. Phys. Chem. C* **2007**, *111*, 10394–10397. [CrossRef]
18. Hughes, R.E.; Kennard, C.; Sullenger, D.B.; Weakliem, H.A.; Hoard, J.L. The structure of β-rhombohedral boron. *J. Am. Chem. Soc.* **1963**, *85*, 361–362. [CrossRef]
19. Fujimori, M.; Nakata, T.; Nakayama, T.; Nishibori, E.; Kimura, K.; Takata, M.; Sakata, M. Peculiar covalent bonds in α-rhombohedral boron. *Phys. Rev. Lett.* **1999**, *82*, 4452–4455. [CrossRef]
20. Shirai, K. Phase diagram of boron crystals. *Jpn. J. Appl. Phys.* **2017**, *56*, 05FA06. [CrossRef]
21. Widom, M.; Mihalkovic, M. Symmetry-broken crystal structure of elemental boron at low temperature. *Phys. Rev. B* **2008**, *77*, 064113. [CrossRef]
22. Bykova, E.; Parakhonskiy, G.; Dubrovinskaia, N.; Chernyshov, D.; Dubrovinsky, L. The crystal structure of aluminum doped β-rhombohedral boron. *J. Solid State Chem.* **2012**, *194*, 188–193. [CrossRef]
23. Sun, Z.X.; Wu, Y.Y.; Han, X.X.; Zhang, G.J.; Liu., X.F. Growth mechanisms of alpha-boron and beta-boron in a copper melt at ambient pressure and its stabilities. *CrystEngComm* **2017**, *19*, 3947–3954. [CrossRef]
24. Frenking, G.; Holzmann, N. Chemistry: A boron-boron triple bond. *Science* **2012**, *336*, 1394–1395. [CrossRef]
25. He, J.L.; Wu, E.D.; Wang, H.T.; Liu, R.P.; Tian, Y.J. Ionicities of boron-boron bonds in B12 icosahedra. *Phys. Rev. Lett.* **2005**, *94*, 015504. [CrossRef]
26. Zhao, S.; Wu, Y.Y.; Sun, Z.X.; Zhou, B.; Liu, X.F. Superhard copper matrix composite reinforced by ultrafine boron for wear-resistant bearings. *ACS Appl. Nano Mater.* **2018**, *1*, 5382–5388. [CrossRef]
27. Fu, X. Uncovering the internal structure of five-fold twinned nanowires through 3D electron diffraction mapping. *Chin. Phys. B* **2020**, *29*, 068101. [CrossRef]

28. Liu, T.; Jiang, P.; You, Q.; Ye, S. Five-fold twinned pentagonal gold nanocrystal structure exclusively bounded by {110} facets. *CrystEngComm* **2013**, *15*, 2350–2353. [CrossRef]
29. Chakrabarti, D.J.; Laughlin, D.E. The B-Cu (Boron-Copper) system. *Bull. Alloy Phase Diagr.* **1982**, *3*, 45–48. [CrossRef]
30. Nie, J.F.; Wu, Y.Y.; Li, P.T.; Li, H.; Liu, X.F. Morphological evolution of TiC from octahedron to cube induced by elemental nickel. *CrystEngComm* **2012**, *14*, 2213–2221. [CrossRef]
31. Werheit, H. Comment on "New ground-state crystal structure of elemental boron". *Phys. Rev. Lett.* **2017**, *118*, 089601. [CrossRef] [PubMed]
32. Zhang, T.; Li, X.J.; Sun, Y.Q.; Liu, D.L.; Li, C.C.; Cai, W.P.; Li, Y. A universal route with fine kinetic control to a family of penta-twinned gold nanocrystals. *Chem. Sci.* **2021**, *12*, 12631–12639. [CrossRef] [PubMed]

Article

The Structural, Electronic, Magnetic and Elastic Properties of Full-Heusler Co$_2$CrAl and Cr$_2$MnSb: An Ab Initio Study

Sara J. Yahya [1], Mohammed S. Abu-Jafar [1,*], Said Al Azar [2], Ahmad A. Mousa [3], Rabah Khenata [4], Doha Abu-Baker [1] and Mahmoud Farout [1]

[1] Department of Physics, An-Najah National University, Nablus P.O. Box 7, Palestine
[2] Department of Physics, Faculty of Science, Zarqa University, Zarqa 13132, Jordan
[3] Department of Basic Sciences, Middle East University, Amman 11831, Jordan
[4] Laboratoire de Physique Quantique et de Modélisation Mathématique de la Matière (LPQ3M), Université de Mascara, Mascara 29000, Algeria
* Correspondence: mabujafar@najah.edu

Abstract: In this paper, the full-potential, linearized augmented plane wave (FP-LAPW) method was employed in investigating full-Heusler Co$_2$CrAl's structural, elastic, magnetic and electronic properties. The FP-LAPW method was employed in computing the structural parameters (bulk modulus, lattice parameters, c/a and first pressure derivatives). The optimized structural parameters were determined by generalized gradient approximation (GGA) for the exchange-correlation potential, V_{xc}. Estimating the energy gaps for these compounds was accomplished through modified Becke–Johnson potential (mBJ). It was found that the conventional Heusler compound Co$_2$CrAl with mBJ and GGA approaches had a half-metallic character, and its spin-down configuration had an energy gap. It was also found that the conventional and inverse Heusler Cr$_2$MnSb and tetragonal (139) (Co$_2$CrAl, Cr$_2$MnSb) compounds with a half-metallic character had direct energy gaps in the spin-down configuration. To a certain degree, the total magnetic moments for the two compounds were compatible with the theoretical and experimental results already attained. Mechanically, we found that the conventional and inverse full-Heusler compound Co$_2$CrAl was stable, but the inverse Cr$_2$MnSb was unstable in the ferromagnetic state. The conventional Heusler compound Cr$_2$MnSb was mechanically stable in the ferromagnetic state.

Keywords: full-Heusler compound; electronic band structure; magnetic order; elastic properties; FP-LAPW

1. Introduction

Since their discovery in 1903, Heusler compounds have found many applications including spintronics [1], shape-memory devices [2] and thermoelectric power generators [3]. Heusler compounds have a type of face-centered cubic (fcc) crystal structure. These compounds can be categorized into two classes: XYZ (half-Heuslers), which consist of three FCC sub-lattices, and X$_2$YZ (full-Heuslers), which have four FCC sub-lattices, where transition elements are represented by X and Y, and the s, p . . . elements are represented by Z [4].

Full-Heusler X$_2$YZ compounds crystallize in two kinds of inverse and conventional forms. Conventional Heusler compounds crystallize in a Cu$_2$MnAl structure with a space group of Fm-3m (space group number 225) having atomic positions of X$_2$ (1/4,1/4,1/4), (3/4,3/4,3/4), Y (1/2,1/2,1/2) and Z (0,0,0). Inverse Heusler compounds crystallize in a Hg$_2$CuTi structure with space group of F-43m (space group number 216) having atomic positions of X$_2$ (1/4,1/4,1/4), (1/2,1/2,1/2), Y (3/4,3/4,3/4) and Z (0,0,0) [5]. Some Heusler compounds have a half-metallic (HM) character [5–18], where only a single conduction spin channel exists for half metals. For one spin channel, the spin-polarized band structure shows metallic behavior. On the other hand, at the Fermi level, the other spin band structure shows a gap. Therefore a 100% spin polarization is exhibited by half-metallic materials.

Certain studies have explored the electronic, magnetic, elastic and structural properties of these compounds using different methods. Zhang et al. [19] focused on the Co_2CrAl Heusler compound's electronic band structure and transport properties. They measured the lattice parameter, magnetic moment and indirect band gap and found those to be 5.74, 3 μ_B and 0.475 eV, respectively. Hakimi et al. [20] conducted an experimental study of the Co_2CrAl compound's magnetic and structural properties. They found that the conventional Co_2CrAl's total magnetic moment was 2 μ_B.

Ozdogan and Galanakis [21] determined half-metallic antiferromagnetic Cr_2MnSb's magnetic and electronic properties for both conventional and inverse structure types. They found that for both structural types, Cr_2MnSb is a half-metallic ferrimagnetic compound for a broad array of lattice constants.

Heusler alloys are well-known for their potential application in the spin-transfer torque (STT) sector. These materials crystallize in multifaceted structures in both cubic and tetragonal symmetries with multiple magnetic sublattices. Galanakis [22] conducted research on the magnetic and electronic properties of both full-Heusler and half-Heusler alloys. The full-Heusler alloys investigations included Co_2MnSi and Co_2MnGe, and the half-Heusler alloys included PtMnSb, CoMnSb and NiMnSb.

Atsufumi Hirohata et al. [23] reviewed the development of anti-ferromagnetic (AFM) Heusler alloys for the replacement of iridium as a critical raw material (CRMs). They established correlations between the crystalline structure of these alloys and the magnetic properties, i.e., antiferromagnetism. This study revealed that the Heusler alloys consisting of elements with moderate magnetic moments require perfectly or partially ordered crystalline structures to exhibit AFM behavior. Using elements with large magnetic moments, a fully disordered structure was found to show either AFM or ferrimagnetic (FIM) behavior. The considered alloys may become useful for device applications due to the additional increase in their anisotropy and grain volume being able to sustain AFM behavior above room temperature.

Recently, Abu Baker et al. [24] investigated the elastic, electronic, magnetic and structural characteristics of half-metallic ferromagnetic full-Heusler alloys, namely conventional Co_2TiSn and inverse Zr_2RhGa, employing the FP-LAPW technique. The lattice parameters for the conventional Co_2TiSn and inverse Zr_2RhGa were found to be 6.094 A^0 and 6.619 A^0, respectively. In addition, the total magnetic moments for these compounds were recorded as 1.9786 μ_B and 1.99 μ_B, respectively. The conventional Co_2TiSn and inverse Zr_2RhGa compounds had indirect energy gaps of 0.482 eV and 0.573 eV, respectively. From their electronic properties, it can be noted that the conventional full-Heusler Co_2TiSn compound and the inverse full-Heusler Zr_2RhGa compound had stability from a mechanical perspective.

Furthermore, Gupta et al. prepared Cr_2MnSb thin films on a MgO (001) substrate using the DC/RF magnetron sputtering method. The XRD analysis of the deposited films revealed that they crystallized in a cubic phase with full B2 and partial $L2_1$ ordering [25]. Previously, Dubowik et al. deposited 100 nm Co_2CrAl films on glass and NaCl substrates using the flash evaporation technique [26].

Paudel and Zhu [27] showed that the full Heusler alloy Co_2ScSb is stable at the ferromagnetic phase with an optimized lattice constant of 6.19. They confirmed the structural stability from the calculations of the negative cohesive, formation energy and real phonon frequency. Paudel and Zhu [28] also showed that the half-metallic ferromagnetic properties of a Fe_2MnP alloy have energy band gaps of 0.34 eV and half-metallic gaps of 0.09 eV at an optimized lattice parameter of 5.56.

In this article, the motivation for investigating the mechanical, electronic and magnetic characteristics of the full-Heusler compounds Co_2CrAl and Cr_2MnSb in both the conventional and inverse form was to study in detail their mechanical and structural stability and preferable magnetic phase, as well as to introduce their elastic properties and behaviors. The article is organized as follows: after the introduction and background review, the computational methods and model are introduced. This is followed by a presentation of the results, discussion and conclusion.

2. Computational Method

In the current study, the calculations were accomplished using the full-potential, linearized augmented plane wave procedure executed in the WIEN2k [29] suite. Generalized gradient approximation (GGA) was used to calculate the structural parameters, i.e., the lattice parameters and bulk modulus. The GGA method depends on the local gradient of the electronic density in addition to the value of the density, giving a more accurate description of variations in the electron–electron interactions. GGA functionals provide a severe underestimation of the energy band gaps, so a modified Becke–Johnson (mBJ-GGA) functional was used to improve the energy band gaps. For the compound Co_2CrAl, the muffin-tin radii (R_{MT}) of the Co, Cr and Al atoms were taken to be 2.1, 2.05 and 1.95 a.u., respectively, and for the compound Cr_2MnSb, the R_{MT} of Cr, Mn and Sb atoms were 2.14, 2.2 and 2.2 a.u., respectively. Moreover, 35 special k-points in the irreducible Brillion Zone (IBZ) with a grid size of $10 \times 10 \times 10$ (equal to 1000 k-points in the Full Brillion Zone (FBZ)) [30] were employed in obtaining self-consistency calculations for the Co_2CrAl and Cr_2MnSb compounds. In addition, the plane waves quantity was limited as $K_{MAX} \times R_{MT} = 8$, and the wave functions' expansions was set by l = 10 inside the muffin-tin spheres. Furthermore, the self-consistent computations were only perceived as well-converged when the computed aggregate crystal energy converged to lower than 10^{-5} Ry. In addition, the cubic phase's elastic constants were computed using the second-order derivatives within the WIEN2k-code contained formalism.

3. Results and Discussion

3.1. Structural Properties

By fitting the total energy to Murnaghan's equation of state (EOS) [31], the optimized lattice constant (a), bulk modulus (B) and its pressure derivative (B') were computed as given below:

$$E(v) = E_0 + \frac{BV}{B'}\left[\frac{\left(\frac{V_0}{V}\right)^{B'}}{B'-1}+1\right] - \frac{B'V_0}{B'-1} \quad (1)$$

where B represents the bulk modulus at the equilibrium volume, B' is the pressure derivative of the bulk modulus at the equilibrium volume and E_0 is the minimum energy. The bulk modulus (B) and the pressure (P) are given by $B = -V\frac{dP}{dV} = V\frac{d^2E}{dV^2}$ and $P = -\frac{dE}{dV}$.

The conventional Heusler Co_2CrAl and Cr_2MnSb compounds had a space group Fm-3m L21 (225) and the inverse Heusler Co_2CrAl and Cr_2MnSb compounds had a space group F-43m Xa (216) [5], while tetragonal crystal lattices were the result of stretching of the cubic lattice along with one of its vectors. This resulted in the cube taking the shape of a rectangular prism whose base was a square (a by a), and the height (c) was different from the base edge a. Therefore, the tetragonal Heusler Co_2CrAl and Cr_2MnSb compounds had space groups of I4/mmm (139) and I-4m2 (119). The full-Heusler structural properties of the Co_2CrAl and Cr_2MnSb compounds were calculated. Figure 1 presents the crystal structures of the full-Heusler Co_2CrAl and Cr_2MnSb compounds. The aggregate energy as a function of the volume for the Heusler Co_2CrAl and Cr_2MnSb compounds are presented in Figures 2 and 3. Moreover, the state (EOS) was used to compute the optimized structural parameters, presented in Tables 1 and 2.

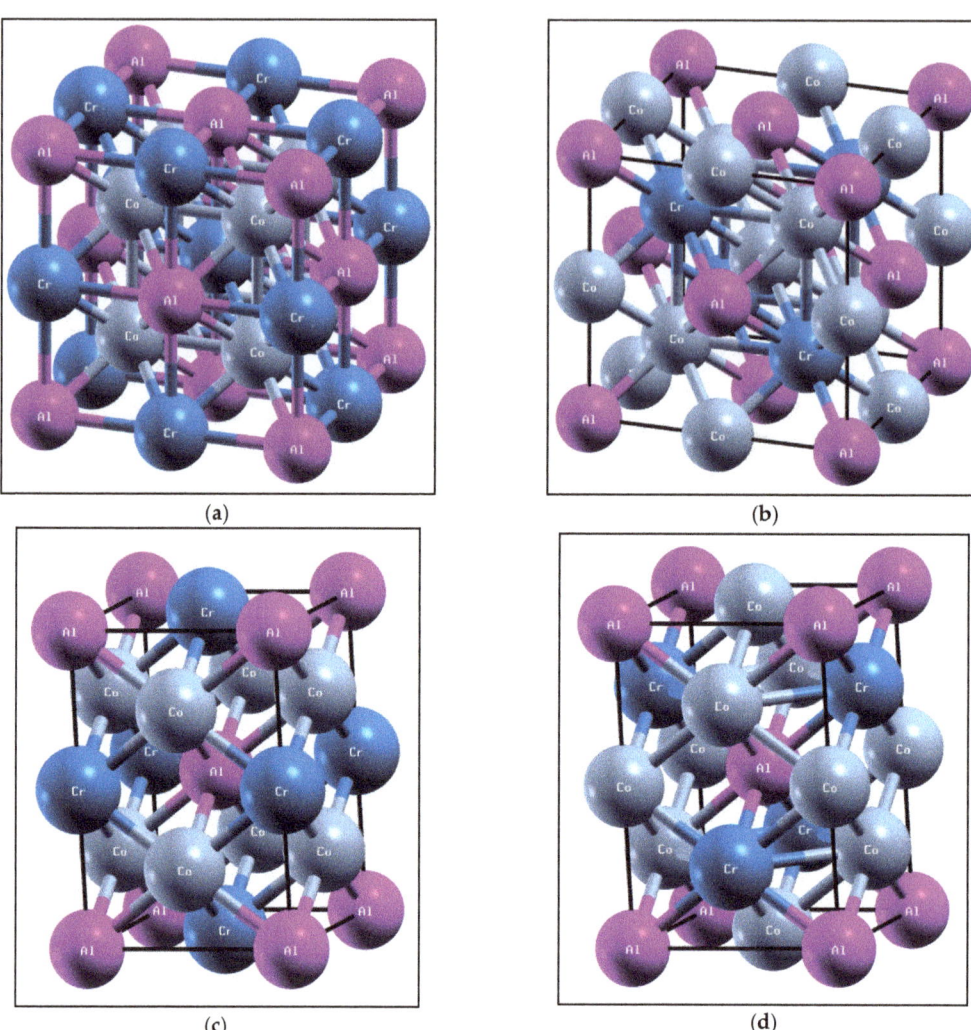

Figure 1. Different crystal structures of Heusler Co_2CrAl. (**a**) Conventional Heusler structure Co_2CrAl Fm-3m L21 (225), (**b**) inverse Heusler Co_2CrAl structure F-43m X (216), (**c**) tetragonal structure I4/mmm (139) and (**d**) tetragonal structure I-4m2 (119).

Tables 1 and 2 show our computed lattice parameters compared with other theoretical and experimental lattice parameters of conventional and inverse Heusler Co_2CrAl and Cr_2MnSb compounds. The calculated lattice parameters for the conventional Heusler Co_2CrAl compound deviated from the measured one within 0.38% [19]. The calculated lattice parameters for the conventional and inverse Heusler Cr_2MnSb compounds perfectly agreed with the theoretical outcomes [21]. As far as we know, comparable experimental results for conventional and inverse Heusler Cr_2MnSb compounds are not available. These results ensured the reliability of the present first-principle computations.

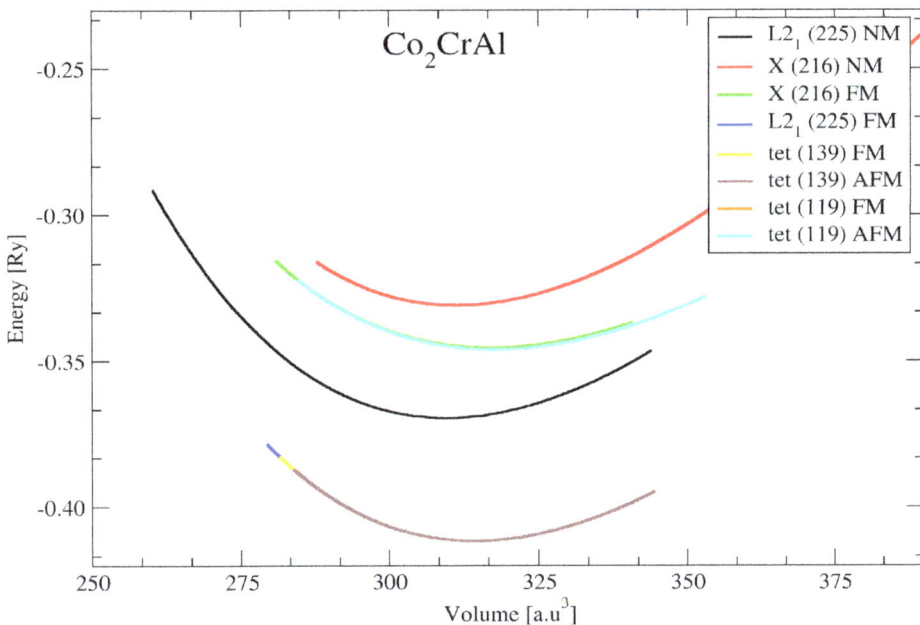

Figure 2. The total energy (Ry) versus volume (a.u.^{3}) for different crystal structures of Heusler Co_2CrAl.

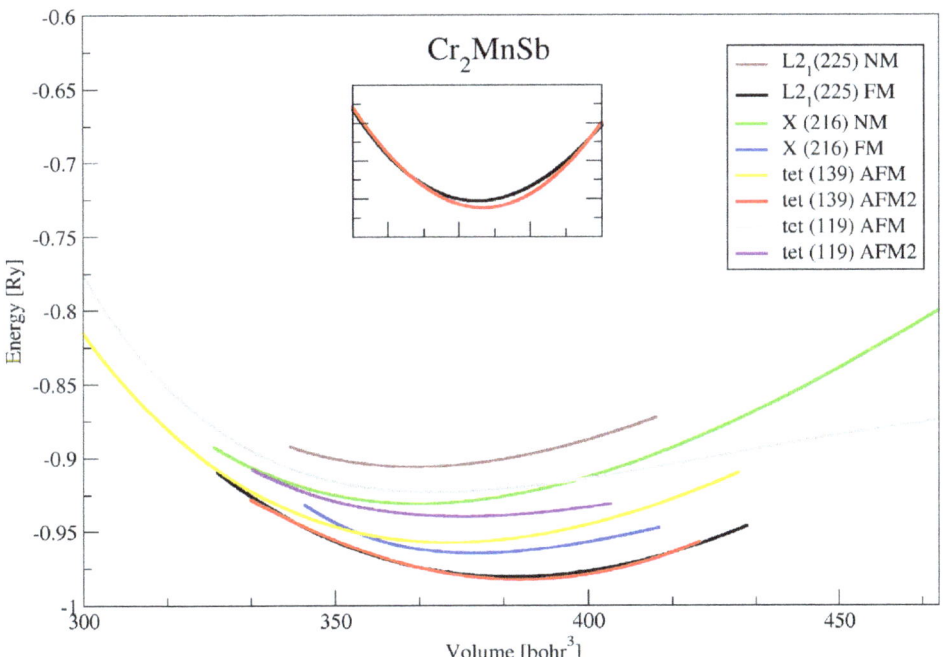

Figure 3. The total energy (Ry) versus volume (a.u.^{3}) for different crystal structures of Heusler Cr_2MnSb.

Table 1. Calculated lattice parameter (a), bulk modulus (B), and total energy (E_{tot}) for Heusler Co_2CrAl compound.

Structure	Space Group	Magnetic Phase	Reference	Lattice Parameter a (Å)	B (GPa)	E_{total} (Ry)/f.u	c/a
Co_2CrAl	Conventional Fm-3m (225)	NM	Present	5.6830	213.468	−8161.3694	1
	Conventional Fm-3m (225)	FM	Present	5.7082	206.811 [19]	−8161.4115 [19]	1
			Experimental	5.74 [19] 5.70 [26]			
			Theoretical	5.73 [20]			
	Inverse F-43m (216)	NM	Present	5.6936	212.056	−8161.3309	1
	Inverse F-43m (216)	FM	Present	5.7398	169.0397	−8161.3454	1
	I4/mmm (139)	FM	Present	4.011	202.24	−8161.41148	1.4175
	I4/mmm (139)	AFM	Present	4.0239	202.133	−8161.4095	1.4175
	I-4m2 (119)	FM	Present	3.9382	166.847	−8161.34608	1.5253
	I-4m2 (119)	AFM	Present	3.9441	163.781	−8161.3458	1.5253

Table 2. Calculated lattice parameter (a), bulk modulus (B), and total energy (E_{tot}) for Heusler Cr_2MnSb compound.

Structure	Space Group	Magnetic Phase	Reference	Lattice Parameter a (Å)	B (GPa)	E_{total} (Ry)/f.u	c/a
Cr_2MnSb	Conventional Fm-3m (225)	NM	Present	6.1116	201.77	−19,487.9803	1
	Conventional Fm-3m (225)	FM	Present	6.1381	248.45	−19,487.98197	1
			Theoretical	6.0 [21]			
			Experimental	5.95 [25]			
	Inverse F-43m (216)	NM	Present	6.0724	220.322	−19,487.9644	1
	Inverse F-43m (216)	FM	Present	6.0571	149.5381	−19,487.9683	1
			Theoretical Result	5.9 [21]			
	I4/mmm (139)	FM	Present	4.3337	296.852	−19,487.982	1.4158
	I4/mmm (139)	AFM	Present	4.2637	386.728	−19,487.9823	1.4158
	I-4m2 (119)	FM	Present	4.0617	142.459	−19,487.9394	1.6513
	I-4m2 (119)	AFM	Present	4.0459	153.8705	−19,487.9394	1.6513

According to the results obtained in this study, our volume optimization results showed that AFM tetragonal distortion (No. 139) was more preferred than FM cubic $L2_1$ for the Cr_2MnSb compound with a slightly small energy difference $\Delta E_{tet-cubic} = 0.0047 \text{R}\frac{\text{Ry}}{\text{f.u}}$ (see Equation (2)). On the other hand, FM cubic $L2_1$ was more preferred than tetragonal distortion for the Co_2CrAl case, with an energy difference $\Delta E_{tet-cubic} = 0.002 \frac{\text{Ry}}{\text{f.u}}$. To make the AFM tetragonal phase stable, the energy difference with a cubic structure should be greater than 0.1 eV/f.u. As reported previously, Cr_2MnSb crystallizes in a cubic $L2_1$ structure with a fully compensated ferrimagnetic configuration, where the magnetic moment of Cr and Mn are dominated by antiparallel exchange [21].

$$\Delta E_{tet-cubic} = E_{tet} - E_{cubic} \qquad (2)$$

3.2. Magnetic Properties

This part involved the calculation of the inverse, conventional and tetragonal I4/mmm (139) Heusler Co_2CrAl and Cr_2MnSb compounds' partial and total magnetic moments. The results obtained were compared with other theoretical values as shown in Tables 3 and 4.

Table 3. Total magnetic moment for inverse, conventional and tetragonal I4/mmm (139) Heusler Co_2CrAl compound.

Compounds		Magnetic Moment in μ_B					
		Co	Co	Cr	Al	Interstitial	Total Magnetic Moment (M^{tot}) in μ_B
Inverse Co_2CrAl	Present	0.96069	1.36311	−1.26906	−0.00841	−0.21497	0.83116
Conventional Co_2CrAl	Present	1.01815	1.01815	1.32570	−0.06082	−0.30118	3
	Theoretical Result	0.650 [19]	0.650 [19]	1.745 [19]	−0.045 [19]	–	3 [19]
	Theoretical Result	–	–	–	–	–	2.96 [22]
Tetragonal I4/mmm (139) Co_2CrAl	Present	−0.04084	−0.04084	0.81717	1.45741	−0.05097	2.99994

Table 4. Total magnetic moment for inverse, conventional and tetragonal I4/mmm (139) Heusler Cr_2MnSb compound.

Compounds		Magnetic Moment in μ_B					
		Cr	Cr	Mn	Sb	Interstitial	Total Magnetic Moment (M^{tot}) in μ_B
Inverse Cr_2MnSb	Present	−1.72053	2.68899	−1.05810	0.04777	0.04187	0
	Theoretical Result	1.96 [21]	−3.18 [21]	1.29 [21]	-	-	0 [21]
Conventional Cr_2MnSb	Present	−1.51854	−1.51854	3.21064	0.06167	−0.23513	0.00011
	Theoretical Result	1.77 [21]	1.77 [21]	−3.44 [21]	-	-	0.01 [21]
Tetragonal I4/mmm (139) Cr_2MnSb	Present	0.05294	0.05294	−1.46141	3.08253	−0.20953	0.00312

We found that the conventional and tetragonal Heusler Co_2CrAl compounds were ferromagnetic compounds. Furthermore, the total magnetic moment for the inverse Co_2CrAl compound was $M^{tot} = 0.83116$ μ_B, while it was $M^{tot} = 3$ μ_B for the conventional Co_2CrAl compound. The physics interpretation behind this huge difference between the total spin magnetic moment of the conventional and inverse Co_2CrAl was due to antiparallel exchange interactions between the Cr atom and Co atom in the case of inverse Co_2CrAl, whereas it was a direct interaction in the case of the conventional phase. Therefore, it can be noted from the results produced here that the conventional Co_2CrAl compound's computed total magnetic moment perfectly matched with the prior theoretical results [19,22], as Table 3 shows. Theoretically, a compound with a total magnetic moment M^{tot} with an integer value means it is a half-metallic material.

Table 4 shows the results for the inverse, conventional and tetragonal I4/mmm (139) Heusler Cr_2MnSb compounds. Table 4 shows that the inverse Heusler Cr_2MnSb had a negative spin moment for one Cr atom and one Mn atom and a positive spin moment for the second Cr atom. The conventional Heusler Cr_2MnSb had a negative spin moment for its Cr atoms and a positive spin moment for its Mn atom. The Sb atom's spin moment was extremely tiny in both structural types of Cr_2MnSb. The electronic configurations in the Mn and Cr atoms were similar, and they had one electron difference. Consequently, their exchange maintained the compound's ferromagnetic character, which led to small variations in the spin moments per site.

We found that the conventional Heusler Cr_2MnSb had a small total magnetic moment (non-zero total magnetization) due to the decrease in the atomic disorder in the Mn–Sb sublattice. This implied that the conventional Heusler Cr_2MnSb compound had a ferrimagnetic order.

On the other hand, we found that the inverse Heusler Cr_2MnSb had a zero total magnetic moment, which meant that this compound had an antiferromagnetic magnetic order. The tetragonal Cr_2MnSb had a small total magnetic moment, which meant that this compound was ferrimagnetic.

3.3. Electronic Properties

In this section, the partial and total density of states and the band structure for the inverse and conventional Heusler (Co_2CrAl, Cr_2MnSb) compounds were investigated. An analysis of the density of states and band structure showed that the conventional Co_2CrAl, conventional Cr_2MnSb and inverse Cr_2MnSb Heusler compounds exhibited a half-metallic conduct in a ferromagnetic state. This implied that the spin-up electrons in the materials had a metallic behavior; when they behaved as semiconducting with a spin-down direction, the materials had a semiconducting behavior. On the other hand, the inverse Heusler Co_2CrAl compound had a metallic behavior. The tetragonal I4/mmm (139) Heusler (Co_2CrAl, Cr_2MnSb) compounds had a half-metallic behavior in the antiferromagnetic state.

Figure 4a,b shows the metallic behavior of the spin up and spin down band structures within the PBE-GGA method for the inverse Heusler Co_2CrAl compound with zero energy gap. Figure 5a,b also shows the metallic behavior of the spin-up and spin-down band structures within the mBJ-GGA method for the inverse Heusler Co_2CrAl compound with zero energy gap.

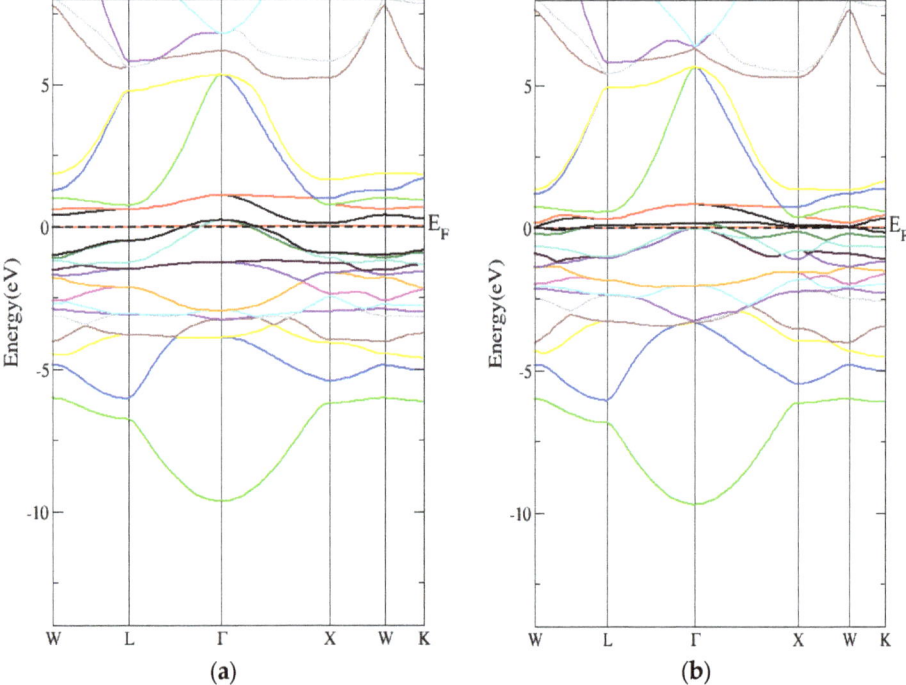

Figure 4. The band structure for the inverse Heusler Co_2CrAl compound by employing the PBE-GGA technique for (**a**) spin-up inverse Heusler Co_2CrAl compound and (**b**) spin-down inverse Heusler Co_2CrAl compound.

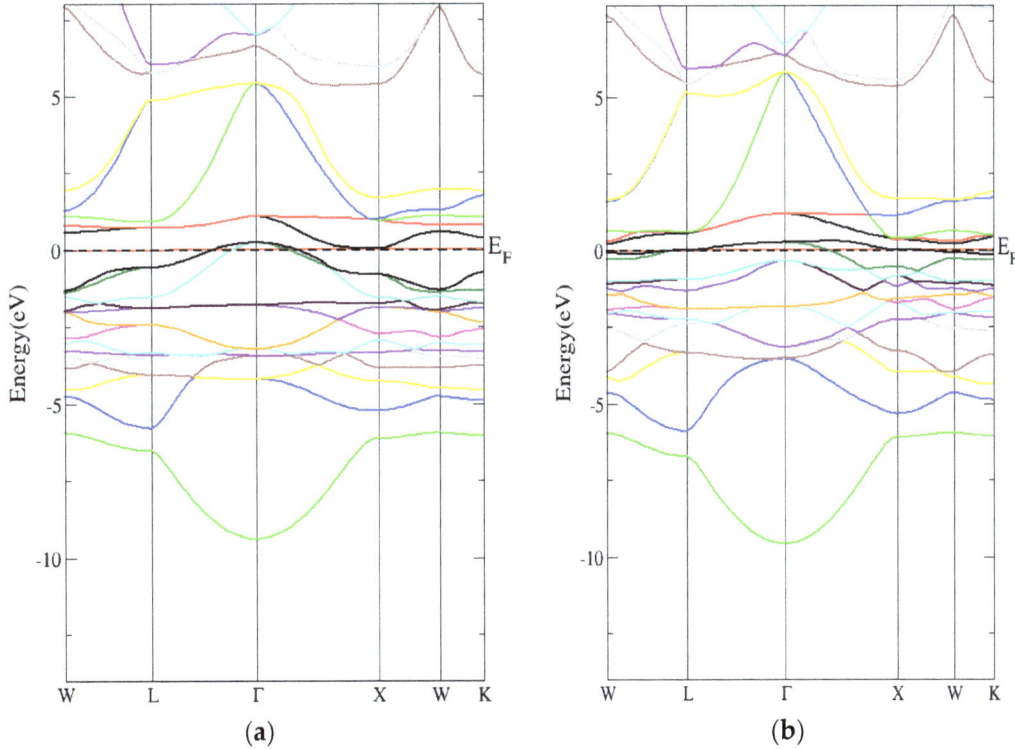

Figure 5. The band structure for the inverse Heusler Co_2CrAl compound by employing the mBJ-GGA technique for (**a**) spin-up inverse Heusler Co_2CrAl compound (**b**) spin-down inverse Heusler Co_2CrAl compound.

From Figure 6a,b, the conventional Heusler Co_2CrAl compound's band structure had an indirect energy band gap (spin down) using the PBE-GGA technique. In addition, Figure 7a,b shows that the conventional Heusler Co_2CrAl compound's band structure had an indirect energy band gap (spin-down) when using the mBJ-GGA technique. As indicated in Table 5, the indirect energy gaps within PBE-GGA and mBJ-GGA were 0.6 eV and 0.9 eV, respectively.

Table 5. The energy band gaps for conventional and inverse Co_2CrAl compound using PBE-GGA and mBJ methods.

Compounds	Band Gap Type	High Symmetry Lines	E_g-PBE-GGA (eV)	E_g-mBJ-GGA (eV)
Conventional $-Co_2CrAl$	Indirect	$\Gamma - X$	0.6	0.9
Inverse $- Co_2CrAl$	Metallic	-	-	-

From Figure 8a,b, the band structure (spin-down) of the inverse Heusler Cr_2MnSb compound had a direct energy band gap using the PBE-GGA technique. In addition, Figure 9a,b shows that the inverse Heusler Cr_2MnSb compound's band structure (spin-down) had a direct energy band gap when the mBJ-GGA technique was used. As can be seen in Table 6, the direct energy gaps within the PBE-GGA and mBJ-GGA methods were 0.8 eV and 0.9 eV, respectively.

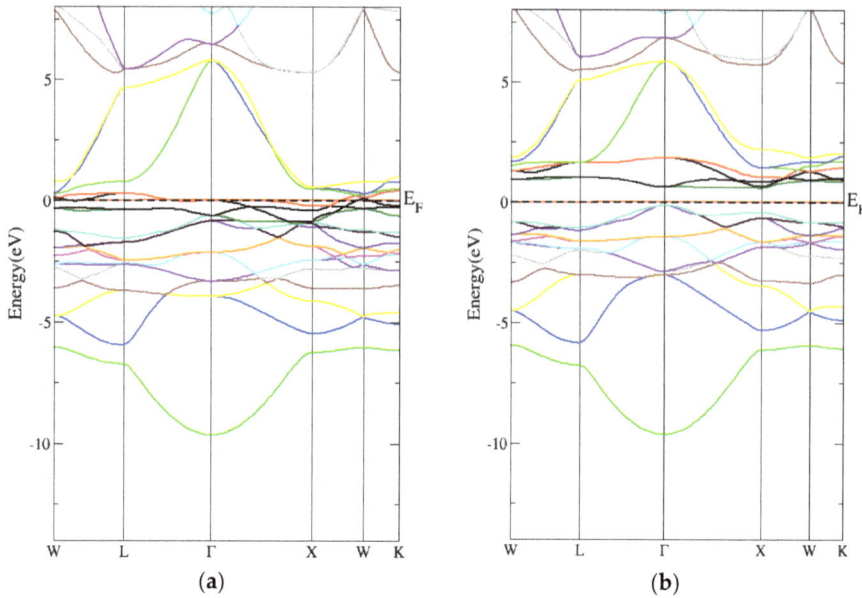

Figure 6. The band structure for the conventional Heusler Co_2CrAl compound by employing the PBE-GGA technique for (**a**) spin-up conventional Heusler Co_2CrAl compound and (**b**) spin-down conventional Heusler Co_2CrAl compound.

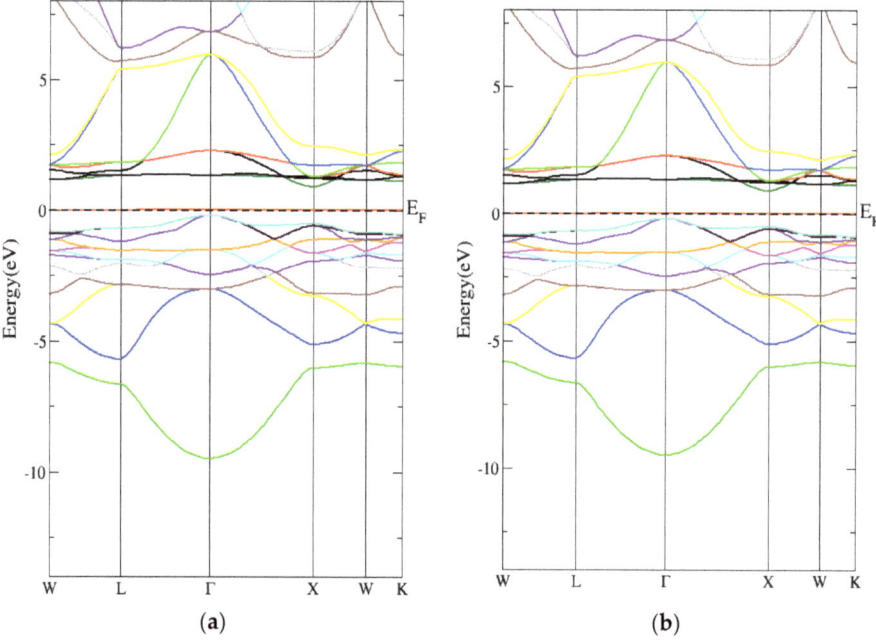

Figure 7. The band structure for the conventional Heusler Co_2CrAl compound by employing the mBJ-GGA technique for (**a**) spin-up conventional Heusler Co_2CrAl compound and (**b**) spin-down conventional Heusler Co_2CrAl compound.

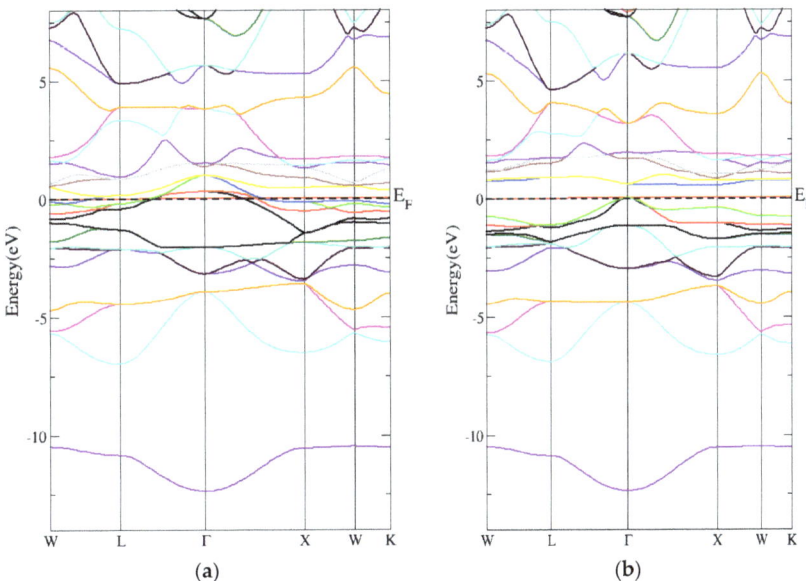

Figure 8. The band structure for the inverse Heusler Cr_2MnSb compound by employing the PBE-GGA technique for (**a**) spin-up inverse Heusler Cr_2MnSb compound and (**b**) spin-down inverse Heusler Cr_2MnSb compound.

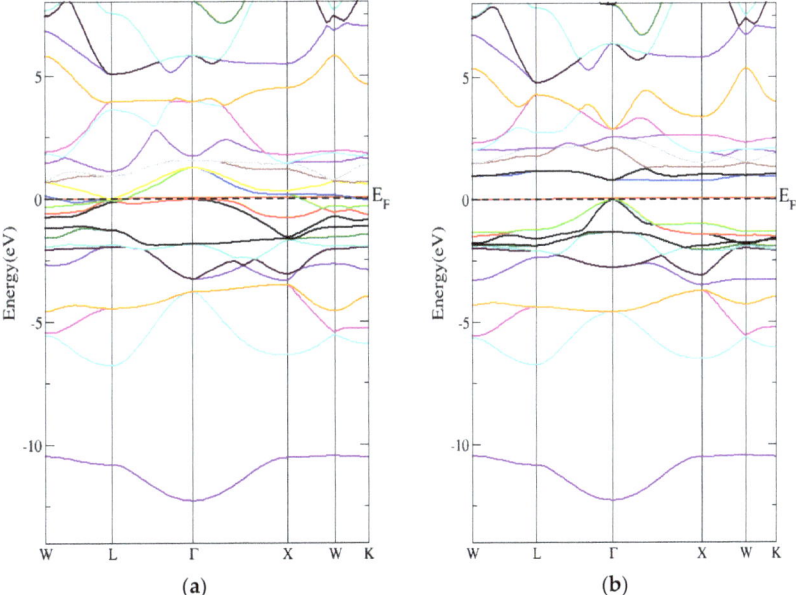

Figure 9. The band structure for the inverse Heusler Cr_2MnSb compound by employing the mBJ-GGA technique for (**a**) spin-up inverse Heusler Cr_2MnSb compound and (**b**) spin-down inverse Heusler Cr_2MnSb compound.

Table 6. The energy band gaps for conventional and inverse Cr$_2$MnSb compound using PBE-GGA and mBJ methods.

Compounds	Band Gap Type	High Symmetry Lines	E_g-PBE-GGA (eV)	E_g-mBJ-GGA (eV)
Conventional-Cr$_2$MnSb	Direct	Γ	0.9	1
Inverse-Cr$_2$MnSb	Direct	Γ	0.8	0.9

Figure 10a,b shows that the band structure of the conventional Heusler Cr$_2$MnSb compound had a direct energy band gap (spin-down) using the PBE-GGA technique. In addition, Figure 11a,b shows that the Heusler Cr$_2$MnSb compound's band structure had a direct energy band gap (spin-down) using the mBJ-GGA technique. As indicated in Table 6, the indirect energy gaps within PBE-GGA and mBJ-GGA methods were 0.9 eV and 1 eV, respectively.

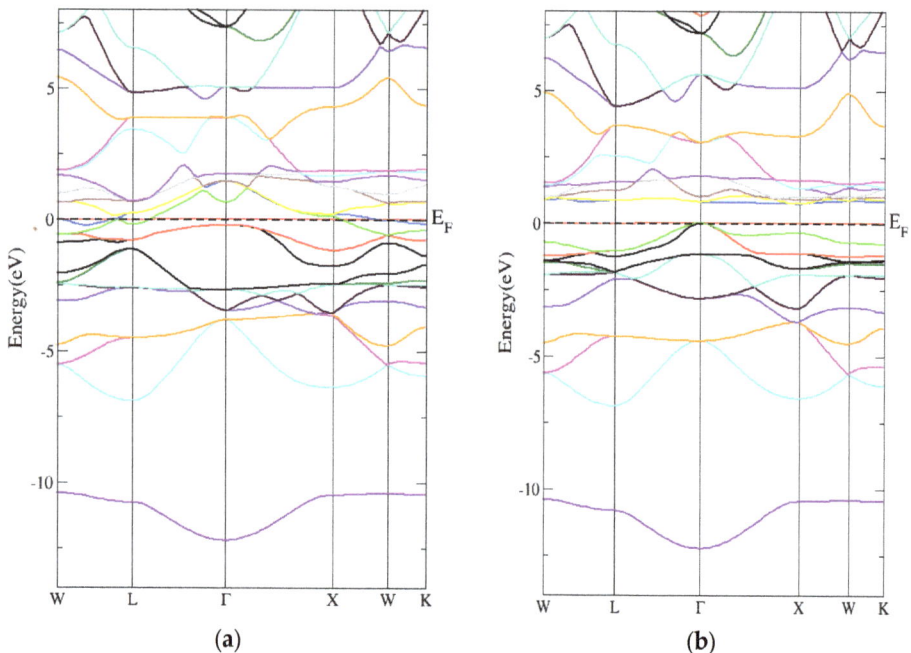

Figure 10. The band structure for the conventional Heusler Cr$_2$MnSb compound by employing the PBE-GGA technique for (**a**) spin-up conventional Heusler Cr$_2$MnSb compound and (**b**) spin-down conventional Heusler Cr$_2$MnSb compound.

Figure 12a,b presents the Heusler Co$_2$CrAl compound band structure for tetragonal I4/mmm (139) in the AFM state. Figure 12a shows that the tetragonal I4/mmm (139) Heusler Co$_2$CrAl compound's spin-up band structure had a metallic character, while Figure 12b indicates that the tetragonal I4/mmm (139) Heusler Co$_2$CrAl compound's spin-down band structure had a direct energy band gap. The direct energy gap was found to be 0.8 eV, as shown in Table 7. Figure 13a,b shows the tetragonal I4/mmm (139) Heusler Cr$_2$MnSb compound's band structure in the AFM state. Figure 13a indicates that the tetragonal I4/mmm (139) Heusler Cr$_2$MnSb compound's spin-up band structure had a metallic nature, while Figure 13b illustrates that the tetragonal I4/mmm (139) Heusler Cr$_2$MnSb compound's spin-down band structure had a direct energy band gap. The direct energy gap was found to be 0.9 eV, as shown in Table 7.

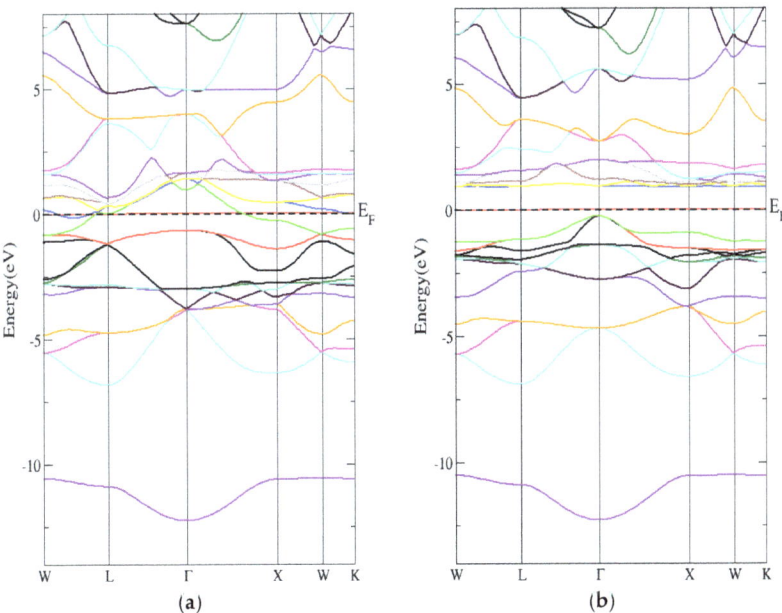

Figure 11. The band structure for the conventional Heusler Cr_2MnSb compound by employing the mBJ-GGA technique for (**a**) spin-up conventional Heusler Cr_2MnSb compound and (**b**) spin-down conventional Heusler Cr_2MnSb compound.

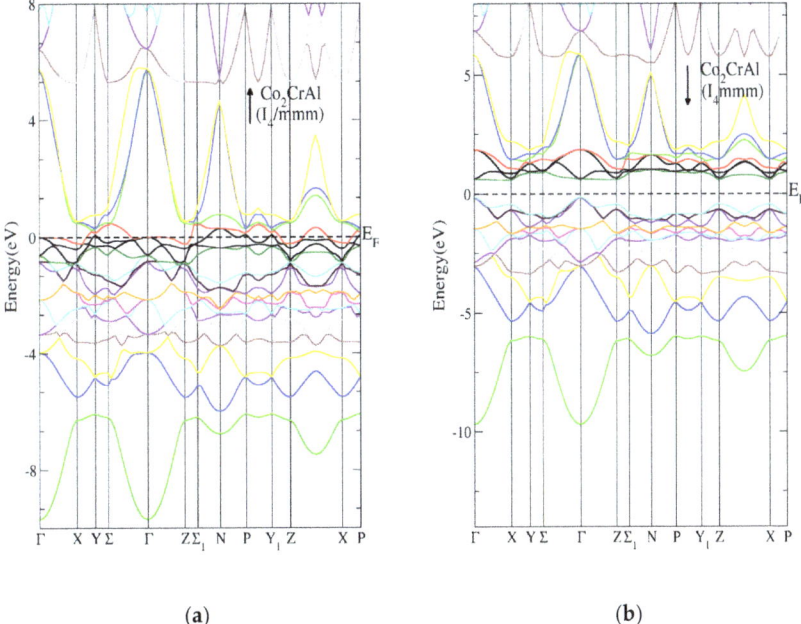

Figure 12. The band structure for the tetragonal I4/mmm (139) Heusler Co_2CrAl compound in AFM state. (**a**) Spin-up tetragonal I4/mmm Heusler (139) Co_2CrAl compound and (**b**) spin-down tetragonal I4/mmm Heusler (139) Co_2CrAl compound.

Table 7. The energy band gaps for the tetragonal I4/mmm (139) Heusler Co_2CrAl and Cr_2MnSb compounds in AFM state.

Compounds	Band Gap Type	High Symmetry Lines	E_g(eV)
Co_2CrAl	Direct	Γ	0.8
Cr_2MnSb	Direct	Γ	0.9

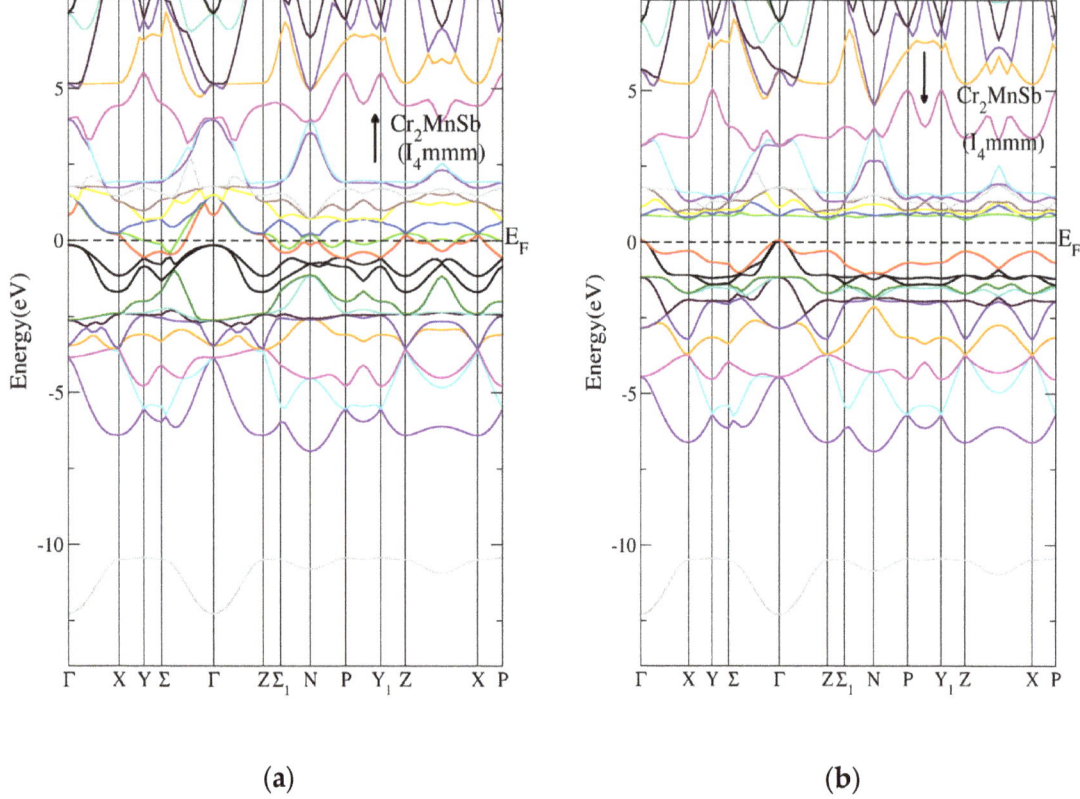

Figure 13. The band structure for the tetragonal I4/mmm (139) Heusler Cr_2MnSb compound in AFM state. (**a**) Spin-up tetragonal I4/mmm Heusler (139) Cr_2MnSb compound and (**b**) spin-down tetragonal I4/mmm Heusler (139) Cr_2MnSb compound.

For the conventional and inverse Heusler Co_2CrAl and Cr_2MnSb compounds, the partial and total density of states for the spin-down, spin-up and inverse Heusler Co_2CrAl and Cr_2MnSb compounds are presented in Figures 14–17. The density of states in Figures 14–17 also show half-metallic behaviors for the conventional Heusler Co_2CrAl and the inverse and conventional Heusler Cr_2MnSb compounds with a minor energy band gap in the spin-down segment. This implied that the behavior of these compounds was half-metallic.

In the conventional Co_2CrAl's (Figure 15) spin-down segment, the valence band resulted from the d-state of Co, the d-state of Cr and the tiny effect of the Al in the s-state. The d-state of Co, the d-state of Cr and the tiny effect of the Al in the s-state were accredited to the conduction band. In the spin-down channel of the conventional Co_2CrAl, the valence band was attributed to the d-state of Co, the d-state of Cr and the minor effect of Al in the s-state. On the other hand, the conduction band resulted from the d-state of Co, the d-state of Cr and the minor effect of the Al in the s-state.

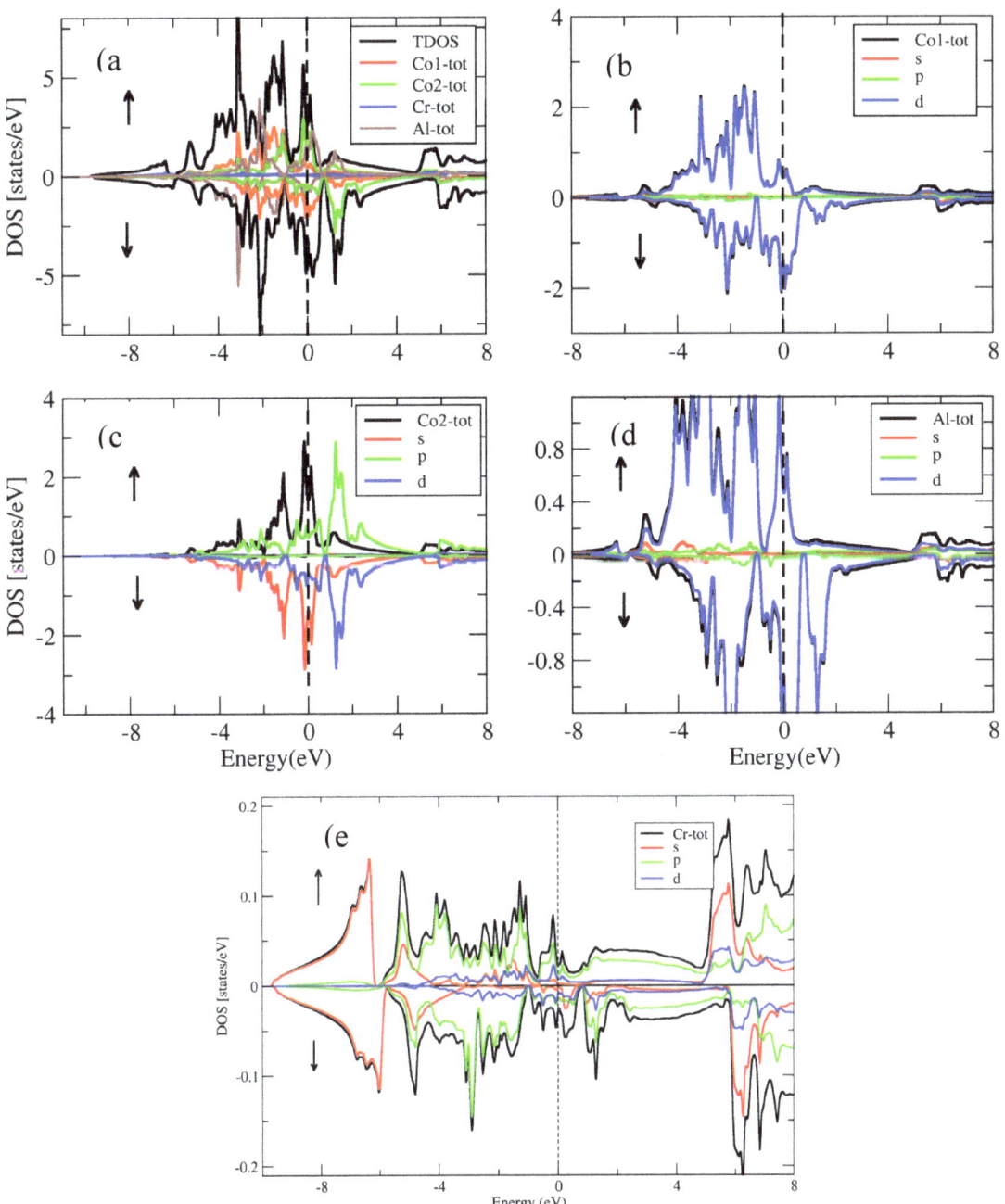

Figure 14. (a) Total density of states for the inverse Co_2CrAl compound and partial density of states for (b) Co1 atom, (c) Co2 atom, (d) Al atom and (e) Cr atom.

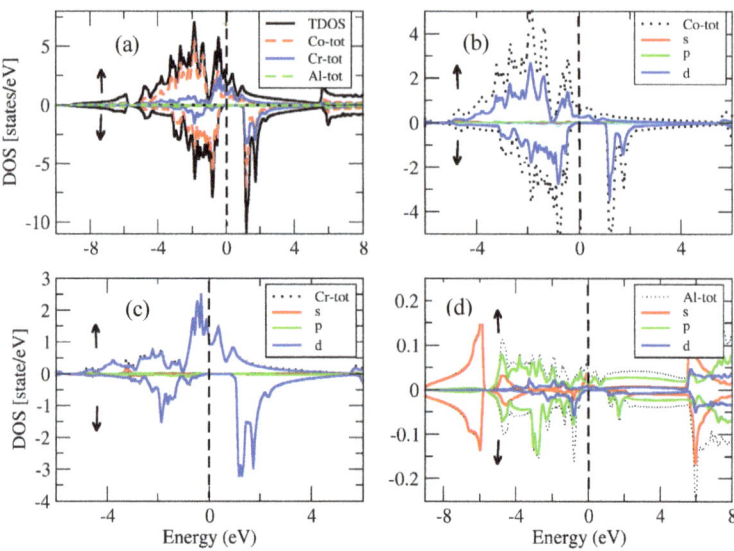

Figure 15. (**a**) Total density of states for the conventional Co_2CrAl compound and the partial density of states for (**b**) Co atom, (**c**) Cr atom and (**d**) Al atom.

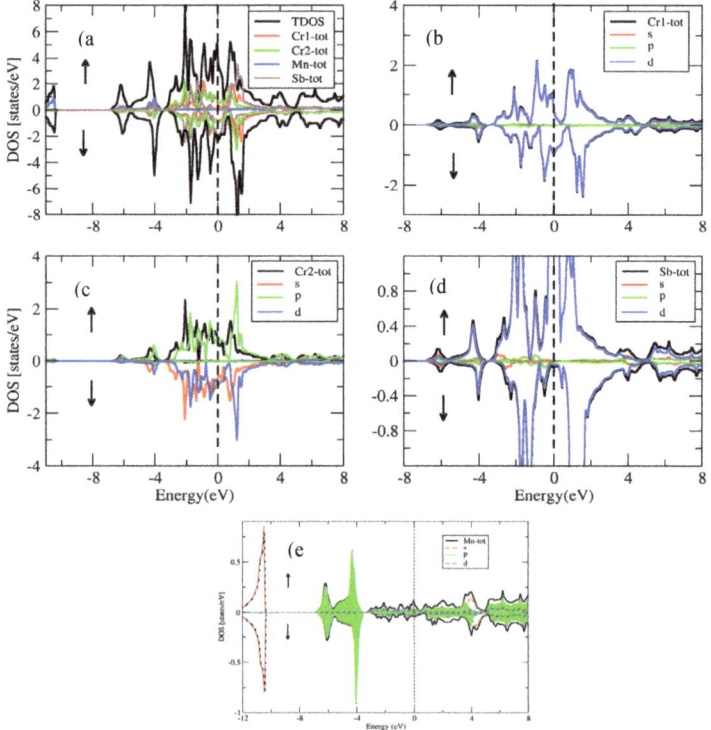

Figure 16. (**a**) Total density of states for the inverse Cr_2MnSb compound and partial density of states for (**b**) Cr1 atom, (**c**) Cr2 atom, (**d**) Sb atom and (**e**) Mn atom.

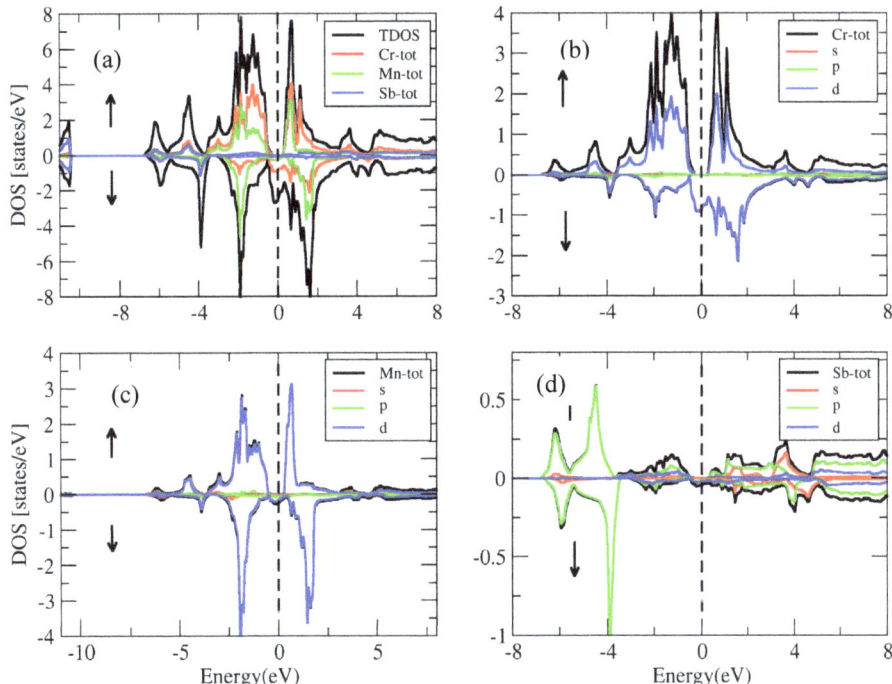

Figure 17. (a) Total density of states for the conventional Cr_2MnSb compound and the partial density of states for (b) Cr atom, (c) Mn atom and (d) Sb atom.

In the spin-up sector of the inverse Cr_2MnSb, (Figure 16), the valence band was attributed to the d-state of Cr, the d-state of Mn, the s-state and the p-state of Sb. On the other hand, the conduction band resulted from the d-state of Cr, the d-state of Mn, the minor effect of the s-state and Sb's p-state. In the spin-down channel of the inverse Cr_2MnSb, the valence band was attributed to the minor contribution of the d-state of Cr, the d-state of Mn, the s-state and the p-state of Sb. Meanwhile, the conduction band was attributed to the d-state of Cr, the d-state of Mn, the minor contribution of the s-state and the p-state of Sb.

In the conventional Cr_2MnSb (Figure 17) spin-up sector, the valence band was attributed to the d-state of Cr, the d-state of Mn, the s-state and the p-state of Sb. On the other hand, the conduction band was attributed to the d-state of Cr, the minor effect of the s-state and the p-state of Sb. In the spin-down channel of the conventional Cr_2MnSb, the valence band was attributed to the d-state of Cr, the d-state of Mn, the s-state and the p-state of Sb, while the conduction band was the result of the d-state of Cr, the minor effect of the s-state and the p-state of Sb.

3.4. Elastic Properties

This part presents the computation of the bulk modulus (B), the shear modulus (S), the elastic constants (C_{ij}), the B/S ratio, Poisson's ratio, Young's modulus (Y) and the anisotropic factor (**A**) of the inverse and conventional Heusler Cr_2MnSb and Co_2CrAl compounds. The standard mechanical stability condition or cubic crystal [32] was $C_{11} > 0$, $C_{11} + 2C_{12} > 0$, $C_{11} - C_{12} > 0$ and $C_{44} > 0$.

Table 8 presents our calculations for the inverse and conventional Heusler Cr_2MnSb and Co_2CrAl compounds. We concluded that the inverse and conventional Heusler Co_2CrAl were mechanically stable. The inverse Cr_2MnSb was found to be mechanically unstable in the ferromagnetic state. On the other hand, the conventional Cr_2MnSb was mechanically stable in the ferromagnetic state.

Table 8. Reuss's bulk modulus (B), shear modulus (S), elastic constants (C_{ij}), B/S ratio, Poisson's ratio (v), Young's modulus (Y) and anisotropic factor (A) of the FM conventional and inverse Heusler (Co_2CrAl, Cr_2MnSb) compounds.

Materials	C_{11} (GPa)	C_{12} (GPa)	C_{44} (GPa)	B (GPa)	S (GPa)	B/S	Y (GPa)	v	A
Conventional Co_2CrAl	268.592	169.462	154.343	202.505	83.628	2.422	220.528	0.319	3.114
Inverse Co_2CrAl	275.6864	234.2043	138.9022	248.032	42.364	5.855	120.246	0.470	6.697
Inverse Cr_2MnSb	197.1538	138.7903	−488.5573	158.245	80.134	1.975	205.683	0.283	−16.7
Conventional Cr_2MnSb	267.3836	224.7424	100.9041	238.956	40.474	5.904	114.933	0.420	4.732

We used the Reuss approximation [33] to calculate the bulk and shear modulus. The following equation can be used to calculate the Reuss shear modulus S_R:

$$S_R = \frac{5C_{44}(C_{11} - C_{12})}{4C_{44} + 3(C_{11} - C_{12})} \tag{3}$$

The following equation gives the cubic structure's bulk modulus:

$$B = \frac{1}{3}(C_{11} + 2C_{12}) \tag{4}$$

The Young modulus (Y) is given by the following:

$$Y = \frac{9BS_R}{(S_R + 3B)} \tag{5}$$

The anisotropic factor and Poisson's ratio are given by the following:

$$A = \frac{2C_{44}}{C_{11} - C_{12}} \tag{6}$$

$$v = \frac{3B - 2S_R}{2(3B + S_R)} \tag{7}$$

Reuss's bulk modulus (B), the shear modulus (S), the elastic constants (C_{ij}), the B/S ratio, Poisson's ratio (v), Young's modulus (Y) and the anisotropic factor (A) of the FM conventional and inverse Heusler Co_2CrAl and Cr_2MnSb compounds are shown in Table 8.

A material's hardness is measured by its shear modulus and bulk modulus [33]. Therefore, the ratio B/S measures a specific material's brittleness and ductility. A material is ductile when $B/S < 1.75$. Otherwise, it is brittle [34]. From the present calculations in Table 8, the B/S ratios of the inverse and conventional Heusler Co_2CrAl compounds were 5.855 and 2.422, respectively. Both the inverse and conventional Heusler compounds were ductile in nature, depending on the B/S ratio values. The B/S ratio values for the inverse and conventional Heusler Co_2CrAl, compounds were found to be 1.975 and 5.904, respectively. Both the inverse and conventional Heusler Cr_2MnSb compounds had a ductile character, depending on the B/S ratio values.

The stiffness of materials is measured using Young's modulus. Materials with a higher Young's modulus (Y) value are stiffer. Poisson's ratio can be employed for understanding the character of bonding and stability of a material. A Poisson's ratio value higher than

0.26 indicates that the material is ductile; otherwise, it is brittle [35]. From the present calculations summarized in Table 8, the Poisson's ratio values of the inverse and conventional Heusler Co_2CrAl compounds were 0.470 and 0.319, respectively. Depending on the Poisson's ratio values, both inverse and conventional Heusler compounds had a ductile nature. The Poisson's ratio values for the inverse and conventional Heusler Cr_2MnSb compounds were found to be 0.283 and 0.420, respectively. Depending on the Poisson's ratio values, both the inverse and conventional Heusler Cr_2MnSb compounds had a ductile nature. Poisson's ratio for compounds with covalent bonds is lower than 0.25, while for compounds with dominating ionic bonds, Poisson's ratio lies between 0.25 to 0.50. From Table 8, it appeared that the conventional and inverse Heusler Co_2CrAl and Cr_2MnSb compounds had prominent ionic bonds. In the same vein, elastic anisotropy is a crucial parameter for measuring the level of material's anisotropy [36]. The value of A is unity for an isotropic material. Otherwise, the elastic anisotropy of the material is elastic [37]. The present values of the anisotropy factor in Table 8 for the inverse and conventional Heusler Co_2CrAl and Cr_2MnSb compounds showed that these compounds were anisotropic elasticity.

4. Conclusions

This study focused on the elastic, magnetic, electronic and structural properties of inverse and conventional Heusler (Co_2CrAl, Cr_2MnSb) compounds. The results showed that the conventional Heusler Co_2CrAl, the conventional and inverse Heusler Cr_2MnSb and the tetragonal (139) Heusler Co_2CrAl and Cr_2MnSb compounds were half-metals. This half-metallic character is a promising characteristic of materials for spintronic applications. The indirect energy gap of the conventional Heusler Co_2CrAl compound was 0.6 eV within the PBE-GGA scheme. The energy band gap within the mBJ-GGA scheme for the conventional Heusler Co_2CrAl compound was computed to be 0.9 eV. Within the PBE-GGA technique, the inverse and conventional Heusler Cr_2MnSb compounds had a direct energy band gap of 0.8 eV and 0.9 eV, respectively. Within the mBJ-GGA method, the energy gaps for the inverse and conventional Heusler Cr_2MnSb compounds were 0.9 eV and 1 eV, respectively. The tetragonal Heusler Co_2CrAl and Cr_2MnSb compounds had direct energy band gaps and were 0.8 eV and 0.9 eV, respectively, within the PBE-GGA method. We discovered that the conventional Heusler Co_2CrAl compound was a ferromagnetic compound with a total magnetic moment of $M^{tot} = 3$ μ_B. On the other hand, the total magnetic moment for the inverse Co_2CrAl compound was $M^{tot} = 0.831$ μ_B. The conventional and tetragonal Heusler Cr_2MnSb compounds had a small total magnetic moment, which meant that these compounds were ferromagnetic. We found that the conventional and inverse Heusler Co_2CrAl compounds were mechanically stable. However, the inverse Cr_2MnSb compound was mechanically unstable in the ferromagnetic state. On the other hand, the conventional Cr_2MnSb is mechanically stable in the ferromagnetic state. The B/S values indicated that the inverse and conventional Heusler compounds for both Cr_2MnSb and Co_2CrAl had ductile characteristics. From the Poisson's ratio values, we found that the conventional and inverse Heusler Co_2CrAl and Cr_2MnSb compounds had dominant ionic bonds. Finally, the conventional and inverse Heusler Co_2CrAl and Cr_2MnSb compounds were anisotropy elastic.

Author Contributions: M.S.A.-J.: conceptualization, methodology, software, investigation, validation, visualization, formal analysis, writing—review & editing, supervision, project administration. S.J.Y.: data curation, methodology, formal analysis, writing—original draft, software. S.A.A.: data curation, formal analysis, writing—review & editing, software, investigation, validation. A.A.M.: data curation, methodology, formal analysis, writing—review & editing, software. D.A.-B.: data curation, methodology, formal analysis, software. M.F.: data curation, methodology, formal analysis, writing—review & editing, software. R.K.: methodology, formal analysis, writing—review & editing. All authors have read and agreed to the published version of the manuscript.

Funding: This research received no external funding.

Data Availability Statement: The data that support the findings of this study are available from the corresponding author upon reasonable request.

Conflicts of Interest: The authors declare that they have no known competing financial interests or personal relationships that could have appeared to influence the work reported in this paper.

References

1. de Groot, R.A.; Mueller, F.M.; van Engen, P.G.; Buschow, K.H.J. New Class of Materials: Half-Metallic Ferromagnets. *Phys. Rev. Lett.* **1983**, *50*, 2024. [CrossRef]
2. Blum, C.G.F.; Ouardi, S.; Fecher, G.H.; Balke, B.; Kozina, X.; Stryganyuk, G.; Ueda, S.; Kobayashi, K.; Felser, C.; Wurmehl, S.; et al. Exploring the details of the martensite–austenite phase transition of the shape memory Heusler compound Mn_2NiGa by hard X-ray photoelectron spectroscopy, magnetic and transport measurements. *Appl. Phys. Lett.* **2011**, *98*, 252501. [CrossRef]
3. Ouardi, S.; Fecher, G.H.; Balke, B.; Kozina, X.; Stryganyuk, G.; Felser, C.; Lowitzer, S.; Ködderitzsch, D.; Ebert, H.; Ikenaga, E. Electronic transport properties of electron- and hole-doped semiconducting C1b Heusler compounds: $NiTi_{1-x}M_xSn$ (M = Sc, V). *Phys. Rev. B* **2010**, *82*, 085108. [CrossRef]
4. Heusler, F. *Über magnetische Manganlegierungen. Verhandlungen der Deutschen Physikalischen Gesellschaft*; German Physical Society: Bad Honnef, Germany, 1903.
5. Ahmad, A.; Srivastava, S.K.; Das, A.K. Effect of L21 and XA ordering on phase stability, half-metallicity and magnetism of Co_2FeAl Heusler alloy: GGA and GGA+U approach. *J. Magn. Magn. Mater.* **2019**, *491*, 165635. [CrossRef]
6. Dowben, P. Half Metallic Ferromagnets. *J. Phys. Condens. Matter* **2007**, *19*, 310301. [CrossRef]
7. Dai, X.F.; Liu, G.D.; Chen, L.J.; Chen, J.L.; Xiao, G.; Wu, G.H. Mn_2CoSb compound: Structural, electronic, transport and magnetic properties. *Solid State Commun.* **2006**, *140*, 533. [CrossRef]
8. Liu, G.D.; Dai, X.F.; Liu, H.Y.; Chen, J.L.; Li, Y.X. Mn_2CoZ (Z = Al, Ga, In, Si, Ge, Sn, Sb) compounds: Structural, electronic, and magnetic properties. *Phys. Rev. B* **2008**, *77*, 014424. [CrossRef]
9. Bayar, E.; Kervan, N.; Kervan, S. Half-metallic ferrimagnetism in the Ti_2CoAl Heusler compound. *J. Magn. Magn. Mater.* **2011**, *323*, 2945. [CrossRef]
10. Fang, Q.L.; Zhang, J.M.; Xu, K.W.; Ji, V. Electronic structure and magnetism of Ti_2FeSi: A first-principles study. *J. Magn. Magn. Mater.* **2013**, *345*, 171. [CrossRef]
11. Kervan, N.; Kervan, S. A first-principle study of half-metallic ferrimagnetism in the Ti_2CoGa eusler compound. *J. Magn. Magn. Mater.* **2012**, *324*, 645. [CrossRef]
12. Jia, H.Y.; Dai, X.F.; Wang, L.Y.; Liu, R.; Wang, X.T.; Li, P.P.; Cui, Y.T.; Liu, G.D. Doping effect on electronic structures and band gap of inverse Heusler compound: Ti_2CoSn. *J. Magn. Magn. Mater.* **2014**, *367*, 33. [CrossRef]
13. Birsan, A.; Palade, P.; Kuncser, V. Prediction of half metallic properties in Ti_2CoSi Heusler alloy based on density functional theory. *J. Magn. Magn. Mater.* **2013**, *331*, 109. [CrossRef]
14. Yamamoto, M.; Marukame, T.; Ishikawa, T.; Matsuda, K.; Uemura, T.; Arita, M. Fabrication of fully epitaxial magnetic tunnel junctions using cobalt-based full- Heusler alloy thin film and their tunnel magnetoresistance characteristics. *J. Phys. D Appl. Phys.* **2006**, *39*, 824. [CrossRef]
15. Kaemmerer, S.; Thomas, A.; Hütten, A.; Reiss, G. Co_2MnSi Heusler alloy as magnetic electrodes in magnetic tunnel junctions. *Appl. Phys. Lett.* **2004**, *85*, 79. [CrossRef]
16. Okamura, S.; Miyazaki, A.; Sugimoto, S.; Tezuka, N.; Inomata, K. Large tunnel magnetoresistance at room temperature with a Co_2FeAl full-Heusler alloy electrode. *Appl. Phys. Lett.* **2005**, *86*, 232503. [CrossRef]
17. Sakuraba, Y.; Hattori, M.; Oogane, M.; Ando, Y.; Kato, H.; Sakuma, A.; Miyazaki, T.; Kubota, H. Electronic structure, magnetism and disorder in the Heusler compound Co_2TiSn. *Appl. Phys. Lett.* **2006**, *88*, 192508. [CrossRef]
18. Sakuraba, Y.; Miyakoshi, T.; Oogane, M.; Ando, Y.; Sakuma, A.; Miyazaki, T.; Kubota, H. Direct observation of half-metallic energy gap in Co_2MnSi by tunnelling conductance spectroscopy. *Appl. Phys. Lett.* **2006**, *89*, 052508. [CrossRef]
19. Zhang, M.; Liu, Z.; Hu, H.; Liu, G.; Cui, Y.; Chen, J.; Wu, G.; Zhang, X.; Xiao, G. Is Heusler compound Co_2CrAl a half-metallic ferromagnet: Electronic band structure, and transport properties. *J. Magn. Magn. Mater.* **2004**, *277*, 130. [CrossRef]
20. Hakimi, M.; Kameli, P.; Salamati, H. Structural and magnetic properties of Co_2CrAl Heusler alloys prepared by mechanical alloying. *J. Magn. Magn. Mater.* **2010**, *322*, 3443. [CrossRef]
21. Ozdogan, K.; Galanakis, I. First-principles electronic and magnetic properties of the half-metallic antiferromagnet Cr_2MnSb. *J. Magn. Magn. Mater.* **2009**, *321*, L34. [CrossRef]
22. Galanakis, I. Surface properties of the half- and full-Heusler alloys. *J. Phys. Condens. Matter* **2002**, *14*, 35615. [CrossRef]
23. Hirohata, A.; Huminiuc, T.; Sinclair, J.; Wu, H.; Samiepour, M.; Vallejo-Fernandez, G.; O'Grady, K.; Balluf, J.; Meinert, M.; Reiss, G. Development of antiferromagnetic Heusler alloys for the replacement of iridium as a critically raw material. *J. Phys. D Appl. Phys.* **2017**, *50*, 443001. [CrossRef]
24. Baker, D.N.A.; Abu-Jafar, M.S.; Mousa, A.A.; Jaradat, R.T.; Ilaiwi, K.F.; Khenata, R. Structural, magnetic, electronic and elastic properties of half-metallic ferromagnetism full-Heusler alloys: Normal-Co_2TiSn and inverse-Zr_2RhGa using FP-LAPW method. *Mater. Chem. Phys.* **2020**, *240*, 122122. [CrossRef]

25. Gupta, S.; Matsukura, F.; Ohno, H. Properties of sputtered full Heusler alloy Cr_2MnSb and its application in a magnetic tunnel junction. *J. Phys. D Appl. Phys.* **2019**, *52*, 495002. [CrossRef]
26. Dubowik, J.; Gościańska, I.; Kudryavtsev, Y.V.; Oksenenko, V.A. Structure and magnetism of Co_2CrAl Heusler alloy films. *Mater. Sci.* **2007**, *25*, 1281.
27. Paudel, R.; Zhu, J. Magnetism and half-metallicity in bulk and (100), (111)-surfaces of Co_2ScSb full Heusler alloy for spintronic applications. *Vacuum* **2019**, *169*, 108931. [CrossRef]
28. Paudel, R.; Zhu, J. Investigation of half-metallicity and magnetism of bulk and (111)-surfaces of Fe_2MnP full Heusler alloy. *Vacuum* **2019**, *164*, 336–342. [CrossRef]
29. Blaha, P.; Schwarz, K.; Tran, F.; Laskowski, R.; Madsen, G.K.H.; Marks, L.D. WIEN2k: An APW+lo program for calculating the properties of solids. *J. Chem. Phys.* **2020**, *152*, 074101. [CrossRef]
30. Monkhorst, H.J.; Pack, I.D. Special points for Brillouin-zone integrations. *Phys. Rev. B* **1976**, *13*, 5188. [CrossRef]
31. Murnaghan, F.D. The Compressibility of Media under Extreme Pressures. *Natl. Acad. Sci. USA* **1944**, *30*, 244. [CrossRef]
32. Born, M.; Huang, K. *Dynamical Theory of Crystal Lattices*; Clarendon Press: Oxford, UK, 1954.
33. Reuss, A. Berechnung der Fließgrenze von Mischkristallen auf Grund der Plastizitätsbedingung für Einkristalle. *Z. Angew. Math. Mech.* **1929**, *9*, 49. [CrossRef]
34. Pugh, S.F. Relations between the elastic moduli and the plastic properties of polycrystalline pure metals. *Philos. Mag.* **1954**, *45*, 823. [CrossRef]
35. Frantsevich, I.N.; Voronov, F.F.; Bokuta, S.A. *Elastic Constants and Elastic Moduli of Metals and Insulators Handbook*; Naukova Dumka: Kiev, Ukraine, 1983; p. 60.
36. Zener, C. Elasticity and Inelasticity of Metals. University of Chicago Press: Chicago, IL, USA, 1948.
37. Ravindran, P.; Fast, L.; Korzhavyi, P.A.; Johansson, B. Density functional theory for calculation of elastic properties of orthorhombic crystals: Application to $TiSi_2$. *J. Appl. Phys.* **1998**, *84*, 4891. [CrossRef]

Article

Field-Induced Transition in (Nd,Dy)$_2$Fe$_{14}$B in Ultrahigh Magnetic Fields

N. V. Kostyuchenko [1,*], I. S. Tereshina [2], A. I. Bykov [3], S. V. Galanova [3], R. V. Kozabaranov [3,4], A. S. Korshunov [3], I. S. Strelkov [3], I. V. Makarov [3], A. V. Filippov [3], Yu. B. Kudasov [3,4], D. A. Maslov [3,4], V. V. Platonov [3,4], O. M. Surdin [3,4], P. B. Repin [3], V. D. Selemir [3,4] and A. K. Zvezdin [5,6]

[1] Moscow Institute of Physics and Technology, National Research University, Dolgoprudny, 141701 Moscow, Russia
[2] Faculty of Physics, Lomonosov Moscow State University, 119234 Moscow, Russia
[3] Russian Federal Nuclear Center—VNIIEF, 607188 Sarov, Russia
[4] Sarov Physics and Technology Institute NRNU MEPhI, 607188 Sarov, Russia
[5] Prokhorov General Physics Institute of the Russian Academy of Sciences, 119991 Moscow, Russia
[6] P. N. Lebedev Physical Institute of the Russian Academy of Sciences, 119991 Moscow, Russia
* Correspondence: nvkost@gmail.com

Abstract: We demonstrate the peculiarities of the magnetization process in the ferrimagnetic intermetallic compound (Nd$_{0.5}$Dy$_{0.5}$)$_2$Fe$_{14}$B, which has been studied theoretically and experimentally using ultrahigh magnetic fields. We observe phase transition induced by external ultrahigh magnetic fields (up to 170 T) and also describe the magnetization process analytically in terms of critical transition fields. In this work, the first and second critical fields of the field-induced magnetic transitions, H_{c1} and H_{c2}, were estimated, and the results were verified against experimental data for H_{c1}. Critical field H_{c2} predicting the place of transition to the forced-ferromagnetic state was estimated for the first time for (Nd$_{0.5}$Dy$_{0.5}$)$_2$Fe$_{14}$B compound. A comparison of the magnetization behavior for (Nd$_{0.5}$Dy$_{0.5}$)$_2$Fe$_{14}$B with the basic systems Nd$_2$Fe$_{14}$B and Dy$_2$Fe$_{14}$B is also performed. We demonstrate that, in the Dy$_2$Fe$_{14}$B compound, the field-induced transition type is changed from the first to the second order due to the replacement of the Nd atom by Dy one.

Keywords: complex modified materials; critical fields; exchange coupling; hard magnetic materials; ultrahigh magnetic fields; R–Fe exchange; rare-earth intermetallics

1. Introduction

In the last few decades, searching for new materials for high-performance permanent magnets is an overriding task for modern physicists and technologists as these magnets are key driving components for electric motors, wind turbines, mobile phones, magnetic memory, and several other products [1–3].

Curie temperature T_c, magnetic anisotropy constant K, and saturation magnetization M_s of such materials are fundamental characteristics used to classify the existing permanent magnets. At Curie temperature, a material loses its ferromagnetic properties; hence, the higher the T_c, the better are the magnets to be used under extreme conditions. High values of saturation magnetization and magnetic anisotropy constants contribute to the creation of high-coercivity magnets that are very important for different practical applications [4–8].

Scientists pay the greatest attention to intermetallic compounds based on rare earth metals (R) and iron, in the fundamental magnetic properties, including exchange interaction parameters of R-Fe compounds, which can be studied most effectively by magnetization measurements in high magnetic fields. Ferrimagnetically ordered compounds are the most interesting because the ferrimagnetic structure is affected by applied external magnetic field, and a sequence of spin–reorientation phase transitions can be observed until the compound reaches magnetic saturation. In order to achieve the full magnetic saturation and

Citation: Kostyuchenko, N.V.; Tereshina, I.S.; Bykov, A.I.; Galanova, S.V.; Kozabaranov, R.V.; Korshunov, A.S.; Strelkov, I.S.; Makarov, I.V.; Filippov, A.V.; Kudasov, Y.B.; et al. Field-Induced Transition in (Nd,Dy)$_2$Fe$_{14}$B in Ultrahigh Magnetic Fields. *Crystals* **2022**, *12*, 1615. https://doi.org/10.3390/cryst12111615

Academic Editor: Xiaoguan Zhang

Received: 26 October 2022
Accepted: 8 November 2022
Published: 11 November 2022

Copyright: © 2022 by the authors. Licensee MDPI, Basel, Switzerland. This article is an open access article distributed under the terms and conditions of the Creative Commons Attribution (CC BY) license (https://creativecommons.org/licenses/by/4.0/).

to maintain the compound in the forced-ferromagnetic state, the ultrahigh magnetic fields are required [9–12]. In addition, interest of magnetic phase transitions study in $(R,R')_2Fe_{14}B$ compounds is increasing due to the recent discovery of skyrmions in $Nd_2Fe_{14}B$ [13].

Obtaining high and ultrahigh fields, as well as obtaining reliable experimental data on the magnetization of samples, is a technically difficult task, which is the subject of great efforts of scientists from different countries as a rule. Generation of magnetic fields involves having an electric current flow through a coil, and the field intensity B is proportional to the current I. The heat dissipation I^2R is proportional to the square of the magnetic field, where R is the coil's resistance. The mechanical pressure is also proportional to B^2, with a proportionality coefficient of approximately 4 atm/T^2. Heating and mechanical forces are the two essential problems for the generation of high fields. The stored energy in magnetic field depends on the volume of the magnet, so the size of the field volume is also an important characteristic of the magnet. Another key parameter is whether a magnet is DC or pulsed and, in the latter case, additional important parameters are the duration, temporal profile of the pulse, and the pulse repetition rate. There are several different approaches to overcome the heat–dissipation and mechanical-stability challenges, which is demonstrated in Figure 1 [14]. In all cases, choice of materials is crucially important, so such development is a task at the intersection of physics, engineering, and materials sciences.

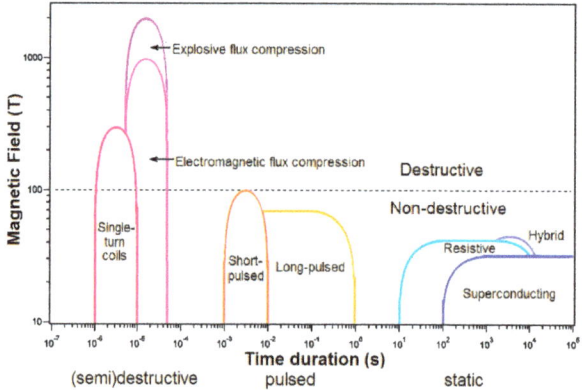

Figure 1. Overview of methods for obtaining high and ultrahigh magnetic fields [14].

It should be noted that the highest-field magnet is not necessarily the best choice for a particular experiment. The figure-of-merit (FOM), depending on the experiment, could be the stored energy B^2V (where V is the field volume), or the effective B^2L (where L is the length of the field), or something else such as the tunability to a desired field value, or the broad operating range of the field values while keeping spatial homogeneity.

There are several laboratories around the world that conduct generation and research of high magnetic fields: the French Laboratoire National des Champs Magnetiques Intenses (LNCMI) with two locations; the German Dresden High Magnetic Field Laboratory (Hochfeld-Magnetlabor Dresden, HLD); and the High Field Magnet Laboratory (HMFL) in the Netherlands. These laboratories operate within the European Magnetic Field Laboratory (EMFL). The US National High Magnetic Field Laboratory (NHMFL) also has three locations. There are two laboratories in China: the High Magnetic Field Laboratory of the Chinese Academy of Sciences (CHMFL) and the Wuhan National High Magnetic Field Center (WHMFC), and four laboratories in Japan: the Tsukuba Magnet Laboratory (TML), the High Field Laboratory for Superconducting Materials, the International Megagauss Science Laboratory (IMGSL), and the Center for Advanced High Magnetic Field Science.

In our work, magnetization measurements were carried out at the Russian Federal Nuclear Center in Sarov in pulsed magnetic fields up to 170 T. This laboratory was one of the first in the world to obtain megagauss magnetic fields [15]. As the object of our study, we used the composition $(Nd_{0.5}Dy_{0.5})_2Fe_{14}B$, which, due to the Dy atoms, is a ferrimagnet. Previously, this composition was studied by us in magnetic fields up to 58 T at the Dresden High Magnetic Field Laboratory. The experiment showed that the fields used were not enough to observe the forced-ferromagnetic state but allowed us to discuss the transition to the non-collinear phase and make a prediction about the place of the transition in a forced-ferromagnetic state. In addition, at the Megagauss Laboratory of Institute for Solid State Physics of University of Tokyo, the transition from a collinear ferrimagnet and a non-collinear spin-flop-like phase was observed in the $Dy_2Fe_{14}B$ compound in magnetic fields up to 120 T [16]. It was also shown that such studies make it possible to compare the obtained experimental data with existing modern theoretical models and obtain the most important information about the main fundamental parameters for the $(Nd,Dy)_2Fe_{14}B$ compound.

2. Materials and Methods

2.1. Sample Preparation

The samples for this study $(Nd_xDy_{1-x})_2Fe_{14}B$ with x = 0, 0.5 and 1 were prepared using an arc furnace. The procedure for the samples obtaining is described in more details in the following works [17,18]. The phase composition of the samples was determined using the standard X-ray powder diffraction (XRD) at room temperature. XRD studies indicated that the $Nd_2Fe_{14}B$ and $Dy_2Fe_{14}B$ alloys were single-phase, while, in the $(Nd,Dy)_2Fe_{14}B$ samples, traces of the second phase were seen (~5%). The investigated alloys have a tetragonal structure of the $Nd_2Fe_{14}B$ type (space group $P4_2/mnm$) at room temperature. The lattice parameters for $Nd_2Fe_{14}B$, $(Nd_xDy_{1-x})_2Fe_{14}B$ and $Dy_2Fe_{14}B$ alloys are a = 0.880 nm, c = 1.219 nm, a = 0.877 nm, c = 1.212 nm, a = 0.873 nm, and c = 1.190 nm, respectively. The decreasing lattice parameter for the Dy-substituted compound is due to the smaller atomic radius of Dy^{3+} as compared to Nd^{3+}. This agrees well with literature data [16].

2.2. Method of Magnetization Measurements

Magnetization measurements were performed at the Russian Federal Nuclear Center in Sarov in pulsed magnetic fields up to 170 T on powder samples. An ultrahigh magnetic field up to 600 T was created in a magnetocumulative generator MC-1 [19]. In a thin-walled wire solenoid, the discharge of a powerful capacitor bank with a stored energy of 2 MJ created a seed magnetic field of about 16 T. During the battery discharge, a converging shock wave was initiated in the annular charge of high explosives surrounding the solenoid. It came out on the surface of the solenoid at the seed field maximum (approximately 80 μs after the start of the discharge). When the shock wave passed through the solenoid, the solenoid wires were welded and formed a homogeneous conducting cylinder shell with the trapped magnetic flux. The ultrahigh magnetic field was generated by explosive magnetic flux compression for about 16 μs. The MC-1 generator has been widely used earlier and proved to be a reliable tool for scientific research [20]. The high uniformity of the magnetic field in large useful volumes (about 10 cm^3) made it possible to install four researched samples in one experiment.

2.3. Registration of the Signal

The registration of the time derivative of the magnetic field was carried out by a set of pick-up coils with different sensitivities (7 coils in total, some of them were duplicated to increase reliability). It allowed measurements of the magnetic field induction with an accuracy of 5% over the entire operating range of the MC-1 generator. The magnetization of the studied samples was measured using compensated pick-up coils [21,22]. A pair of two identical coils were of diameter d = 2.8 mm and had number of turns N = 20. A special winding of the sensor was performed, which provided a significant reduction of the total electrical voltage between the coils of the sensor [21]. The signal induced in the compensation coils consists of a "useful" part and background signal, which stemmed from the coil decompensation and was proportional to the time derivative of the magnetic field. The degree of decompensation for all sensors was less than 2%. To take into account a slight attenuation of the signals of the compensation sensors in cable lines, the whole system was pre-calibrated immediately before the experiment. The compensated pick-up coils were proven to be a reliable technique for magnetization measurements at pulsed magnetic fields [20–22]. The compensation sensors were placed in a glass helium cryostat, into which liquid helium was raised from a transport Dewar vessel before the experiment.

The absolute values of the magnetization were calibrated by measuring the magnetization curves up to 14 T in static fields using a PPMS 14T magnetometer (Quantum Design, San Diego, CA, USA).

3. Results

Figure 2 displays the magnetization curves of the $Dy_2Fe_{14}B$ and $Nd_2Fe_{14}B$ single crystals measured at 1.8 and 10 K, respectively. Measurements have been performed along the main crystallographic axes and compared with the data given in the works [18,23]. $Nd_2Fe_{14}B$ samples display an easy-cone anisotropy, in contrast to the uniaxial $Dy_2Fe_{14}B$ with an anisotropy field of 27 T. In the inset in Figure 2, we show the magnetization curve of aligned polycrystal sample $Dy_2Fe_{14}B$ in magnetic fields up to 120 T obtained at 10 K [16]. The anomaly near 100 T (at 105 and 101 T for increasing and decreasing fields, respectively, which correspond to intermittent changes in the M(H) curve) arises from the first-order transition between a collinear ferrimagnet and a non-collinear spin-flop-like phase. In the $Nd_2Fe_{14}B$ single crystal, a jump in the M(H) magnetization curve is observed along the [100] direction in magnetic field 17 T.

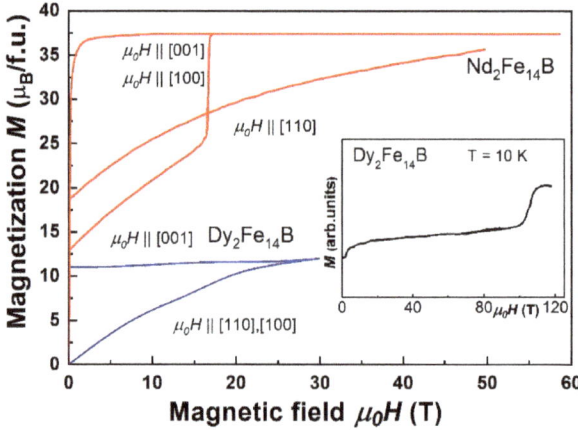

Figure 2. Magnetization curves of $Dy_2Fe_{14}B$ and $Nd_2Fe_{14}B$ single crystals applied along the main crystallographic directions at 1.8 and 10 K, respectively [18,23]. Inset: The data for an aligned polycrystal $Dy_2Fe_{14}B$ at 10 K (for increasing fields [16]) are shown for comparison.

Figure 3 demonstrates the magnetization curves of the $(Nd_{0.5}Dy_{0.5})_2Fe_{14}B$ compound along the perpendicular [001] easy axis. Magnetization measurements were performed by us at the Dresden High Magnetic Field Laboratory in pulsed magnetic fields up to 58 T previously [18]. The measurements up to 170 T obtained for the first time at the Russian Federal Nuclear Center in Sarov. It should be mentioned that a compensated pick-up sensor, in fact, measured a time derivative of magnetization. This is why the flat segments of the magnetization curve in Figure 3 were beyond the sensitivity and shown by the dashed line. A partial substitution of Dy by Nd atom in the latter compound results in an increase of the anisotropy field where the easy- and hard-axis magnetization curves intersect. Such new results, containing features on the magnetization curve, can also be used for modern numerical calculations, including for obtaining or refining the exchange and crystal-field parameters in the quantum model of the crystal electric field [16–18].

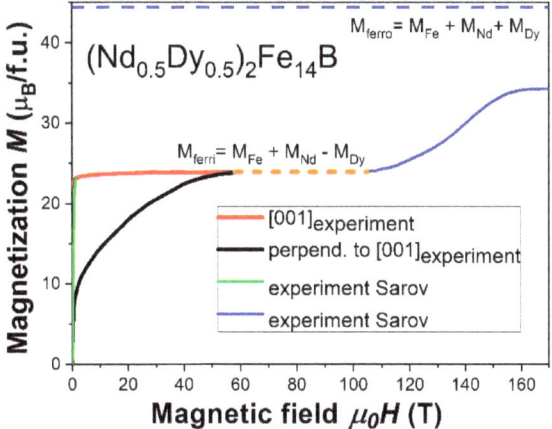

Figure 3. Magnetization curves of $(Nd_{0.5}Dy_{0.5})_2Fe_{14}B$ at 1.8 K in magnetic fields up to 58 T applied along the perpendicular easy axis c (red and black lines) [18] and at 4.2 K in fields up to 170 T (for increasing fields) applied along the c-axis (green and blue lines).

Figure 3 shows that the experimental data obtained in different laboratories are in good agreement with each other. Moreover, it can be stated that ultrahigh magnetic fields in the $(Nd_{0.5}Dy_{0.5})_2Fe_{14}B$ compound induce a magnetic phase transition similar to that observed earlier in the $Dy_2Fe_{14}B$ compound. These transitions can be interpreted as intermediate processes from ferrimagnetism to forced ferromagnetism. A significant difference here is the fact that the magnetic phase transition in the substituted composition $(Nd_{0.5}Dy_{0.5})_2Fe_{14}B$ occurs smoothly, covering a quite large range of magnetic fields from 105 T to 160 T, exhibiting features characteristic of second-order phase transitions.

4. Discussion

In order to analyze obtained experimental data, it is important to understand which sections of the magnetization curves correspond to which phase: ferrimagnetic, non-collinear, or ferromagnetic (field-induced forced ferromagnetic state). This can be understood using a quite simple analytical approach in terms of the first and second critical fields. The first critical field H_{c1} corresponds to the transition from the ferrimagnetic to the non-collinear phase, and the second H_{c2} to the ferromagnetic one. In order to estimate them, we applied the analytical approach (previously described in detail in Refs. [24–27]) well proven for the $(R,R')_2Fe_{14}B$ intermetallic compounds.

For the $R_2Fe_{14}B$ compounds, both critical fields H_{c1} and H_{c2} are equal to [24,26,27]:

$$H_{c1} = \lambda \left(M_{Fe} - 2M_R\right) + \frac{H_a \, 2M_R}{\lambda \, (M_{Fe} - 2M_R)},$$
$$H_{c2} = \lambda \left(M_{Fe} + 2M_R\right) - \frac{H_a \, 2M_R}{\lambda \, (M_{Fe} + 2M_R)}$$
(1)

where λ is the R-Fe intersublattice exchange parameter, $H_a = \frac{2K_1}{M_{Fe}}$ is the magnetic anisotropy field, and K_1 is magnetic anisotropy constant [5]. Here, the second term describing the anisotropy has been added in order to make a more accurate critical fields estimation [27]. Thus, the accuracy of analytical evaluations of the critical fields reaches several Tesla.

For the $(Nd_{0.5}Dy_{0.5})_2Fe_{14}B$ compound with two different rare-earth ions, the critical fields H_{c1} and H_{c2} have the form [27]

$$H_{c1} = \lambda_{Dy}\left(M_{Fe} - 2M_{Dy}\zeta_1\right) - \frac{2M_{Dy}H_a\zeta_1^2}{(M_{Fe} - 2M_{Dy}\zeta_1)},$$

$$H_{c2} = \lambda_{Dy}\left(M_{Fe} + 2M_{Dy}\zeta_2\right) + \frac{2M_{Dy}H_a\zeta_2^2}{(M_{Fe} + 2M_{Dy}\zeta_2)}, \quad (2)$$

$$\zeta_i(H_{ci}) = \frac{1}{1 + \lambda_{Nd}\chi_{Nd}(H_{ci})}; \quad \chi_{Nd}(H_{ci}) = \frac{2M_{Nd}}{\lambda_{Nd}M_{Fe} + H_{ci}}; i = 1, 2$$

λ_{Nd} and χ_{Nd} are the exchange parameter and susceptibility of the Nd sublattice, respectively. In accordance with our previous estimates, $\zeta_i \approx 0.9$ [24]. Formula (2) with anisotropic correction is universal and could be used for various rare-earth intermetallics. H_{c1} and H_{c2} values obtained for $(Nd_{0.5}Dy_{0.5})_2Fe_{14}B$ and $Dy_2Fe_{14}B$ by analyzing high-field experimental data using Formulas (1) and (2) are given in Table 1.

Table 1. Magnetic critical fields' parameters H_{c1} and H_{c2} for $(Nd_{0.5}Dy_{0.5})_2Fe_{14}B$ and $Dy_2Fe_{14}B$.

Compound	H_{c1} (T)	H_{c2} (T)
$(Nd_{0.5}Dy_{0.5})_2Fe_{14}B$	105	240
$Dy_2Fe_{14}B$	105	250

H_{c2} value provides important information, which the magnitudes of external magnetic fields require to reach the ferromagnetic state, and allow us to plan new ultrahigh magnetization experiments for studying similar types of compounds. It can be seen that, in order to experimentally observe the transition to the forced-ferromagnetic phase, both compounds (($Nd_{0.5}Dy_{0.5})_2Fe_{14}B$ and $Dy_2Fe_{14}B$) require magnetic fields greater than 250 T. This is confirmed, among other things, by the value of the magnetization in the $(Nd_{0.5}Dy_{0.5})_2Fe_{14}B$ compound at 170 T. It reaches magnitude 34 μ_B/f.u., while the magnitude of the magnetization in a forced-ferromagnetic state is 44.1 μ_B/f.u. (see Table 2), that is, in the magnetic field 170 T, saturation has not yet been obtained.

Table 2. Values of the rare-earth and iron magnetic moments for $(Nd_{0.5}Dy_{0.5})_2Fe_{14}B$ and $Dy_2Fe_{14}B$ intermetallic compounds.

Compound	M_{Fe} (μ_B)	M_R (μ_B)
$(Nd_{0.5}Dy_{0.5})_2Fe_{14}B$	31.4	3 (Nd) and 10 (Dy)
$Dy_2Fe_{14}B$	31.4	10

5. Conclusions

Magnetization measurements were performed for the intermetallic ferrimagnetic compound $(Nd_{0.5}Dy_{0.5})_2Fe_{14}B$ using ultrahigh magnetic fields up to 170 T. Such high magnetic fields make it possible to observe the experimentally field-induced phase transition from the initial ferrimagnetic state to the non-collinear one. The results obtained were compared with the literature high-field magnetization data for the basic $Dy_2Fe_{14}B$ compound. We have determined two critical fields (H_{c1} and H_{c2}) of field-induced transitions for the compounds under study. We demonstrate that, in order to observe the transition to the forced-ferromagnetic state, both compounds (($Nd_{0.5}Dy_{0.5})_2Fe_{14}B$ and $Dy_2Fe_{14}B$) require magnetic fields greater than 250 T. It is shown that the replacement of one rare-earth atom by another is a powerful tool for controlling the properties of rare-earth $R_2Fe_{14}B$-type compounds, on the basis of which modern highly efficient permanent magnets are manufactured.

Author Contributions: Methodology, Y.B.K., D.A.M., V.V.P. and O.M.S.; investigation, N.V.K., A.K.Z., A.I.B., S.V.G., R.V.K., A.S.K., I.S.S., I.V.M. and A.V.F.; writing—original draft preparation, N.V.K.; writing—review and editing, I.S.T. and Y.B.K.; supervision, I.S.T. and Y.B.K.; project administration, A.K.Z., P.B.R. and V.D.S. All authors have read and agreed to the published version of the manuscript.

Funding: This work was supported by the Ministry of Science and Higher Education of the Russian Federation (No. FSMG-2021-0005). Research in ultrahigh magnetic fields was carried out within the framework of the scientific program of the National Center for Physics and Mathematics (Project "Research in high and ultrahigh magnetic fields").

Institutional Review Board Statement: Not applicable.

Informed Consent Statement: Not applicable.

Data Availability Statement: Not applicable.

Conflicts of Interest: The authors declare no conflict of interest.

References

1. Kostyuchenko, N.V.; Tereshina, I.S.; Andreev, A.V.; Doerr, M.; Tereshina-Chitrova, E.A.; Paukov, M.A.; Gorbunov, D.I.; Politova, G.A.; Pyatakov, A.P.; Miyata, A.; et al. Investigation of the field-induced phase transitions in the (R, R′)$_2$Fe$_{14}$B rare-earth intermetallics in ultra-high magnetic fields. *IEEE Trans. Magn.* **2021**, *57*, 2101105. [CrossRef]
2. Coey, J.M.D. Hard Magnetic Materials: A Perspective. *IEEE Trans. Magn.* **2011**, *47*, 4671–4681. [CrossRef]
3. Coey, J.M.D. Perspective and Prospects for Rare Earth Permanent Magnets. *Engineering* **2020**, *6*, 119–131. [CrossRef]
4. Gutfleisch, O.; Willard, M.A.; Brück, E.; Chen, C.H.; Sankar, S.G.; Liu, J.P. Magnetic Materials and Devices for the 21st Century: Stronger, Lighter, and More Energy Efficient. *Adv. Mater.* **2011**, *23*, 821–842. [CrossRef] [PubMed]
5. Sepehri-Amin, H.; Hirosawa, S.; Hono, K. Chapter 4—Advances in Nd-Fe-B Based Permanent Magnets. In *Handbook of Magnetic Materials*; Elsevier: Amsterdam, The Netherlands, 2018; Volume 27, pp. 269–372.
6. Herbst, J.F. R$_2$Fe$_{14}$B materials: Intrinsic properties and technological aspects. *Rev. Mod. Phys.* **1991**, *63*, 819–898. [CrossRef]
7. Li, H.-S.; Coey, J.M.D. Chapter 3—Magnetic properties of ternary rare-earth transition-metal compounds. In *Handbook of Magnetic Materials*; Elsevier: Amsterdam, The Netherlands, 2019; Volume 28, pp. 87–196.
8. Gabay, A.M.; Marinescu, M.; Li, W.F.; Liu, J.F.; Hadjipanayis, G.C. Dysprosium-saving improvement of coercivity in Nd-Fe-B sintered magnets by Dy$_2$S$_3$ additions. *J. Appl. Phys.* **2011**, *109*, 083916. [CrossRef]
9. Liu, W.-Q.; Chang, C.; Yue, M.; Yang, J.-S.; Zhang, D.-T.; Zhang, J.-X.; Liu, Y.-Q. Coercivity, microstructure, and thermal stability of sintered Nd–Fe–B magnets by grain boundary diffusion with TbH$_3$ nanoparticles. *Rare Met.* **2017**, *36*, 718–722. [CrossRef]
10. Poenaru, I.; Lixandru, A.; Güth, K.; Malfliet, A.; Yoon, S.; Škulj, I.; Gutfleisch, O. HDDR treatment of Ce-substituted Nd$_2$Fe$_{14}$B-based permanent magnet alloys—Phase structure evolution, intergranular processes and magnetic property development. *J. Alloys Compd.* **2020**, *814*, 152215. [CrossRef]
11. Sözen, H.İ.; Ener, S.; Maccari, F.; Skokov, K.P.; Gutfleisch, O.; Körmann, F.; Neugebauer, J.; Hickel, T. Ab initio phase stabilities of Ce-based hard magnetic materials and comparison with experimental phase diagrams. *Phys. Rev. Mater.* **2019**, *3*, 084407. [CrossRef]
12. Furlani, E.P. *Permanent Magnet and Electromechanical Devices: Materials, Analysis, and Applications*; Academic Press: Cambridge, MA, USA, 2001.
13. Chaboy, J.; Piquer, C.; Plugaru, N.; Bartolomé, F.; Laguna-Marco, M.A.; Plazaola, F. ^{57}Fe Mössbauer and x-ray magnetic circular dichroism study of magnetic compensation of the rare-earth sublattice in Nd$_{2-x}$Ho$_x$Fe$_{14}$B compounds. *Phys. Rev. B* **2007**, *76*, 134408. [CrossRef]
14. Xiao, Y.; Morvan, F.J.; He, A.N.; Wang, M.K.; Luo, H.B.; Jiao, R.B.; Xia, W.X.; Zhao, G.P.; Liu, J.P. Spin-reorientation transition induced magnetic skyrmion in Nd$_2$Fe$_{14}$B magnet. *Appl. Phys. Lett.* **2020**, *117*, 132402. [CrossRef]
15. Battesti, R.; Beard, J.; Böser, S.; Bruyant, N.; Budker, D.; Crooker, S.A.; Daw, E.J.; Flambaum, V.V.; Inada, T.; Irastorza, I.G.; et al. High magnetic fields for fundamental physics. *Phys. Rep.* **2018**, *765*, 1–39. [CrossRef]
16. Younger, S.; Lindemuth, I.; Reinovsky, R.; Fowler, C.M.; Goforth, J.; Ekdahl, C. Scientific Collaborations Between Los Alamos and Arzamas-16 Using Explosive-Driven Flux Compression Generators. *Los Alamos Sci.* **1996**, *24*, 48–67.
17. Lim, D.W.; Kato, H.; Yamada, M.; Kido, G.; Nakagawa, Y. High-field magnetization process and spin reorientation in (Nd$_{1-x}$Dy$_x$)$_2$Fe$_{14}$B single crystals. *Phys. Rev. B* **1991**, *44*, 10014. [CrossRef] [PubMed]
18. Takeyama, S.; Amaya, K.; Nakagawa, T.; Ishizuka, M.; Nakao, K.; Sakakibara, T.; Goto, T.; Miura, N.; Ajiro, Y.; Kikuchi, H. Magnetisation measurements in ultra-high magnetic fields produced by a single-turn coil system. *J. Phys. E Sci. Instrum.* **1988**, *21*, 1025. [CrossRef]
19. Kostyuchenko, N.V.; Tereshina, I.S.; Gorbunov, D.I.; Tereshina-Chitrova, E.A.; Andreev, A.V.; Doerr, M.; Politova, G.A.; Zvezdin, A.K. Features of magnetization behavior in the rare-earth intermetallic compound (Nd$_{0.5}$Ho$_{0.5}$)$_2$Fe$_{14}$B. *Intermetallics* **2018**, *98*, 139–142. [CrossRef]

20. Shneerson, G.A.; Dolotenko, M.I.; Krivosheev, S.I. *Strong and Superstrong Magnetic Fields Generation*; Walter de Gruter GmbH: Berlin, Germany, 2014.
21. Boriskov, G.V.; Bykov, A.I.; Dolotenko, M.I.; Egorov, N.I.; Kudasov, Y.B.; Platonov, V.V.; Selemir, V.D.; Tatsenko, O.M. Research in ultrahigh magnetic field physics. *Phys. Uspekhi* **2011**, *54*, 421–427. [CrossRef]
22. Kudasov, Y.B. Megagauss magnetization measurements. *Phys. B Condens. Matter* **2001**, *294*, 684–690. [CrossRef]
23. Kostyuchenko, N.V.; Tereshina, I.S.; Gorbunov, D.I.; Tereshina-Chitrova, E.A.; Rogacki, K.; Andreev, A.V.; Doerr, M.; Politova, G.A.; Zvezdin, A.K. High-field magnetization study of $(Nd,Dy)_2Fe_{14}B$: Intrinsic properties and promising compositions. *Intermetallics* **2020**, *124*, 106840. [CrossRef]
24. Eslava, G.G.; Fayyazi, B.; Skokov, K.; Skourski, Y.; Gorbunov, D.; Gutfleisch, O.; Dempsey, N.M.; Givord, D. A two-sublattice model for extracting rare-earth anisotropy constants from measurements on $(Nd,Ce)_2(Fe,Co)_{14}B$ single crystals. *J. Magn. Magn. Mater.* **2021**, *520*, 167470. [CrossRef]
25. Tereshina, I.S.; Pyatakov, A.P.; Tereshina-Chitrova, E.A.; Gorbunov, D.I.; Skourski, Y.; Law, J.M.; Paukov, M.A.; Havela, L.; Doerr, M.; Zvezdin, A.K.; et al. Probing the exchange coupling in the complex modified Ho-Fe-B compounds by high-field magnetization measurements. *AIP Adv.* **2018**, *8*, 125223. [CrossRef]
26. Zvezdin, A.K.; Lubashevskii, I.A.; Levitin, R.Z.; Platonov, V.V.; Tatsenko, O.M. Phase transitions in megagauss magnetic fields. *Phys. Uspekhi* **1998**, *41*, 1037–1042. [CrossRef]
27. Zvezdin, A.K. Magnetic Phase Transitions: Field-induced (Order-to-order). In *Encyclopedia of Materials: Science and Technology*; Buschow, K.H.J., Cahn, R.W., Flemings, M.C., Ilschner, B., Kramer, E.J., Mahajan, S., Veyssière, P., Eds.; Elsevier: Amsterdam, The Netherlands, 2001; pp. 4841–4847.

Article

Hydrogen-Induced Order–Disorder Effects in FePd$_3$

André Götze [1], Siobhan Christina Stevenson [2], Thomas Christian Hansen [3] and Holger Kohlmann [1,*]

[1] Institute of Inorganic Chemistry, Leipzig University, Johannisallee 29, 04103 Leipzig, Germany
[2] School of Chemistry, University of Glasgow, Glasgow G12 8QQ, UK
[3] Institut Laue-Langevin, 71 Avenue des Martyrs, CS 20156, CEDEX 9, 38042 Grenoble, France
* Correspondence: holger.kohlmann@uni-leipzig.de; Tel.: +49-341-9736201

Abstract: Binary intermetallic compounds, such as FePd$_3$, attract interests due to their physical, magnetic and catalytic properties. For a better understanding of their hydrogenation properties, both ordered FePd$_3$ and disordered Fe$_{0.25}$Pd$_{0.75}$ are studied by several *in situ* methods, such as thermal analysis, X-ray powder diffraction and neutron powder diffraction, at moderate hydrogen pressures up to 8.0 MPa. FePd$_3$ absorbs small amounts of hydrogen at room temperature and follows Sieverts' law of hydrogen solubility in metals. [Pd$_6$] octahedral voids are filled up to 4.7(9)% in a statistical manner at 8.00(2) MPa, yielding the hydride FePd$_3$H$_{0.047(9)}$. This is accompanied by decreasing long-range order of Fe and Pd atoms (site occupancy factor of Fe at Wyckoff position 1a decreasing from 0.875(3) to 0.794(4)). This trend is also observed during heating, while the ordered magnetic moment decreases up to the Curie temperature of 495(8) K. The temperature dependences of the magnetic moments of iron atoms in FePd$_3$ under isobaric conditions ($p(D_2) = 8.2(2)$ MPa) are consistent with a 3D Ising or Heisenberg model (critical parameter $\beta = 0.28(5)$). The atomic and magnetic order and hydrogen content of FePd$_3$ show a complex interplay.

Keywords: intermetallics; metal hydrides; neutron diffraction; in situ diffraction; order–disorder effects; interstitial hydrides; deuterides; magnetism

Citation: Götze, A.; Stevenson, S.C.; Hansen, T.C.; Kohlmann, H. Hydrogen-Induced Order–Disorder Effects in FePd$_3$. *Crystals* **2022**, *12*, 1704. https://doi.org/10.3390/cryst12121704

Academic Editors: Jacek Ćwik and Wojciech Polkowski

Received: 14 October 2022
Accepted: 19 November 2022
Published: 24 November 2022

Copyright: © 2022 by the authors. Licensee MDPI, Basel, Switzerland. This article is an open access article distributed under the terms and conditions of the Creative Commons Attribution (CC BY) license (https://creativecommons.org/licenses/by/4.0/).

1. Introduction

The incorporation of hydrogen is a well-known tool for influencing the structural, electric, magnetic and optic properties of intermetallic compounds [1,2]. The interplay between hydrogen uptake and magnetism is often quite complex. Upon hydrogen uptake, intermetallic compounds may lose ferro-(LaCo$_5$) or ferrimagnetism (Y$_6$Mn$_{23}$) or become ferromagnets (CeNi$_3$, Hf$_2$Fe and Th$_7$Fe$_3$) [1]. The influence of hydrogen incorporation on the atomic order in crystal structures of intermetallics is also widely studied. In an extreme case, it may lead to amorphous hydrides in a process known as hydrogen-induced amorphization (HIA) [2]. The process involves short-range diffusion of metallic atoms, and the driving force seems to be the different hydrogen occupation sites in crystalline and amorphous states of the alloy. In some cases, such as Laves phases, the latter can be predicted by geometric factors, such as the atomic size ratio of constituting atoms [2]. Obviously, hydrogen uptake, atomic order (crystal structure) and magnetic order (cooperative phenomena) influence each other. In most studies, however, only the interaction between two of these factors is investigated and the third neglected or assumed not to play a role, which might be an oversimplification in some cases. Clearly, more in-depth investigations are needed to reveal the complex interplay between hydrogen uptake, atomic order and magnetic order in intermetallics. In this respect, we studied the crystal and magnetic structure of FePd$_3$ and its solid solution with hydrogen (also called hydride throughout this text) in detail in order to thoroughly characterize the interesting system FePd$_3$-H$_2$ and to reveal the potential cross-links between these factors. In order to obtain a high depth of knowledge, the hydrogenation of FePd$_3$ was analyzed by time-resolved *in situ* neutron diffraction, mapping the crystal structure, including hydrogen atom positions and magnetic moments.

FePd$_3$ has attracted interest as a functional material due to its diverse physical properties. It has a characteristic pressure-induced invar behavior by means of an anomalously low thermal expansion at high applied pressures [3–5], and it is ferromagnetic with a Curie temperature of 499 K [6]. A soft ferromagnetic behavior was found in carbon-based materials by encapsulation of FePd$_3$ [7–9]. FePd$_3$ can be used in electrocatalysis to enhance the cycle stability of hybrid Li–air batteries [10] or as an electrocatalyst to oxidize formic acid [11]. In addition, a higher attraction between 2-methylfuran and hydrogen compared to palladium was found in hydrogenation catalysis [12]. The use of bimetallic catalysts changes the electronic structure at the surface [13] and decreases the Pd-Pd coordination number; this can hinder the formation of unfavorable surface species, thus avoiding unwanted side reactions, e.g., decarbonylation, in the solvent-free hydrodeoxygenation of furan compounds for a Pd-FeO$_x$/SiO$_2$ catalyst [14]. Knowledge of the crystal structure is of great importance for catalytic applications, since it has a distinct influence on catalytic properties. The phase diagram of the Fe-Pd system shows a solid solution with a large phase width of about ±10% around an Fe:Pd atomic ratio of 1:3 [15]. In addition to the disordered phase Fe$_{0.25}$Pd$_{0.75}$ (Cu type, $Fm\overline{3}m$), an ordered phase is known. FePd$_3$ crystallizes in an ordered variant of a cubic close packing (AuCu$_3$ type, $Pm\overline{3}m$, Figure 1). The annealing times for the ordering process are long due to similar electronic and geometric properties of the constituting atoms [16]. FePd$_3$, with a high degree of crystallographic order, shows a higher hydrogen incorporation at high hydrogen pressures compared to (partially) disordered samples [17]. Disordered Fe$_{0.25}$Pd$_{0.75}$ needs more than two orders of magnitude higher hydrogen pressure to obtain the same electrical resistivity as found in ordered FePd$_3$ [18]. The position of hydrogen atoms in FePd$_3$H$_x$ is not known yet.

In this work, we employ *in situ* studies to show the influence of moderate hydrogen pressures on the order–disorder transition in FePd$_3$, thus complementing studies of hydrogen absorption of FePd$_3$ at high hydrogen pressures [17]. *In situ* neutron powder diffraction, as an established method [19,20], was used to determine the level of atomic disorder, the magnetic moment and the amount of incorporated hydrogen.

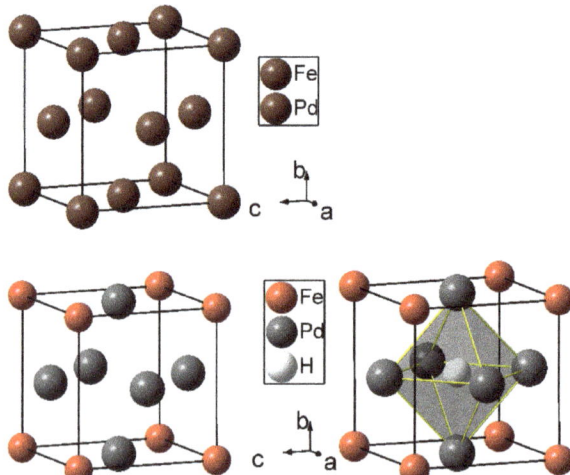

Figure 1. The crystal structures of FePd$_3$ in Cu type (left, disordered, Fe$_{0.25}$Pd$_{0.75}$, $Fm\overline{3}m$ [15]) and in AuCu$_3$ type (middle, ordered, FePd$_3$, $Pm\overline{3}m$ [15]) and the crystal structure of FePd$_3$H$_{0.05}$ in the cubic anti-perovskite type ($Pm\overline{3}m$ (this work)) with hydrogen atoms surrounded by six palladium atoms in an octahedral arrangement.

2. Materials and Methods

Synthesis and Chemical Analysis: Due to air sensitivity, iron was handled in an argon-filled glove box. Disordered $Fe_{0.25}Pd_{0.75}$ was synthesized from stoichiometric amounts of palladium powder (99.95%, ≤150 μm, Goodfellow) and iron granules (99.98%, 1–2 mm, abcr) in sealed silica glass ampoules under argon atmosphere. The mixture was heated to 1423 K (100 K h^{-1} heating rate) for 48 h and afterward quenched in air. The ordered compound was synthesized analogously; however, one small crystal of iodine (resublimed, Merck) was added as a mineralizing agent. This mixture was heated to 923 K (100 K h^{-1} heating rate) for 7 d in a sealed silica ampoule. A further sample was afterward annealed for one month at 773 K. The products were ground in a mortar after cooling.

Iodine was removed by sublimation to the opposite side of the ampoule. Chemical analyses were performed by an EDX INCA SYSTEM from Oxford Instruments, mounted on a Zeiss LEO 1530 scanning electron microscope, with an acceleration voltage of 20 kV and a working distance of 15 mm.

Thermal Analysis: Differential scanning calorimetry (DSC) was performed under hydrogen pressure on a DSC HP 2+ (Mettler Toledo) equipped with a gas pressure chamber. An amount of 20 mg of the powdered sample was put in an aluminum crucible, which was closed with an aluminum lid. This was placed inside the pressure chamber, which was then purged several times with hydrogen gas, before filling to the final hydrogen gas pressure of 5.0 MPa. The sample was heated to 723 K with 10 K min^{-1}, held at this temperature for a minimum of 1 h and cooled to 300 K. Two runs were performed; afterward, the hydrogen pressure was released, the sample removed and structural characterization undertaken by XRPD.

Ex situ **X-ray Powder Diffraction (XRPD):** X-ray powder diffraction data from flat transmission samples were collected on a G670 diffractometer (Huber, Rimsting, Germany) with Mo-K$_{\alpha 1}$ radiation (70.926 pm) and from flat reflection samples on a SmartLab powder high-resolution X-ray powder diffractometer (Rigaku, Tokyo, Japan) with a HyPix-3000 two-dimensional semiconductor detector using Co-K$_\alpha$ radiation with parallel beam. The instrumental resolution function and the wavelength distribution were determined using a measurement on an external silicon NIST640d standard sample. The wavelengths were found to be 178.9789(4) pm and 179.3625(4) pm, close to the usual values for Co-K$_{\alpha 1}$ and Co-K$_{\alpha 2}$; the difference was caused by optical components of the diffractometer affecting the wavelength distribution.

In situ **X-ray Powder Diffraction:** *In situ* X-ray powder diffraction was performed on a SmartLab powder high-resolution X-ray powder diffractometer (Rigaku, Tokyo, Japan) in an Anton Paar XRK 900 reaction chamber (Graz, Austria) with 0.5 MPa hydrogen pressure (H$_2$, Air Liquide, 99.9%) and Co-K$_\alpha$ radiation with parallel beam on a flat FePd$_3$ specimen on top of an Al$_2$O$_3$ layer.

Ex situ **Neutron Powder Diffraction (NPD):** Neutron powder diffraction was carried out at the Institute Laue-Langevin in Grenoble, France, with a high-flux diffractometer D20 in high-resolution mode. Powdered samples (≈1 cm^{-3}) were held in air-tight vanadium containers with 6 mm inner diameter and were each measured for 15 min. The wavelength λ = 186.80(2) pm was calibrated using an external silicon NIST640b standard sample in a 5 mm vanadium container. Deuterides rather than hydrides were used to avoid the high incoherent scattering of ^{1}H.

In situ **Neutron Powder Diffraction:** *In situ* neutron powder diffraction (NPD) was performed on a high-intensity two-axis diffractometer D20 at the Institute Laue-Langevin (ILL), Grenoble, France. Time-resolved neutron diffraction data were collected under deuterium pressure and heating by two lasers. These *in situ* experiments were carried out in (leuco-)sapphire single-crystal cells with 6 mm inner diameter connected to a gas supply system. The details are given elsewhere [19,20]. The sample cell was filled with FePd$_3$ and attached to the gas supply system, which was subsequently evacuated. The reactions were performed under various deuterium pressures (D$_2$, Air Liquide, 99.8% isotope purity). Data sets were obtained with 2 min time resolution. They are presented with an additional internal raw label (NUMOR), referring to proposal 5-24-613 [21]. For the *in situ* studies, NUMORs 131613–131859 were used.

Rietveld Refinement: Rietveld refinements [22,23] were performed using FullProf [24] and Topas [25]. Deuterium atoms were located by difference Fourier analysis. Simultaneous refinements of FePd$_3$ based on XRPD data and neutron powder diffraction data were performed with constrained mixed occupation parameters to reduce correlation with the ordered magnetic moment of the iron atoms. Further details of the crystal structure investigations may be obtained from FIZ Karlsruhe, 76344 Eggenstein-Leopoldshafen, Germany (fax: (+49)7247-808-666; e-mail: crysdata@fiz-karlsruhe.de), on quoting the deposition number CSD-2163436.

3. Results

For reasons of clarity and simplification, partially disordered FePd$_3$ (typically with between 10% and 20% palladium atoms on iron sites and vice versa, *vide infra*) is referred to as ordered, and the completely disordered Fe$_{0.25}$Pd$_{0.75}$, with statistical distribution of iron and palladium atoms as disordered in the following text. The term hydride is used to include all hydrogen isotopes, unless indicated otherwise, e.g., for deuterides used in neutron diffraction experiments. Because of the small amount of dissolved hydrogen (deuterium), these phases may also be seen as solid solutions of hydrogen (deuterium) in FePd$_3$.

3.1. Synthesis and Chemical Analysis

The intermetallic compound FePd$_3$ was synthesized from the elements. To facilitate the ordering of the metal atoms, iodine was added as a mineralizing agent. The annealed and quenched samples were gray powders with a metallic luster. Based on chemical analysis, the empirical formulae Fe$_{0.97(13)}$Pd$_{3.03(13)}$ for the ordered and Fe$_{1.0(2)}$Pd$_{3.0(2)}$ for the disordered phase were determined, with values averaged from at least 20 energy dispersive X-ray (EDX) spectra of each. Based on these results, we assign the same sum formula FePd$_3$ for both the ordered and the disordered phase. We distinguish them by nomenclature, i.e., FePd$_3$ for the (partially) ordered and Fe$_{0.25}$Pd$_{0.75}$ for the disordered phase. The products are stable in air. The powder particles align with the magnetic field of a permanent magnet.

3.2. X-ray Diffraction and Thermal Analysis

X-ray diffraction is well suited to tracking down unit cell volume changes due to hydrogen uptake and to distinguishing between iron and palladium atoms, i.e., to investigating the atomic order. X-ray powder diffraction patterns of ordered FePd$_3$ exhibit anisotropic reflection broadening (Figure A1). Reflections common to both disordered (Cu type, $Fm\bar{3}m$) and ordered (AuCu$_3$ type, $Pm\bar{3}m$) are sharp (*hkl* all even or all odd), whereas those extinct by F centering and seen only in the ordered phase are considerably broader (*hkl* mixed even and odd), e.g., 100 with a half width at full maximum of 0.785°, 111 with 0.188°. This broadening can be attributed to small ordered domains joined by anti-phase boundaries to larger crystallites. This effect is also present in cubic MnPd$_3$ [26]. In the Rietveld refinement, the anisotropic reflection broadening was modeled by dividing the diffraction data set into two patterns with different regions of the 2θ range—one containing only reflections with *hkl* all even or all odd, and one containing all the others, i.e., those with *hkl* mixed even and odd (Figure A1 and Table A1). Each of the two patterns was treated independently in terms of profile parameters but constrained in terms of crystal structure parameters and scale factors. The refinement of occupation parameters (site occupation factors, *SOF*) for the 1*a* and 3*c* sites, with a stoichiometric constraint on the sum formula FePd$_3$, yielded *SOF*(Fe) = 0.876(2) at Wyckoff position 1*a*, i.e., about 12% palladium at iron sites (Figures 2 and A1, and Tables 1 and A1). The long-time annealed sample yielded *SOF*(Fe) = 0.99(2) instead, i.e., it was fully ordered (Figure A2 and Table A2).

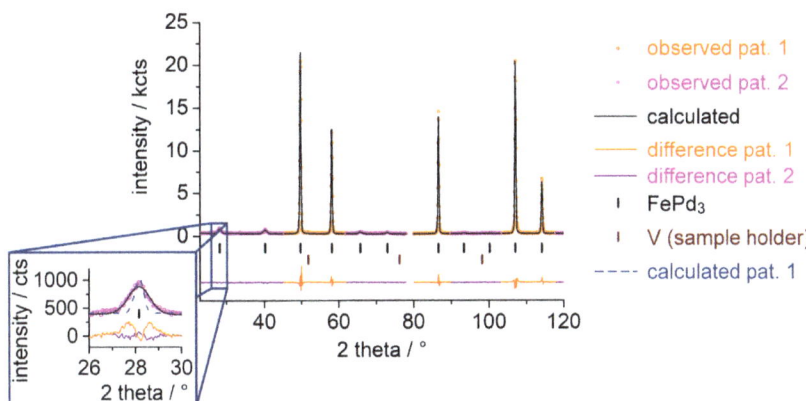

Figure 2. Simultaneous Rietveld refinement ($X^2 = 1.97$) of the crystal and magnetic structure of FePd$_3$ (AuCu$_3$ type, $Pm\bar{3}m$, $a = 385.204(6)$ pm, R_{Bragg}(pat.1) = 2.13, R_{Bragg}(pat.2) = 6.73, R_{magn}(pat.1) = 2.26, R_{magn}(pat.2) = 6.12; for further details, see Tables 1 and A5 and the text) at 296(1) K based on neutron powder diffraction data (NUMOR 131401 [21], λ =186.80(2) pm, D20 ILL, Grenoble, R_p(pat.1) = 0.057, R_p(pat.2) = 0.069, R_{wp}(pat.1) = 0.080, R_{wp}(pat.2) = 0.087, background corrected: R'_p(pat.1) = 0.083, R'_p(pat.2) = 0.667, R'_{wp}(pat.1) = 0.100, R'_{wp}(pat.2) = 0.410) and XRPD data (Figure A1, Co-Kα radiation, Smart Lab, R_p(pat.3) = 0.105, R_p(pat.4) = 0.102, R_{wp}(pat.3) = 0.150, R_{wp}(pat.4) = 0.140, R'_p(pat.3) = 0.160, R'_p(pat.4) = 1.90, R'_{wp}(pat.3) = 0.192, R'_{wp}(pat.4) = 0.804) using FullProf [24]. The inset shows the broadening of pattern 2 reflections through a comparison of intensities for 100 reflections calculated with the reflection width of pattern 1 (blue dashed line) and the reflection width of pattern 2 (black line) with the respective difference plot.

Table 1. Crystal structure parameters of FePd$_3$ ($Pm\bar{3}m$) based on neutron powder diffraction ($a = 385.204(6)$ pm, $\mu_{Fe} = 2.3(2)$ μ_B; see Figure 2) and XRPD ($a = 385.390(3)$ pm; compare Figure A1) at 296(1) K.

Atom	Wyckoff Position	x	y	z	$B_{iso1}/10^{-4}$ pm^2 NPD	$B_{iso2}/10^{-4}$ pm^2 XRPD	SOF
Fe1	1a	0	0	0	2.17(8)	0.19(4)	0.876(2)
Fe2	3c	0	$\frac{1}{2}$	$\frac{1}{2}$	1.11(12)	0.88(4)	(1 − SOF(Fe1))/3
Pd1	1a	0	0	0	B_{iso1}(Fe1)	B_{iso2}(Fe1)	1 − SOF(Fe1)
Pd2	3c	0	$\frac{1}{2}$	$\frac{1}{2}$	B_{iso1}(Fe2)	B_{iso2}(Fe2)	1 − (1 − SOF(Fe1))/3

The thermal analysis (DSC, p_{start}(H$_2$) = 5.00(2) MPa, T_{max} = 723 K) of the FePd$_3$ phase shows no thermal signal (Figure A3). However, the unit cell increases by 0.08% (a(FePd$_3$) = 385.330(5) pm, a(FePd$_3$H$_x$) = 385.433(3) pm, according to Rietveld analysis of XRPD data), and the disorder increases as well (SOF(Fe, 1a, FePd$_3$) = 0.909(10), SOF(Fe, 1a, FePd$_3$H$_x$) = 0.841(12)) based on X-ray powder diffraction (Mo-K$_{\alpha 1}$ radiation) before and after DSC (Figure A4).

In situ X-ray diffraction at 0.50(5) MPa hydrogen pressure on the long-time annealed FePd$_3$ sample shows a decrease in the level of atomic order. After the *in situ* chamber was flushed with hydrogen to a pressure of 0.50(5) MPa, SOF(Fe) decreased from 0.99(2) to 0.877(2). Furthermore, the cell volume increased by 0.2% (a(FePd$_3$) = 385.412(4) pm; a(FePd$_3$H$_x$) = 385.631(4) pm), indicating hydrogen uptake at room temperature. In the subsequent heating and cooling steps of 50 K to a maximum temperature of 550 K, SOF stays constant, and the lattice parameters change reversibly with temperature (Figures A5 and A6, and Tables A3 and A4), indicating that the hydrogen-induced disorder occurs at room temperature and does not proceed further at elevated temperatures.

3.3. Neutron Diffraction

3.3.1. *Ex Situ* Neutron Diffraction

The disordered $Fe_{0.25}Pd_{0.75}$ sample shows no significant cell volume increase (Figure A7, a_{before} = 385.12(4) pm, a_{after} = 385.18(2) pm) when subjected to deuterium gas ($p_{max}(D_2)$ = 8.3 MPa, T_{max} = 558 K). Furthermore, no deuterium atoms can be found in the crystal structure (zero occupation at Wyckoff position $4b$, $\frac{1}{2}, \frac{1}{2}, \frac{1}{2}$; analogous to PdH [27]), and the iron and palladium atoms remain disordered. Therefore, we conclude that disordered $Fe_{0.25}Pd_{0.75}$ does not take up hydrogen under the given conditions. This result is in accordance with hydrogenation studies under higher pressures [17].

The neutron diffraction pattern of ordered $FePd_3$ shows the same anisotropic reflection broadening (Figure 2) as observed with XRPD (*vide supra* and Figure A1). For modeling in Rietveld refinements, the same strategy of dividing into two patterns as described above for XRPD was used. Each of the patterns was constrained to the respective XRPD pattern in a simultaneous refinement on X-ray and neutron diffraction data. Please note that here, and for the following refinements, the residual values for pattern 2 are quite high because of low intensities and broad reflections (purple line in Figures 2, A1, A2, A4 and A7–A27). Approximately 12% palladium atoms on iron sites were found with a fixed composition of $FePd_3$ as a constraint (Table 1). The refined value for the magnetic moment of the iron atoms of 2.3(2) μ_B along [001] is in accordance with the literature data (μ_{Fe} = 2.73(13) μ_B [28]) within two standard uncertainties. Palladium atoms were not included in the magnetic structure, since the refinements did not converge, indicating small μ_{Pd} values. This is in line with the small magnetic moments of palladium atoms in $FePd_3$ found earlier (μ_{Pd} = 0.35 μ_B [29]).

The unit cell volume of ordered $FePd_3$ expanded by 0.20% upon deuterium uptake ($a(FePd_3)$ = 385.204(3) pm, $a(FePd_3D_{0.047(9)})$ = 385.372(2) pm). Deuterium atoms were localized by difference Fourier analysis. For the refinement of deuterium occupation, the magnetic moment of iron was fixed at 2.344 μ_B due to convergence problems. Two possible deuterium sites were tested by Rietveld refinement. The Wyckoff position $3d$ (0 0 $\frac{1}{2}$) yields a negative deuterium occupation and is thus considered to be empty. The occupation of deuterium in [Pd_6] octahedral voids at Wyckoff position $1b$ ($\frac{1}{2} \frac{1}{2} \frac{1}{2}$) was refined to 0.047(9). We therefore conclude the deuteride to be $FePd_3D_{0.047(9)}$ with an anti-perovskite type structure (Figure 1).

3.3.2. *In Situ* Neutron Diffraction

To study the influence of structural order in $FePd_3$ on hydrogenation properties, *in situ* neutron powder diffraction on an ordered sample using deuterium in a sapphire single-crystal cell was performed (Figure 3). Deuterium gas pressure was slowly increased up to 8.0 MPa before raising the temperature to about 550 K. After maintaining the temperature for one hour, the cell was cooled to room temperature under deuterium pressure. As seen in the 2D plot, the intensities of reflections with *hkl* mixed even and odd (e.g., at 2θ = 28°, 40°, 66°, 73°) decrease with increasing temperature. This observation indicates an increasing level of disorder and decreasing ordered magnetic moment. Reflections (2θ = 42.4°; 45.6°; 56.1°) at high temperatures are single-crystal reflections from the sapphire cell that shift into the range of the detector due to thermal expansion.

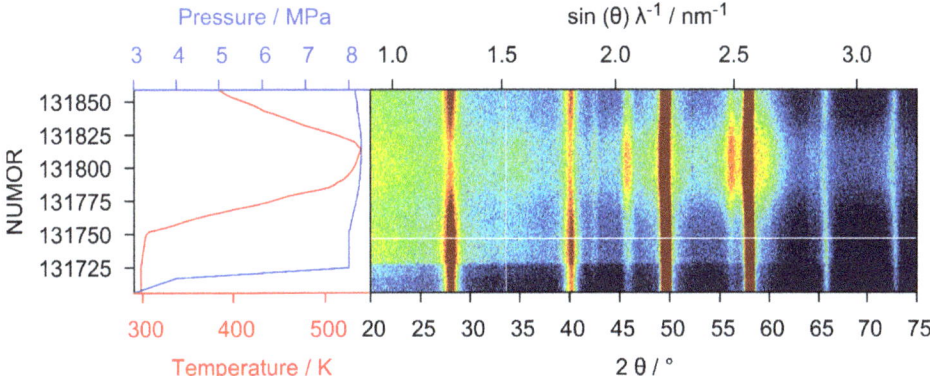

Figure 3. *In situ* neutron powder diffraction data (NUMOR 13,1706–13,1859 [21]) of the deuteration of ordered $FePd_3$ taken with diffractometer D20 (Grenoble, France) at λ = 186.80(2) Å in a single-crystal sapphire cell [19,20] under various temperature and deuterium pressure conditions. Intensities are in false colors.

The Rietveld refinements of selected NUMORs during the isothermal deuterium pressure increase were performed with fixed magnetic moments. Magnetic investigations of isotypic $FePd_3B_x$ show that only the ordered magnetic moment of the immediate boron environment (only Pd atoms) is changed upon boron incorporation [30], while the magnetic moment of the Fe atoms is not affected. Therefore, we consider fixing the magnetic moments of iron atoms to be an appropriate approximation. The refinements during the quasi-isobaric temperature variation were performed with fixed mixed occupancy. This is because *in situ* XRPD at 0.5 MPa hydrogen pressure showed no significant change, and a simultaneous refinement of the magnetic moment, the hydrogen occupation, and the mixed occupancy of the metal atoms did not converge. The lattice parameter a of $FePd_3D_x$ increases with rising deuterium pressure (Figure 4, left), which is reflected in the increasing deuterium content. At 8 MPa deuterium pressure, the composition $FePd_3D_{0.047(9)}$ is reached. The disorder increases with deuterium uptake, as seen from the iron occupation at Wyckoff position 1a decreasing from 0.875(3) to 0.794(3). In the heating step (T_{max} = 539 K) at 8.2(2) MPa deuterium pressure, the lattice parameter a increases almost linearly, and the deuterium content does not change significantly (Figure 4). The ordered magnetic moment decreases with increasing temperature to zero. The data were fitted with a function for a second-order transition based on the Landau theory with the magnetic moment as the order parameter:

$$\mu_{ord.} = A \left(\frac{T_C - T}{T_C} \right)^{\beta}$$

The resulting Curie temperature, T_C, is 495(8) K, and the critical exponent β equals 0.28(5) (Figure 4). The ordered magnetic moments at temperatures above the Curie temperature are not significantly different from zero or are set to zero at the maximum temperature of 539 K. All observed trends are fully reversible during the cooling step.

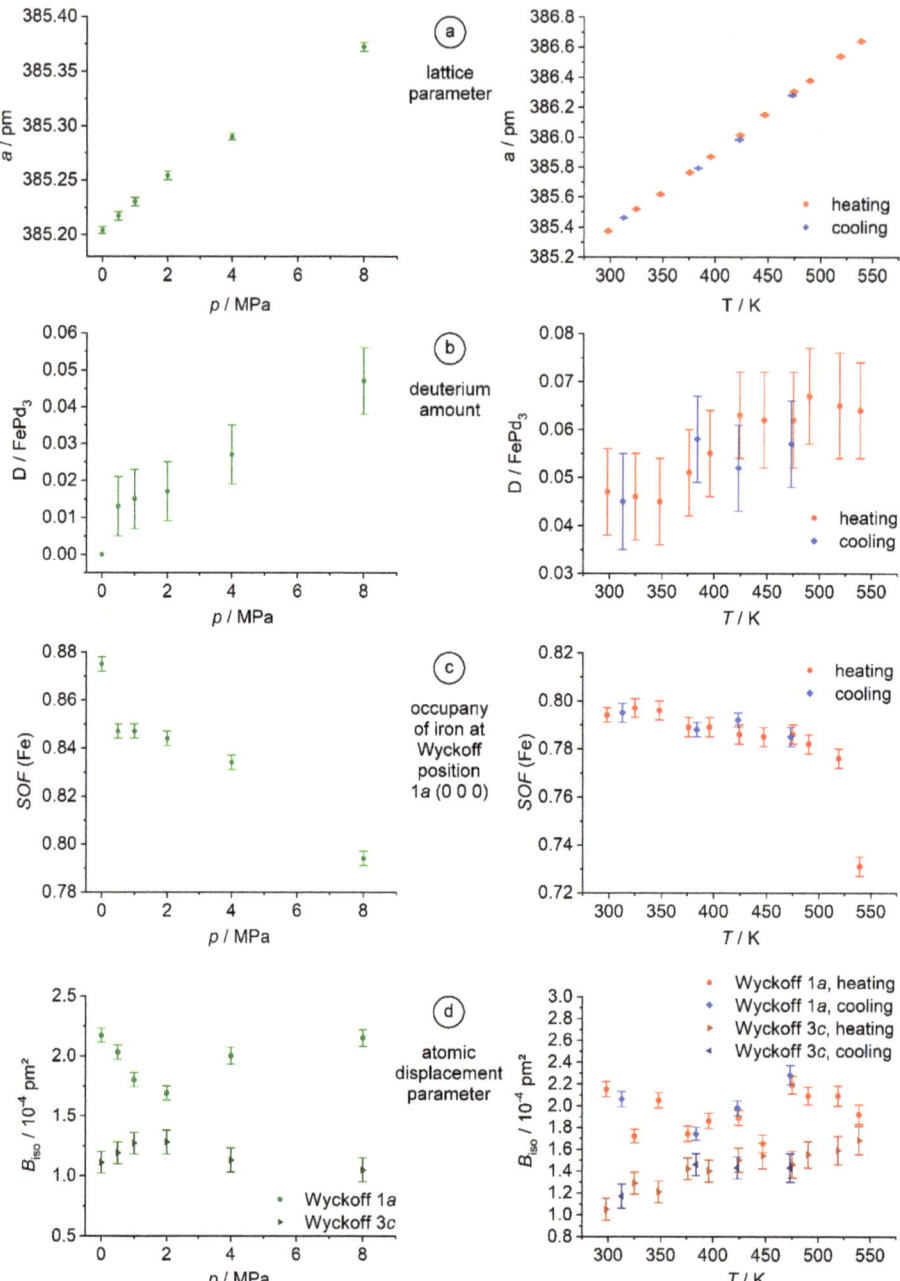

Figure 4. Refined structural parameters of ordered $FePd_3$ during isothermal increase in deuterium pressure (left, NUMOR 131613–131746 [21], T = 298(2) K) and isobaric variation of the temperature (right, NUMOR 131746–131894 [21], $p(D_2)$ = 8.2(2) MPa) based on *in situ* neutron powder diffraction data: lattice parameter (**a**), deuterium content per formula unit (**b**), SOF of iron at Wyckoff position 1a (**c**) and atomic displacement parameters of the metal atoms (**d**). Error bars represent $\pm\sigma$.

4. Discussion

The atomic order in $FePd_3$ can be controlled via the synthesis protocol, most easily by using iodine as a mineralizing agent. Its use allows full atomic order in one month as compared to 91% order in two months [16]. This method is well known for its potential to promote single-crystal growth [31,32], the synthesis of metastable compounds [32] or single-phase ordered compounds with shorter annealing times [31].

The magnetic moment of 2.3(2) μ_B of $FePd_3$ determined by refinement of neutron diffraction data differs somewhat from the literature values (μ_{Fe} = 2.73(13) μ_B [28]); however, the difference is less than two combined standard uncertainties. The difference may also be caused by varying disorder in $FePd_3$, which was not taken into account in early studies [28,33]. The Curie temperature of 495(8) K, determined by a second-order transition fit (Figure 5), is in accordance with the literature (T_c = 499 K [6]). The determined critical exponent of this fit (β = 0.28(5)) is close (less than two standard uncertainties apart) to the expected values of a 3D Ising model (β = 0.325) and a 3D Heisenberg model (β = 0.365) but far from the mean-field model (β = 0.5) [34]. This is in perfect agreement with a short-range interaction as typical for a magnetic exchange. The disparity between this and previous investigations reporting on a Heisenberg magnet with a critical exponent of 0.371 [6] may be explained by the method of determination. The ordered magnetic moment in this study is only refined in the 001 direction, resulting in a bias toward the 3D Ising model. Furthermore, the uncertainties are relatively high due to the correlation of the hydrogen occupation, the mixed occupation and the ordered magnetic moment.

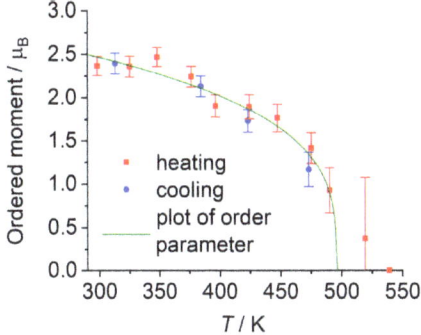

Figure 5. Temperature dependence of refined magnetic moment with fixed occupation (SOF(Fe,1a) = 0.798) in ordered $FePd_3$ under isobaric conditions ($p(D_2)$ = 8.2(2) MPa) based on *in situ* neutron powder diffraction data (see Figure 3). Green line shows the fitted function of the model of second-order phase transition ($\mu_{ord.} = 3.2(2) \mu_B \left(\frac{495(8)\ K - T}{495(8)\ K} \right)^{0.28(5)}$).

Deuterium occupies exclusively [Pd_6] octahedral voids in a statistical manner with small occupation parameters. The small deuterium contents are in accordance with studies at high gas pressures [17] (Figure 6, left). The Pd-D distances are between 192.644(2) pm and 192.686(2) pm for $FePd_3D_{0.013(8)}$ and $FePd_3D_{0.047(9)}$, respectively, and comparable to known hydrides of MPd_3 compounds, such as $MnPd_3H_{0.61}$ (190.0 pm $\leq d$(Pd-H) $\leq$ 197.6 pm) [35].

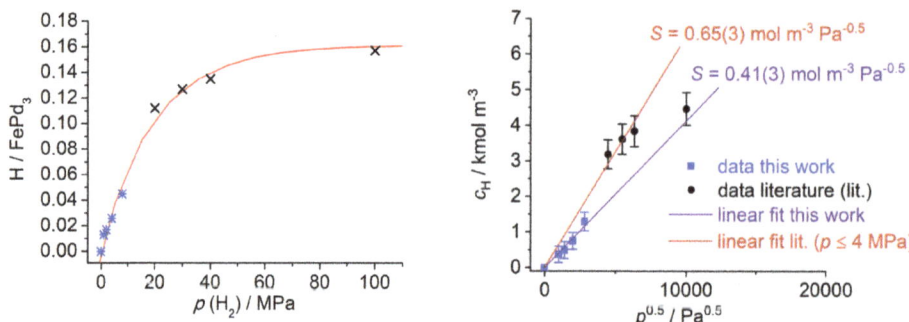

Figure 6. Plot of hydrogen pressure vs. hydrogen content (left) and Sieverts' law (right, $c_H = S \sqrt{p}$ [36]) of FePd$_3$H$_x$. Black symbols mark data from the literature [17], blue symbols from this work. Red line on the left-hand side shows an empirical fitted function ($x = -0.161(5)\mathrm{e}^{-0.050(5)p} + 0.161(4)$) and red and purple lines a linear fit based on Sieverts' law.

FePd$_3$ takes up much less hydrogen than, i.e., MgPd$_3$ [20], MnPd$_3$ [35,37] or InPd$_3$ [38]. This is in accordance with a proposed structure map correlating electronegativity and atomic radius of the metal M with the hydrogen content of MPd$_3$H$_x$ [39]. Iron, with quite a small atomic radius and an electronegativity of 1.6, is predicted to take up small amounts of hydrogen, which is confirmed in this study. Furthermore, the density of states at the Fermi level of FePd$_3$ (6.2 [40] or 11.1 states eV^{-1} atom^{-1} [41]) is remarkably high compared to other MPd$_3$ compounds crystallizing in the AuCu$_3$ type, such as MgPd$_3$ (1.13 states eV^{-1} atom^{-1} [42]), MnPd$_3$ (2.18 [43] or 2.84 states eV^{-1} atom^{-1} [44]) and InPd$_3$ (3.3 states eV^{-1} atom^{-1} [45]), which might also have an impact on the hydrogen uptake capacity. At room temperature, the hydrogenation follows Sieverts' law [36] (Figure 6, right), with low pressure (this work) and high pressure data [17] differing somewhat (Figure 6). The moderate fit and higher error in fit parameters for the high pressure data may be caused by increasing hydrogen–hydrogen interaction in the solid solution and larger differences between fugacity and pressure, the latter of which was used here as an approximation.

To compare the possible hydrogenation of ordered and disordered FePd$_3$, it is useful to look at the maximum hydrogen content of both structure types. The unit cell of MPd$_3$ compounds in the AuCu$_3$ type contains one [Pd$_6$] and three [M_2Pd$_4$] octahedral voids. Therefore, the probability of a [Pd$_6$] octahedral void is 0.25, yielding the formula MPd$_3$H for a maximum occupation of hydrogen in [Pd$_6$] sites. For disordered MPd$_3$ compounds crystallizing in the Cu type, the probability of a [Pd$_6$] octahedral void is 0.178 (=0.75^6), assuming a 75% probability of finding a palladium atom at any position in the crystal structure. A maximum hydrogen occupation using only [Pd$_6$] sites yields the formula MPd$_3$H$_{0.712}$. According to this consideration, the ordered compound may absorb more hydrogen if only [Pd$_6$] sites are involved. It is well known that hydrogen uptake in disordered FePd$_3$ is indeed less than that for ordered FePd$_3$ [17]. In this regard, it is remarkable that the disorder in FePd$_3$H$_x$ increases with hydrogen uptake despite the statistical decrease in [Pd$_6$] octahedral sites. Furthermore, the disorder reversibly increases with temperature, with near constant hydrogen content. The preference of [Pd$_6$] sites for hydrogen atoms can be inferred from many examples of hydrogenation reactions of MPd$_3$ compounds structurally related to cubic close packing. The thermodynamic driving force for hydrogen-induced rearrangements (from TiAl$_3$ type or ZrAl$_3$ type to AuCu$_3$ type) arises from the preference for Pd-H bonding and, consequently, an increase in the number of [Pd$_6$] octahedral voids [46]. Furthermore, other interstitial FePd$_3$ compounds, such as FePd$_3$B$_x$, also prefer [Pd$_6$] octahedral sites for the interstitial atoms [30]. The immediate environment of absorbed hydrogen is responsible for the hydrogen solubility in metals and intermetallic compounds [47]. Under the assumption of a preference of hydrogen atoms for [Pd_6] octahedra, the unexpected increasing disorder upon hydrogenation may be understood in terms of local short-range order of formed HPd_6 octahedra and a lack of long-range crystallographic order.

The findings on the magnetic and structural details of $FePd_3$ and its hydrides show the complex interplay between hydrogen uptake, atomic and magnetic order in intermetallics. This raises questions on the validity of the assumption of constant atomic order, which is often made for investigations on the effects of hydrogen incorporation on magnetic properties, and calls for further investigations on this fascinating subject.

5. Conclusions

The use of iodine as a mineralizing agent decreases the annealing time and enables higher ordering of metal atoms in the synthesis of $FePd_3$. At moderate hydrogen pressures ($p \leq 8$ MPa), disordered $Fe_{0.25}Pd_{0.75}$ absorbs a negligible amount of hydrogen, and ordered $FePd_3$ forms the hydride $FePd_3H_{0.047(9)}$. Hydrogen is incorporated at [Pd_6] octahedral voids, and the hydrogenation follows Sieverts' law. During heating, the ordered magnetic moment decreases, and $FePd_3H_x$ behaves like a 3D Ising or Heisenberg magnet. Simultaneously, the disorder of the metal atoms increases slightly. All temperature-dependent effects are fully reversible. Hydride formation in $FePd_3$ influences crystallographic and magnetic order alike. The citation of Carl G. Jung "In all chaos there is a cosmos, in all disorder a secret order" [48] can be understood as order only arising from chaos. In contrast to that, hydrogen seems to enhance the metal diffusion in $FePd_3$, resulting in long-range disorder arising from local order of the immediate hydrogen environment. This induced metal diffusion and the resulting change in the arrangement of the metal atoms at the surface might raise interest in catalysis.

Author Contributions: Conceptualization, A.G. and H.K.; methodology, A.G., S.C.S., T.C.H. and H.K.; software, A.G. and T.C.H.; validation, A.G., S.C.S., T.C.H. and H.K.; formal analysis, A.G.; investigation, A.G., S.C.S., T.C.H. and H.K.; resources, H.K. and T.C.H.; data curation, A.G. and T.C.H.; writing—original draft preparation, A.G.; writing—review and editing, A.G., S.C.S., T.C.H. and H.K.; visualization, A.G.; supervision, H.K.; project administration, A.G. and H.K.; funding acquisition, H.K. All authors have read and agreed to the published version of the manuscript.

Funding: This work was funded by the Deutsche Forschungsgemeinschaft (Grant 448675425 (KO1803/15-1) and INST 268/379/1 FUGG).

Acknowledgments: We acknowledge the Institut Laue Langevin for provision of beam time with a high-intensity powder diffractometer D20. We thank Simon Keilholz for support with the *in situ* XRPD measurements.

Conflicts of Interest: The authors declare no conflict of interest. The funders had no role in the design of the study; in the collection, analyses, or interpretation of data; in the writing of the manuscript, or in the decision to publish the results.

Appendix A

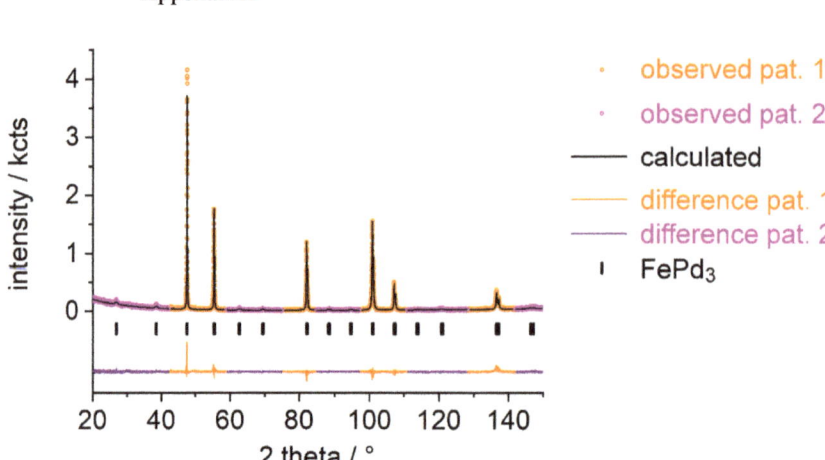

Figure A1. Rietveld refinement of the crystal structure of ordered FePd$_3$ (AuCu$_3$ type, $Pm\bar{3}m$, a = 385.380(5) pm, R_{Bragg}(pat.1) = 5.45, R_{Bragg}(pat.2) = 49.4; for further details, see Table A1) at 296(1) K based on XRPD powder diffraction data (Co-Kα radiation, Smart Lab, R_p(pat.1) = 0.105, R_p(pat.2) = 0.102, R_{wp}(pat.1) = 0.150, R_{wp}(pat.2) = 0.140, X^2 = 2.24, background corrected: R'_p(pat.1) = 0.124, R'_p(pat.2) = 1.90, R'_{wp}(pat.1) = 0.163, R'_{wp}(pat.2) = 0.804) using FullProf [24].

Table A1. Crystal structure parameters of FePd$_3$ ($Pm\bar{3}m$, a = 385.380(5) pm) based on XRPD data (see Figure A1) at 296(1) K.

Atom	Wyckoff Position	x	y	z	$B_{iso}/10^{-4}$ pm^2	SOF
Fe1	1a	0	0	0	0.18(4)	0.87(2)
Fe2	3c	0	$\frac{1}{2}$	$\frac{1}{2}$	0.88(1)	(1 − SOF(Fe1))/3
Pd1	1a	0	0	0	B_{iso}(Fe1)	1 − SOF(Fe1)
Pd2	3c	0	$\frac{1}{2}$	$\frac{1}{2}$	B_{iso}(Fe2)	1 − (1 − SOF(Fe1))/3

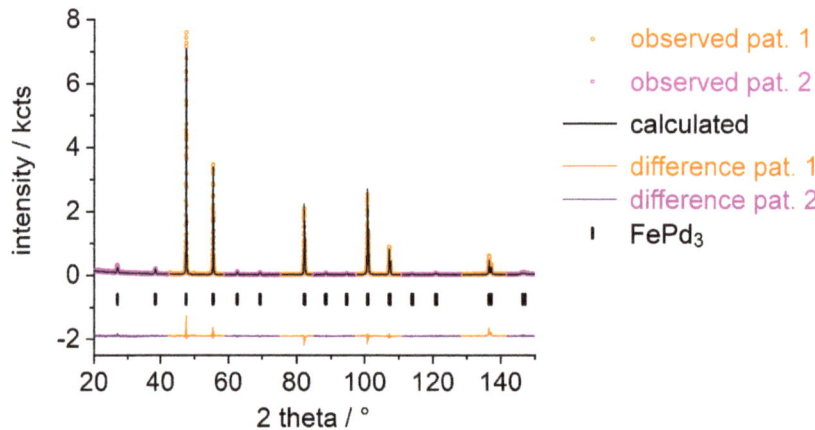

Figure A2. Rietveld refinement of the crystal structure of ordered FePd$_3$ (long-time annealed sample, AuCu$_3$ type, $Pm\bar{3}m$, a = 385.412(4) pm, R_{Bragg}(pat.1) = 8.01, R_{Bragg}(pat.2) = 27.1; for further details, see Table A2) at 296(1) K based on XRPD powder diffraction data (Co-Kα radiation, Smart Lab, R_p(pat.1) = 0.109, R_p(pat.2) = 0.125, R_{wp}(pat.1) = 0.161, R_{wp}(pat.2) = 0.173, X^2 = 3.79, background corrected: R'_p(pat.1) = 0.147, R'_p(pat.2) = 1.23, R'_{wp}(pat.1) = 0.161, R'_{wp}(pat.2) = 0.553) using FullProf [24].

Table A2. Crystal structure parameters of FePd$_3$ (long-time annealed sample, $Pm\bar{3}m$, a = 385.412(4) pm) based on XRPD data (see Figure A2) at 296(1) K.

Atom	Wyckoff Position	x	y	z	$B_{iso}/10^{-4}$ pm^2	SOF [1]
Fe1	1a	0	0	0	1.5(2)	0.99(2)
Fe2	3c	0	$\frac{1}{2}$	$\frac{1}{2}$	1.88(4)	$(1 - SOF(Fe1))/3$
Pd1	1a	0	0	0	B_{iso}(Fe1)	$1 - SOF(Fe1)$
Pd2	3c	0	$\frac{1}{2}$	$\frac{1}{2}$	B_{iso}(Fe2)	$1 - (1 - SOF(Fe1))/3$

[1] The stoichiometric ratio of Fe to Pd atoms was fixed at 1:3.

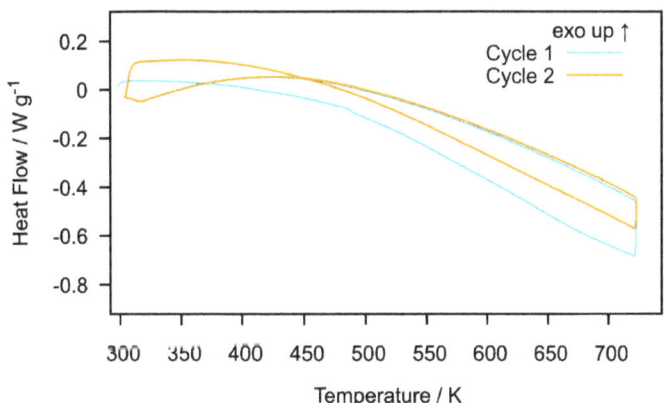

Figure A3. Differential scanning calorimetry (DSC) of ordered FePd$_3$ at 5.0 MPa hydrogen pressure.

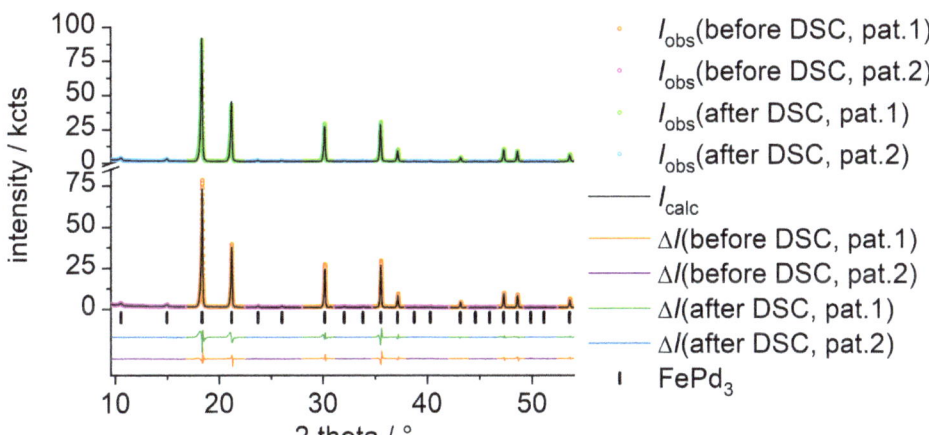

Figure A4. Rietveld refinement of the crystal structure of ordered FePd$_3$ before DSC (bottom, AuCu$_3$ type, $Pm\bar{3}m$, a = 385.330(5) pm, SOF(Fe, 1a) = 0.909(10), B_{iso}(1a) = 0.35(9) 10^{-4} pm^2, B_{iso}(3c) = 0.91(2) 10^{-4} pm^2, R_{Bragg}(pat.1) = 2.83, R_{Bragg}(pat.2) = 2192, R_P(pat.1) = 0.089, R_P(pat.2) = 0.020, R_{wp}(pat.1) = 0.142, R_{wp}(pat.2) = 0.025, GOF = 5.09) and after DSC (top, AuCu$_3$ type, $Pm\bar{3}m$, a = 385.433(3) pm, SOF(Fe, 1a) = 0.841(12), B_{iso}(1a) = 0.96(12) 10^{-4} pm^2, B_{iso}(3c) = 0.69(3) 10^{-4} pm^2, R_{Bragg}(pat.1) = 4.10, R_{Bragg}(pat.2) = 2150, R_P(pat.1) = 0.131, R_P(pat.2) = 0.024, R_{wp}(pat.1) = 0.189, R_{wp}(pat.2) = 0.030, GOF = 5.09) at 297(1) K based on XRPD powder diffraction data (Mo-K$_{\alpha 1}$ radiation, Huber G670) using Topas [25].

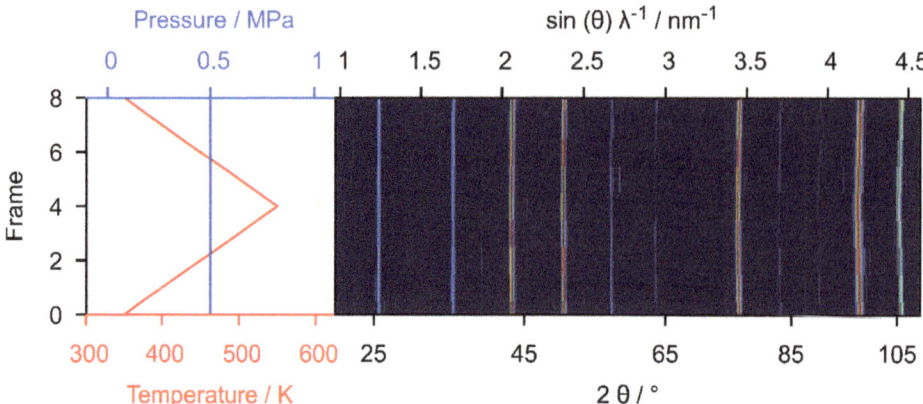

Figure A5. *In situ* X-ray powder diffraction data of the hydrogenation of ordered $FePd_3$ (long-time annealed sample) taken with SmartLab diffractometer in an XRK 900 reaction chamber with Co-Kα radiation under various temperatures and 0.5 MPa hydrogen pressure. λ_1 is used for sin (θ) λ^{-1} scale. Intensities are in false colors.

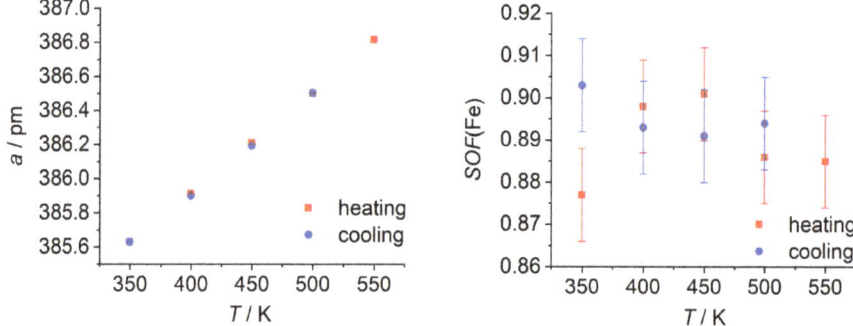

Figure A6. Refined structural parameters of ordered $FePd_3$ (long-time annealed sample, see Figure A5 and Tables A3 and A4) during isobaric variation of the temperature ($p(H_2)$ = 0.50(5) MPa) based on *in situ* XRPD data: lattice parameter (left) and SOF of iron at Wyckoff position 1a (right). Error bars represent $\pm \sigma$.

Table A3. Conditions and refined parameters of the Rietveld refinement of $FePd_3$ (long-time annealed sample, $Pm\bar{3}m$) based on *in situ* XRPD data (see Figures A5 and A6, and Table A4) using Topas; X-ray absorption modeled with an overall B value of -4×10^{-4} pm^2 [25].

Frame	T/K	p/MPa	a/pm	V/10^6 pm^3	$B_{iso}(1a)/10^{-4}$ pm^2	$B_{iso}(3c)/10^{-4}$ pm^2	SOF(Fe, 1a) [1]
0	350 (2)	0.50 (5)	385.631 (4)	57.348 (2)	1.9 (2)	2.54 (4)	0.877 (11)
1	400 (2)	0.50 (5)	385.914 (4)	57.474 (2)	1.8 (2)	2.67 (4)	0.898 (11)
2	450 (2)	0.50 (5)	386.211 (5)	57.607 (2)	1.8 (2)	2.76 (5)	0.901 (11)
3	500 (2)	0.50 (5)	386.503 (5)	57.738 (2)	1.9 (2)	2.79 (5)	0.886 (11)
4	550 (2)	0.50 (5)	386.818 (5)	57.879 (2)	2.2 (2)	2.97 (5)	0.885 (11)
5	500 (2)	0.50 (5)	386.506 (5)	57.739 (2)	2.0 (2)	2.80 (5)	0.894 (11)
6	450 (2)	0.50 (5)	386.194 (4)	57.599 (2)	2.2 (2)	2.68 (4)	0.891 (11)
7	400 (2)	0.50 (5)	385.902 (4)	57.469 (2)	1.8 (2)	2.71 (4)	0.893 (11)
8	350 (2)	0.50 (5)	385.632 (4)	57.348 (2)	1.5 (2)	2.54 (4)	0.903 (11)

[1] The stoichiometric ratio of Fe to Pd atoms was fixed at 1:3.

Table A4. Residual parameters of the Rietveld refinement of FePd$_3$ (long-time annealed sample, $Pm\bar{3}m$) based on *in situ* XRPD data (see Figures A5 and A6, and Table A3) using Topas [25].

Frame	R_{p1}	R_{p2}	R_{wp1}	R_{wp2}	χ^2	R_{Bragg}
0	0.155	0.41	0.273	0.51	1.06	5.34
1	0.158	0.41	0.275	0.51	1.08	5.78
2	0.153	0.42	0.266	0.52	0.98	5.7
3	0.154	0.42	0.263	0.52	0.96	5.58
4	0.161	0.42	0.270	0.53	1.02	5.84
5	0.156	0.41	0.268	0.52	1.00	5.74
6	0.150	0.41	0.262	0.52	0.96	5.83
7	0.151	0.42	0.266	0.52	1.00	5.44
8	0.152	0.42	0.265	0.52	1.00	5.38

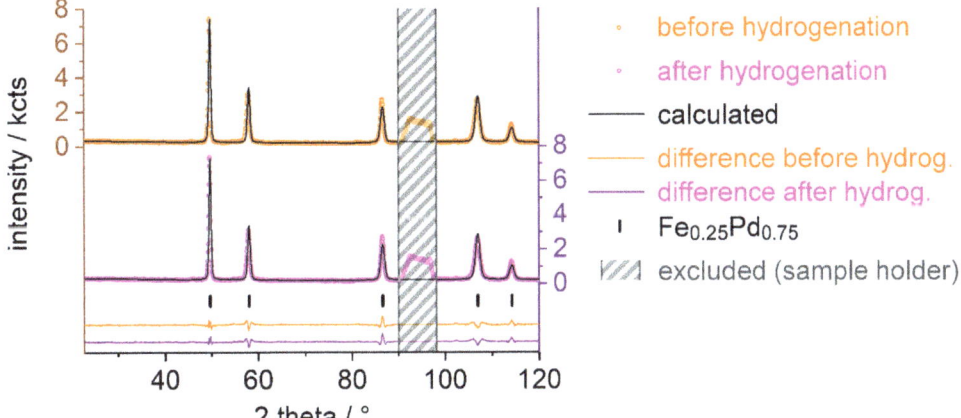

Figure A7. Rietveld refinement of the crystal structure of disordered Fe$_{0.25}$Pd$_{0.75}$ before hydrogenation (top, Cu type, $Fm\bar{3}m$, a = 385.12(4) pm, $B_{iso}(1a)$ = 1.52(5) 10^{-4} pm^2, R_{Bragg}(pat.1) = 3.39, R_p(pat.1) = 0.073, R_{wp}(pat.1) = 0.093, GOF = 2.54) and after hydrogenation (p_{max}(D$_2$) = 8.3 MPa, T_{max} = 558 K) at 8.00(2) MPa deuterium pressure (bottom, Cu type, $Fm\bar{3}m$, a = 385.18(5) pm, $B_{iso}(1a)$ = 1.47(5) 10^{-4} pm^2, R_{Bragg}(pat.1) = 3.47, R_p(pat.1) = 0.065, R_{wp}(pat.1) = 0.084, GOF = 3.61) at 297(1) K based on neutron powder diffraction data (NUMORs 132451 and 132636 [21], λ = 186.80(2) pm, D20 ILL, Grenoble) using FullProf [24].

Table A5. Set constraints of the simultaneous Rietveld refinement of FePd$_3$ ($Pm\bar{3}m$) based on NPD (Figure 2) and XRPD data (Figure A1) using FullProf [24].

	Phase1	Phase2 (Magnetic Phase)	Phase3
contribution to pattern [1]	pattern 1 NPD pattern 2 NPD	pattern 1 NPD pattern 2 NPD	pattern 1 XRPD pattern 2 XRPD
a	a1	a1	a3
So(Fe)	SOF(Fe)	SOF(Fe)	SOF(Fe)
$B_{iso}(1a)$	$B_{iso}1(1a)$	$B_{iso}1(1a)$	$B_{iso}3(1a)$
$B_{iso}(3c)$	$B_{iso}1(3c)$	$B_{iso}1(3c)$	$B_{iso}3(3c)$

[1] Scale factors of patterns 1 and 2 of NPD and XRPD data are constrained, respectively; the profile parameters of each pattern are refined separately.

Table A6. Conditions and refined parameters of the Rietveld refinement of FePd$_3$ ($Pm\bar{3}m$) based on *in situ* NPD data (see Figures 2–4 and A9–A14, and Table A7) using FullProf [24].

p/MPa	T/K	a/pm	$V/10^6$ pm^3	$B_{iso}(1a)/10^{-4}$ pm^2	$B_{iso}(3c)/10^{-4}$ pm^2	SOF(Fe, 1a)$_{1,2}$	SOF(D)
0.0001 (1)	296 (1)	385.204 (3)	57.157 (1)	2.17 (6)	1.11	0.875 (3)	–[2]
0.50 (1)	296 (1)	385.217 (4)	57.163 (1)	2.03 (6)	1.19	0.847 (4)	0.013 (8)
1.00 (1)	296 (1)	385.230 (3)	57.169 (1)	1.80 (6)	1.27	0.847 (4)	0.015 (8)
2.00 (1)	296 (1)	385.254 (4)	57.180 (1)	1.69 (6)	1.28	0.844 (4)	0.017 (8)
4.00 (1)	296 (1)	385.290 (3)	57.196 (1)	2.00 (7)	1.13	0.834 (4)	0.027 (8)
8.00 (2)	296 (1)	385.372 (4)	57.232 (1)	2.15 (7)	1.05	0.794 (4)	0.047 (9)

[1] The stoichiometric ratio of Fe to Pd atoms was fixed at 1:3; [2] the Wyckoff position 1b is not occupied.

Table A7. Residual parameters of the Rietveld refinement of FePd$_3$ ($Pm\bar{3}m$) based on *in situ* NPD data (see Figures 2–4 and A9–A14, and Table A6) using FullProf [24].

p/MPa	R_{p1}	R_{p2}	R_{wp1}	R_{wp2}	χ^2	R_{Bragg1}	R_{Bragg2}
0.0001 (10)	0.057	0.069	0.080	0.087	2.38	2.15	7.03
0.50 (1)	0.058	0.073	0.084	0.090	2.49	2.59	15.1
1.00 (1)	0.057	0.068	0.080	0.086	2.42	2.32	11.0
2.00 (1)	0.057	0.065	0.079	0.083	2.37	2.14	10.3
4.00 (1)	0.056	0.064	0.077	0.081	2.32	1.97	13.2
8.00 (1)	0.058	0.062	0.077	0.081	2.52	2.74	18.3

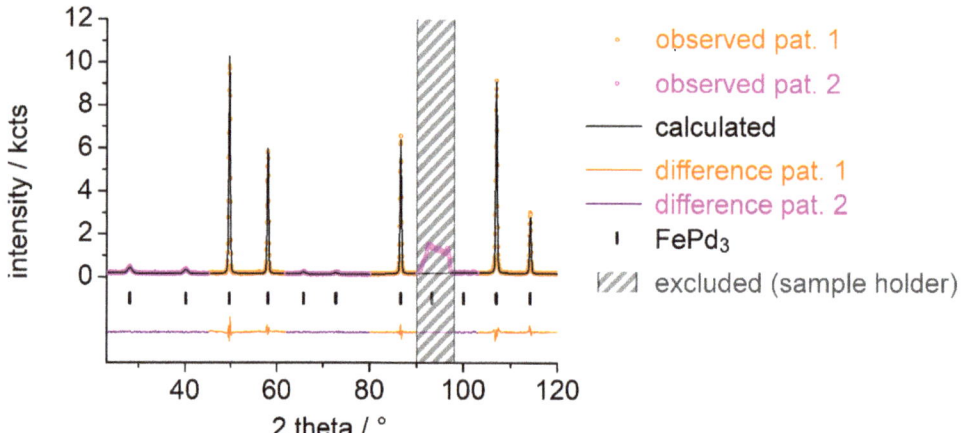

Figure A8. Rietveld refinement of the crystal structure of FePd$_3$ (for details, see first row in Tables A6 and A7) based on neutron powder diffraction data (NUMOR 131613 [21], λ = 186.80(2) pm, D20 ILL, Grenoble) in a single-crystal sapphire cell at 296(1) K and applied vacuum (0.0001(10) MPa) using FullProf [24].

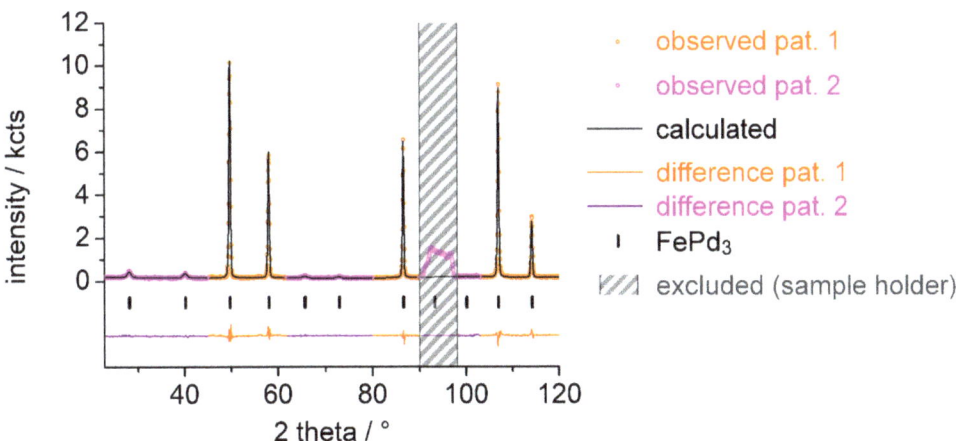

Figure A9. Rietveld refinement of the crystal structure of FePd$_3$D$_{0.013(8)}$ (for details, see second row in Tables A6 and A7) based on neutron powder diffraction data (NUMOR 131629 [21], λ = 186.80(2) pm, D20 ILL, Grenoble) in a single-crystal sapphire cell at 296(1) K and 0.50(1) MPa deuterium pressure using FullProf [24].

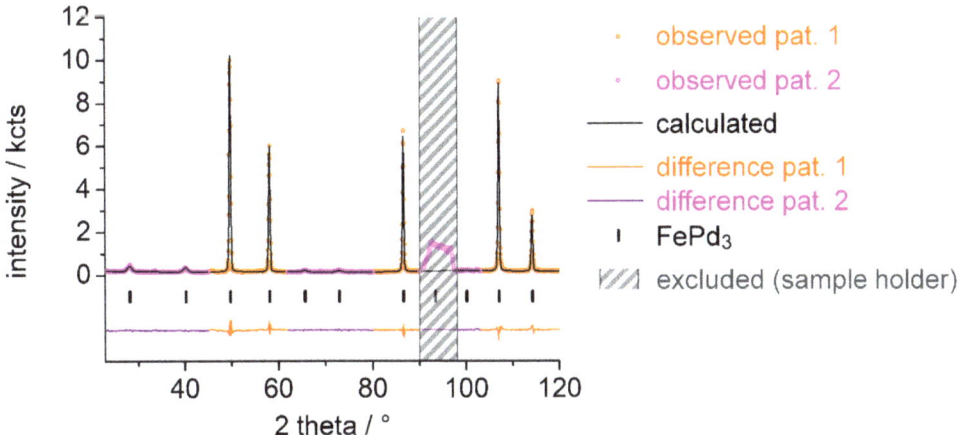

Figure A10. Rietveld refinement of the crystal structure of FePd$_3$D$_{0.015(8)}$ (for details, see third row in Tables A6 and A7) based on neutron powder diffraction data (NUMOR 131634 [21], λ = 186.80(2) pm, D20 ILL, Grenoble) in a single-crystal sapphire cell at 296(1) K and 1.00(1) MPa deuterium pressure using FullProf [24].

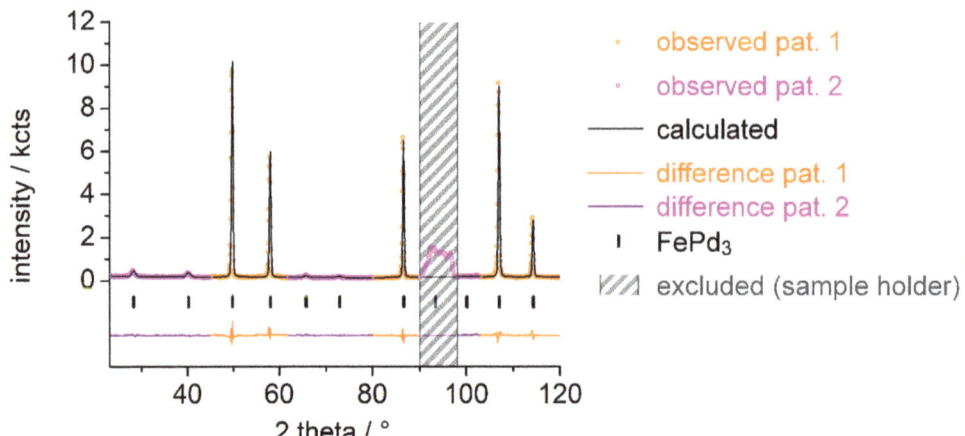

Figure A11. Rietveld refinement of the crystal structure of FePd$_3$D$_{0.017(8)}$ (for details, see fourth row in Tables A6 and A7) based on neutron powder diffraction data (NUMOR 131642 [21], λ = 186.80(2) pm, D20 ILL, Grenoble) in a single-crystal sapphire cell at 296(1) K and 2.00(1) MPa deuterium pressure using FullProf [24].

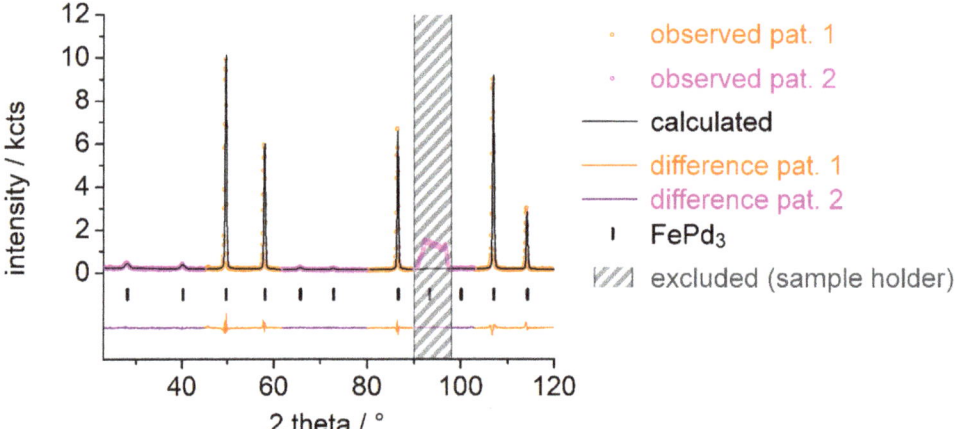

Figure A12. Rietveld refinement of the crystal structure of FePd$_3$D$_{0.027(8)}$ (for details, see fifth row in Tables A6 and A7) based on neutron powder diffraction data (NUMOR 131717 [21], λ = 186.80(2) pm, D20 ILL, Grenoble) in a single-crystal sapphire cell at 296(1) K and 4.00(1) MPa deuterium pressure using FullProf [24].

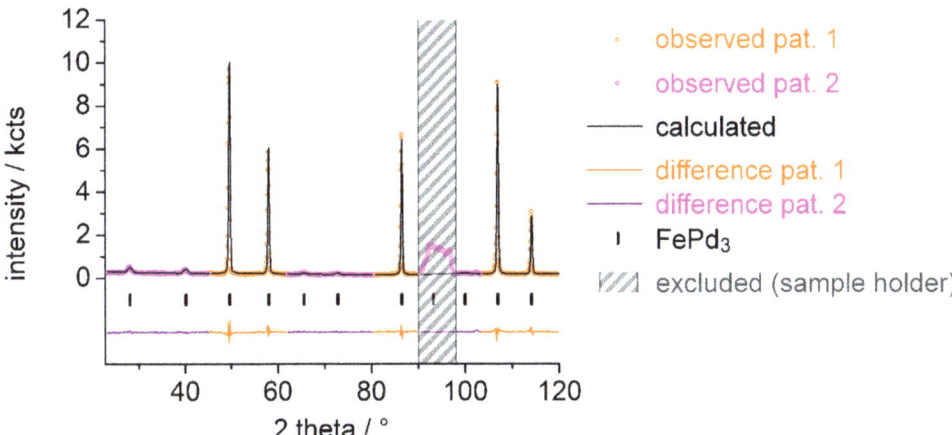

Figure A13. Rietveld refinement of the crystal structure of FePd$_3$D$_{0.047(9)}$ (for details, see sixth row in Tables A6 and A7, or first row in Tables A8 and A9) based on neutron powder diffraction data (NUMOR 131746 [21], λ = 186.80(2) pm, D20 ILL, Grenoble) in a single-crystal sapphire cell at 296(1) K and 8.00(1) MPa deuterium pressure using FullProf [24].

Table A8. Conditions and refined parameters of the Rietveld refinement of FePd$_3$ ($Pm\bar{3}m$) based on *in situ* NPD data (see Figures 2–4 and Table A9) using FullProf [24].

T/K	p/MPa	a/pm	V/10^6 pm^3	B_{iso}(1a)/10^{-4} pm^2	B_{iso}(3c)/10^{-4} pm^2	SOF(Fe, 1a) [1,2]	SOF(D)	R_z(Fe)/μ_B [1]
298 (1)	8.00 (1)	385.372 (4)	57.232 (1)	2.15 (7)	1.05	0.794 (3)	0.047 (9)	2.37 (11)
325 (3)	8.03 (1)	385.518 (4)	57.297 (1)	1.72 (6)	1.29	0.797 (4)	0.046 (9)	2.36 (12)
348 (3)	8.06 (1)	385.618 (3)	57.342 (1)	2.05 (7)	1.21	0.796 (4)	0.045 (9)	2.47 (11)
376 (3)	8.10 (1)	385.762 (4)	57.406 (1)	1.74 (7)	1.42	0.789 (4)	0.051 (9)	2.24 (12)
396 (3)	8.12 (1)	385.868 (4)	57.453 (1)	1.86 (7)	1.4	0.789 (4)	0.055 (9)	1.90 (13)
424 (3)	8.15 (1)	386.013 (4)	57.518 (1)	1.89 (7)	1.5	0.786 (4)	0.063 (9)	1.89 (14)
447 (3)	8.18 (1)	386.149 (4)	57.579 (1)	1.65 (8)	1.54	0.785 (4)	0.062 (10)	1.8 (2)
475 (3)	8.21 (1)	386.304 (4)	57.649 (1)	2.19 (8)	1.46	0.786 (4)	0.062 (10)	1.4 (2)
490 (3)	8.22 (1)	386.377 (4)	57.681 (1)	2.09 (8)	1.55	0.782 (4)	0.067 (10)	0.9 (3)
519 (2)	8.25 (1)	386.539 (4)	57.754 (1)	2.09 (9)	1.59	0.776 (4)	0.065 (11)	0.4 (7)
539 (1)	8.29 (1)	386.639 (4)	57.799 (1)	1.92 (9)	1.68	0.731 (4)	0.064 (10)	0 [3]
473 (3)	8.25 (1)	386.280 (4)	57.638 (1)	2.28 (9)	1.43	0.785 (4)	0.057 (9)	1.2 (2)
423 (2)	8.20 (1)	385.983 (4)	57.505 (1)	1.98 (7)	1.43	0.792 (3)	0.052 (9)	1.73 (13)
384 (1)	8.14 (1)	385.792 (3)	57.420 (1)	1.74 (6)	1.46	0.788 (3)	0.058 (9)	2.13 (12)
313 (1)	8.00 (1)	385.461 (4)	57.272 (1)	2.06 (7)	1.17	0.795 (4)	0.045 (10)	2.39 (12)

[1] R_z was refined with fixed occupation from coupled refinement (Table 1); then, SOF was refined with fixed R_z;
[2] The stoichiometric ratio of Fe to Pd atoms was fixed at 1:3; [3] R_z was fixed at 0.

Table A9. Residual parameters of the Rietveld refinement of FePd$_3$ ($Pm\bar{3}m$) based on *in situ* NPD data (see Figures 2–4 and Table A8) using FullProf [24].

T/K	R_{p1}	R_{p2}	R_{wp1}	R_{wp2}	χ^2	R_{Bragg1}	R_{Bragg2}	R_{Magn1}	R_{Magn2}
298 (1)	0.058	0.062	0.077	0.081	2.52	2.74	18.3	2.60	20.5
325 (3)	0.059	0.058	0.081	0.074	2.54	3.41	21.3	3.97	14.5
348 (3)	0.056	0.059	0.075	0.073	2.29	2.03	9.46	2.38	9.68
376 (3)	0.054	0.057	0.075	0.073	2.31	2.07	10.5	2.28	10.3
396 (3)	0.054	0.058	0.075	0.074	2.33	2.74	13.6	3.28	10.2
424 (3)	0.057	0.055	0.079	0.070	2.45	3.38	13.5	3.59	12.7
447 (3)	0.058	0.057	0.080	0.072	2.54	3.57	9.79	3.31	6.46

Table A9. Cont.

T/K	R_{p1}	R_{p2}	R_{wp1}	R_{wp2}	χ^2	R_{Bragg1}	R_{Bragg2}	R_{Magn1}	R_{Magn2}
475 (3)	0.060	0.055	0.083	0.071	2.71	4.00	17.0	4.14	15.9
490 (3)	0.058	0.057	0.082	0.073	2.70	3.69	26.1	3.86	22.7
519 (2)	0.060	0.055	0.087	0.069	2.95	4.56	27.0	5.32	25.0
539 (1)	0.058	0.056	0.085	0.070	2.86	3.65	31.6	-[1]	-[1]
473 (2)	0.059	0.058	0.082	0.074	2.78	4.38	36.9	4.51	24.7
423 (2)	0.057	0.055	0.079	0.070	2.42	3.38	28.0	3.73	12.7
384 (1)	0.053	0.059	0.073	0.074	2.24	2.44	9.15	2.55	10.3
313 (1)	0.062	0.062	0.085	0.078	2.79	3.16	31.5	3.98	15.1

[1] R_z was fixed at 0.

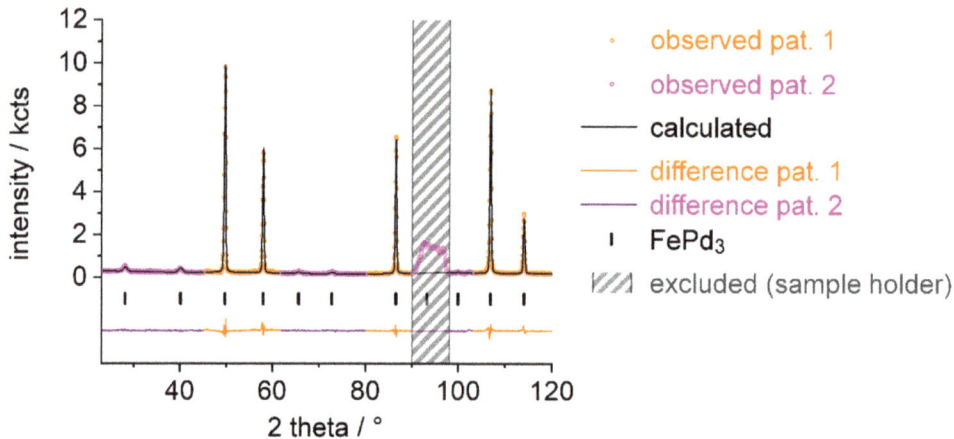

Figure A14. Rietveld refinement of the crystal structure of FePd$_3$D$_{0.046(9)}$ (for details, see second row in Tables A8 and A9) based on neutron powder diffraction data (NUMOR 131755 [21], λ = 186.80(2) pm, D20 ILL, Grenoble) in a single-crystal sapphire cell at 325(3) K and 8.03(1) MPa deuterium pressure using FullProf [24].

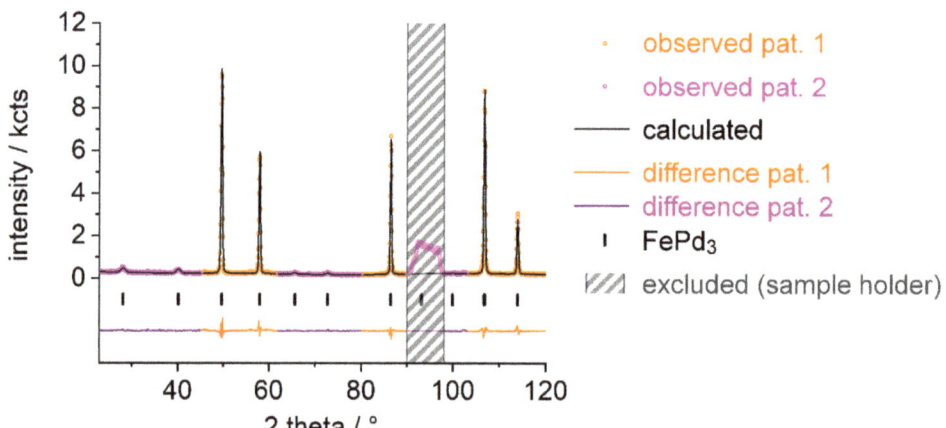

Figure A15. Rietveld refinement of the crystal structure of FePd$_3$D$_{0.045(9)}$ (for details, see third row in Tables A8 and A9) based on neutron powder diffraction data (NUMOR 131760 [21], λ = 186.80(2) pm, D20 ILL, Grenoble) in a single-crystal sapphire cell at 348(3) K and 8.06(1) MPa deuterium pressure using FullProf [24].

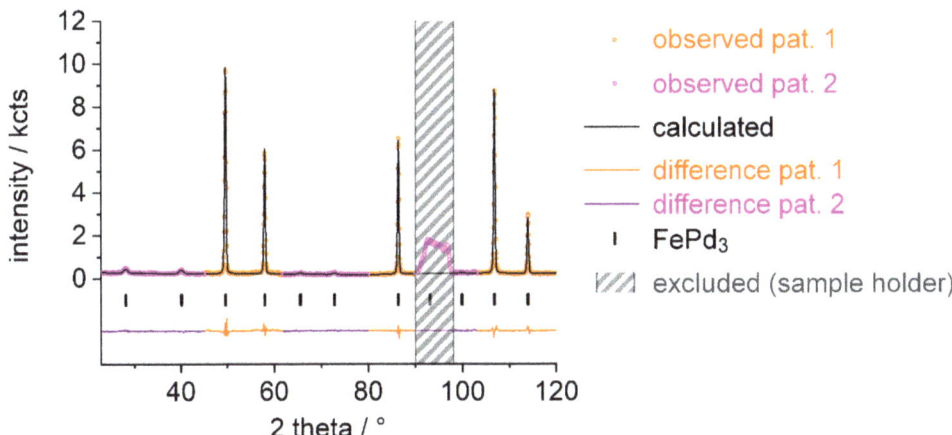

Figure A16. Rietveld refinement of the crystal structure of FePd$_3$D$_{0.051(9)}$ (for details, see fourth row in Tables A8 and A9) based on neutron powder diffraction data (NUMOR 131765 [21], λ = 186.80(2) pm, D20 ILL, Grenoble) in a single-crystal sapphire cell at 376(3) K and 8.10(1) MPa deuterium pressure using FullProf [24].

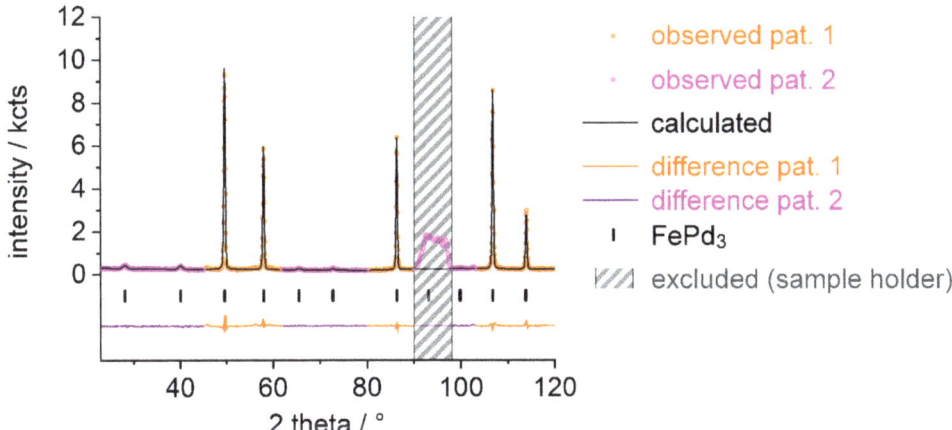

Figure A17. Rietveld refinement of the crystal structure of FePd$_3$D$_{0.055(9)}$ (for details, see fifth row in Tables A8 and A9) based on neutron powder diffraction data (NUMOR 131768 [21], λ = 186.80(2) pm, D20 ILL, Grenoble) in a single-crystal sapphire cell at 396(3) K and 8.12(1) MPa deuterium pressure using FullProf [24].

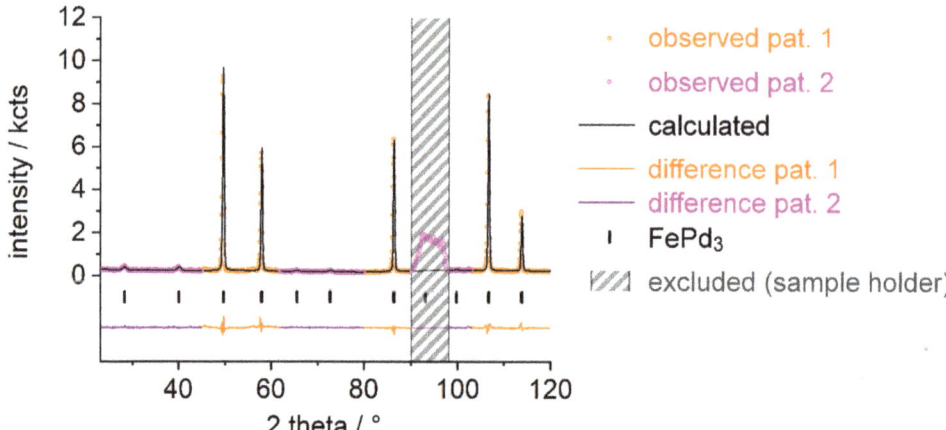

Figure A18. Rietveld refinement of the crystal structure of FePd$_3$D$_{0.063(9)}$ (for details, see sixth row in Tables A8 and A9) based on neutron powder diffraction data (NUMOR 131772 [21], λ = 186.80(2) pm, D20 ILL, Grenoble) in a single-crystal sapphire cell at 424(3) K and 8.15(1) MPa deuterium pressure using FullProf [24].

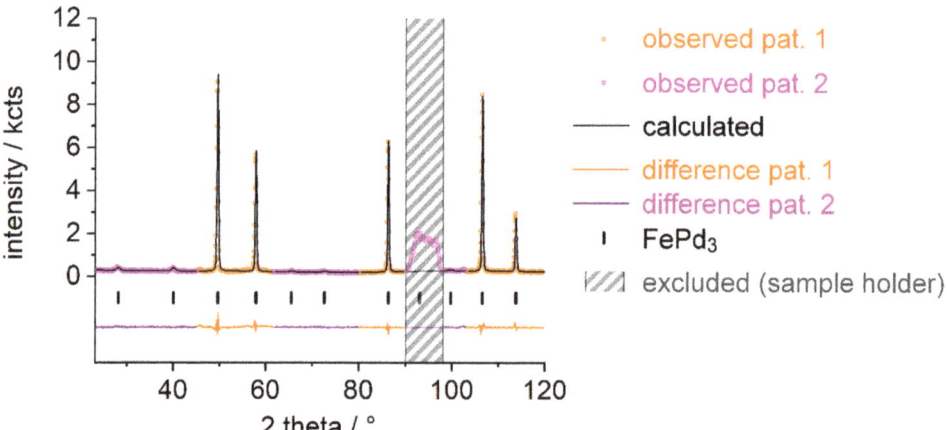

Figure A19. Rietveld refinement of the crystal structure of FePd$_3$D$_{0.062(10)}$ (for details, see seventh row in Tables A8 and A9) based on neutron powder diffraction data (NUMOR 131776 [21], λ = 186.80(2) pm, D20 ILL, Grenoble) in a single-crystal sapphire cell at 447(3) K and 8.18(1) MPa deuterium pressure using FullProf [24].

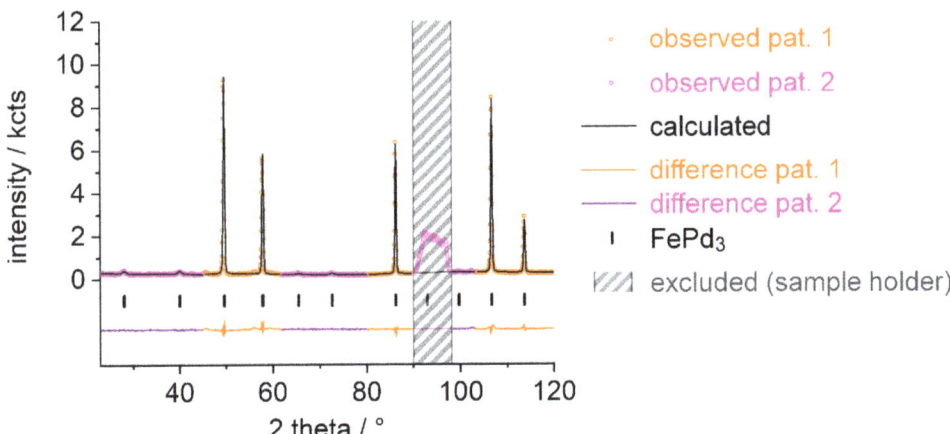

Figure A20. Rietveld refinement of the crystal structure of FePd$_3$D$_{0.062(10)}$ (for details, see eighth row in Tables A8 and A9) based on neutron powder diffraction data (NUMOR 131781 [21], λ = 186.80(2) pm, D20 ILL, Grenoble) in a single-crystal sapphire cell at 475(3) K and 8.21(1) MPa deuterium pressure using FullProf [24].

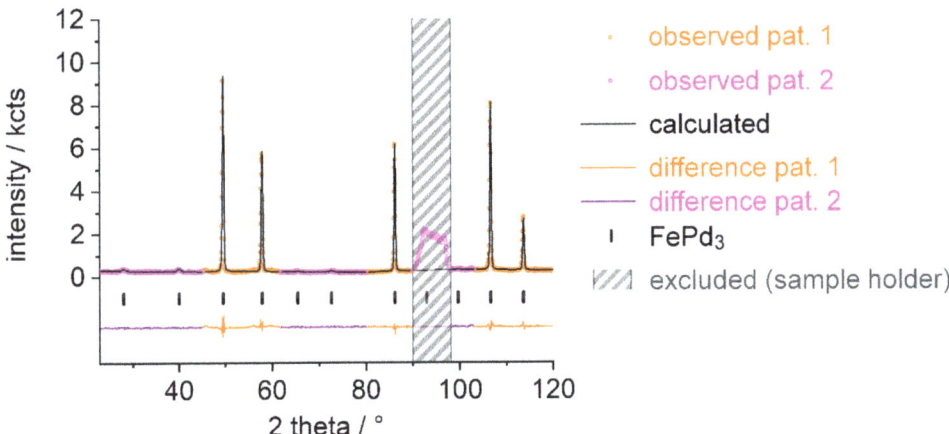

Figure A21. Rietveld refinement of the crystal structure of FePd$_3$D$_{0.067(10)}$ (for details, see ninth row in Tables A8 and A9) based on neutron powder diffraction data (NUMOR 131783 [21], λ = 186.80(2) pm, D20 ILL, Grenoble) in a single-crystal sapphire cell at 490(3) K and 8.22(1) MPa deuterium pressure using FullProf [24].

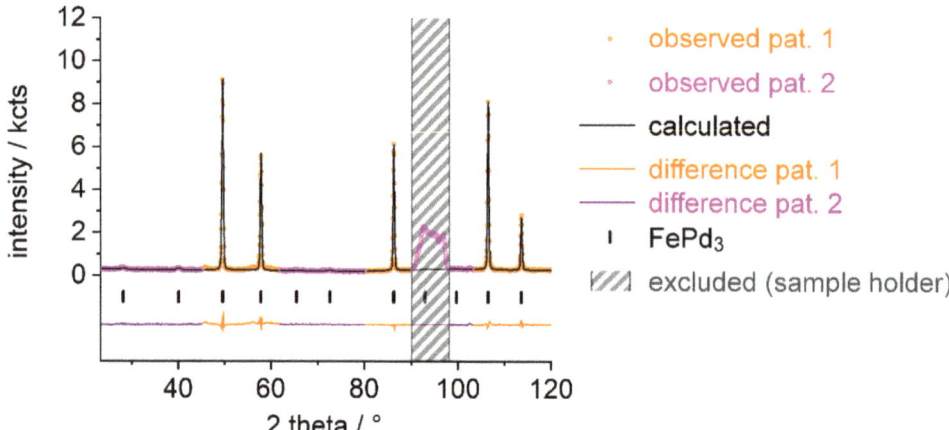

Figure A22. Rietveld refinement of the crystal structure of FePd$_3$D$_{0.065(11)}$ (for details, see tenth row in Tables A8 and A9) based on neutron powder diffraction data (NUMOR 131791 [21], λ = 186.80(2) pm, D20 ILL, Grenoble) in a single-crystal sapphire cell at 519(2) K and 8.25(1) MPa deuterium pressure using FullProf [24].

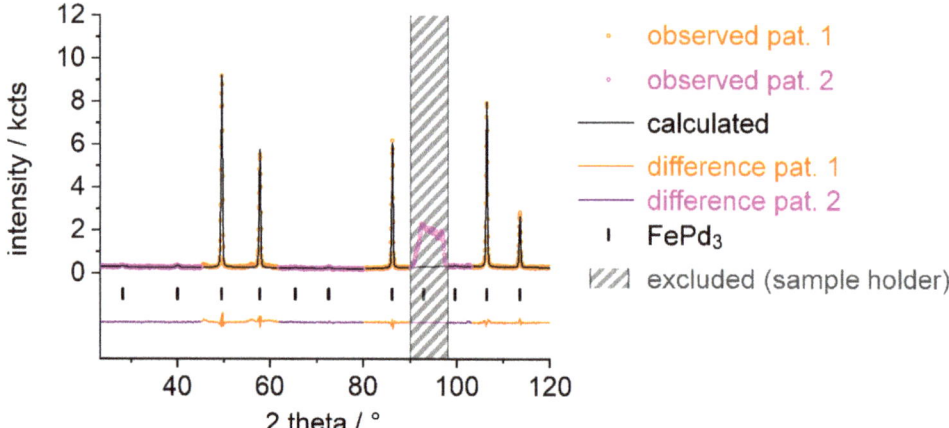

Figure A23. Rietveld refinement of the crystal structure of FePd$_3$D$_{0.064(10)}$ (for details, see eleventh row in Tables A8 and A9) based on neutron powder diffraction data (NUMOR 131815 [21], λ = 186.80(2) pm, D20 ILL, Grenoble) in a single-crystal sapphire cell at 539(1) K and 8.29(1) MPa deuterium pressure using FullProf [24].

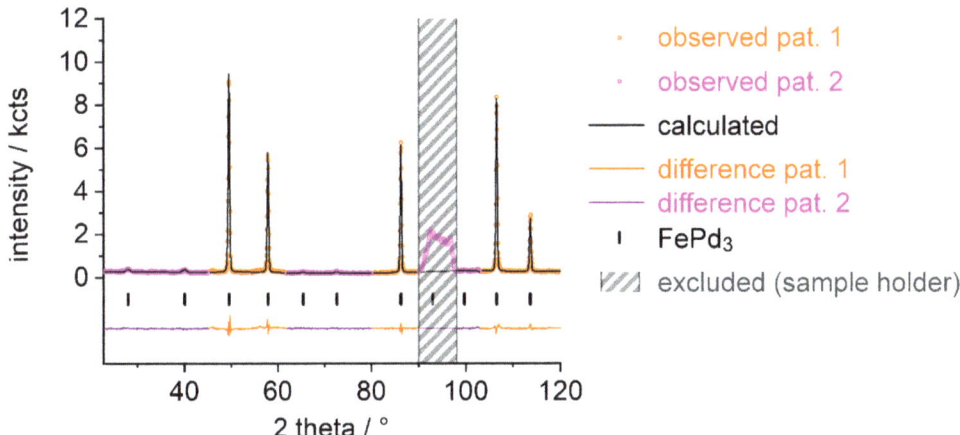

Figure A24. Rietveld refinement of the crystal structure of FePd$_3$D$_{0.057(9)}$ (for details, see twelfth row in Tables A8 and A9) based on neutron powder diffraction data (NUMOR 131833 [21], λ = 186.80(2) pm, D20 ILL, Grenoble) in a single-crystal sapphire cell at 473(2) K and 8.25(1) MPa deuterium pressure using FullProf [24].

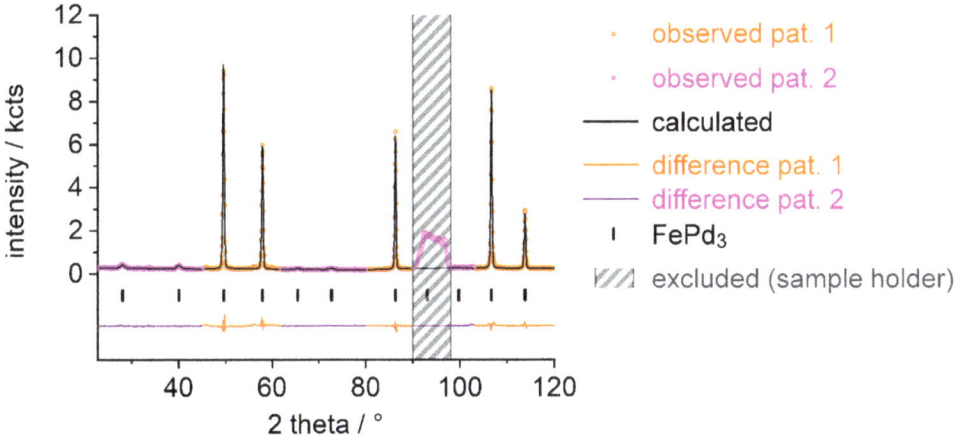

Figure A25. Rietveld refinement of the crystal structure of FePd$_3$D$_{0.052(9)}$ (for details, see thirteenth row in Tables A8 and A9) based on neutron powder diffraction data (NUMOR 131847 [21], λ = 186.80(2) pm, D20 ILL, Grenoble) in a single crystal sapphire cell at 423(2) K and 8.20(1) MPa deuterium pressure using FullProf [24].

6. Longworth, G. Temperature dependence of the Fe^{57} hfs in the ordered alloys FePd$_3$ and FePd near the Curie temperature. *Phys. Rev.* **1968**, *172*, 572–576. [CrossRef]
7. Guo, J.; Ye, Q.; Lan, M.; Wang, S.; Yu, T.; Gao, F.; Hu, D.; Wang, P.; He, Y.; Boi, F.S.; et al. Cl-assisted highly efficient synthesis of FePd$_3$ alloys encapsulated in graphite papers: A two stage CVD approach. *RSC Adv.* **2016**, *6*, 40676–40682. [CrossRef]
8. Boi, F.S.; Guo, J.; Lan, M.; Xiang, G.; He, Y.; Wang, S.; Chen, H. In Situ encapsulation of Pd crystals inside foam-like carbon films continuously filled with α-Fe: Investigating the nucleation of FePd$_3$ alloys. *RSC Adv.* **2016**, *6*, 54189–54192. [CrossRef]
9. Boi, F.S.; Du, Y.; Ivaturi, S.; He, Y.; Wang, S. New insights on the magnetic properties of ferromagnetic FePd$_3$ single-crystals encapsulated inside carbon nanomaterials. *Mater. Res. Express* **2017**, *4*, 35021. [CrossRef]
10. Cui, Z.; Li, L.; Manthiram, A.; Goodenough, J.B. Enhanced cycling stability of hybrid Li–air batteries enabled by ordered Pd$_3$Fe intermetallic electrocatalyst. *J. Am. Chem. Soc.* **2015**, *137*, 7278–7281. [CrossRef]
11. Liu, Z.; Fu, G.; Li, J.; Liu, Z.; Xu, L.; Sun, D.; Tang, Y. Facile synthesis based on novel carbon-supported cyanogel of structurally ordered Pd$_3$Fe/C as electrocatalyst for formic acid oxidation. *Nano Res.* **2018**, *11*, 4686–4696. [CrossRef]
12. Dimas-Rivera, G.L.; La Rivera De Rosa, J.; Lucio-Ortiz, C.J.; Martínez-Vargas, D.X.; Sandoval-Rangel, L.; García Gutiérrez, D.I.; Solis Maldonado, C. Bimetallic Pd-Fe supported on γ-Al$_2$O$_3$ catalyst used in the ring opening of 2-methylfuran to selective formation of alcohols. *Appl. Catal. A* **2017**, *543*, 133–140. [CrossRef]
13. Yu, W.; Porosoff, M.D.; Chen, J.G. Review of Pt-based bimetallic catalysis: From model surfaces to supported catalysts. *Chem. Rev.* **2012**, *112*, 5780–5817. [CrossRef] [PubMed]
14. Yang, J.; Li, S.; Zhang, L.; Liu, X.; Wang, J.; Pan, X.; Li, N.; Wang, A.; Cong, Y.; Wang, X.; et al. Hydrodeoxygenation of furans over Pd-FeO$_x$/SiO$_2$ catalyst under atmospheric pressure. *Appl. Catal. B* **2017**, *201*, 266–277. [CrossRef]
15. von Goldbeck, O.K. *Iron—Binary Phase Diagrams*; Springer Berlin Heidelberg: Berlin/Heidelberg, Germany, 1982; pp. 88–91. [CrossRef]
16. Konovalova, E.V.; Perevalova, O.B.; Koneva, N.A.; Veselov, S.V.; Kozlov, E.V. Effect of ordering kinetics on the degree of far atomic order in Pd$_3$Fe alloy. *Bull. Russ. Acad. Sci. Phys.* **2013**, *77*, 288–291. [CrossRef]
17. Flanagan, T.B.; Majchrzak, S.; Baranowski, B. A chemical reaction strongly dependent upon the degree of order of an alloy: The absorption of hydrogen by Pd$_3$Fe. *Philos. Mag.* **1972**, *25*, 257–262. [CrossRef]
18. Lewis, F.A. The Palladium-Hydrogen System, Part III: Alloy Systems and Hydrogen Permeation. *Platin. Met. Rev.* **1982**, *26*, 121–128.
19. Finger, R.; Kurtzemann, N.; Hansen, T.C.; Kohlmann, H. Design and use of a sapphire single-crystal gas-pressure cell for *in situ* neutron powder diffraction. *J. Appl. Crystallogr.* **2021**, *54*, 839–846. [CrossRef] [PubMed]
20. Götze, A.; Auer, H.; Finger, R.; Hansen, T.C.; Kohlmann, H. A sapphire single-crystal cell for *in situ* neutron powder diffraction of solid-gas reactions. *Physica B* **2018**, *551*, 395–400. [CrossRef]
21. Kohlmann, H.; Finger, R.; Götze, A.; Hansen, T.C.; Pflug, C.; Werwein, A. *Reaction Pathways of Hydrogenation-Dehydrogenation of InPd$_3$ and FePd$_3$ by In Situ Neutron Diffraction*; Institut Laue-Langevin: Grenoble, France, 2018. [CrossRef]
22. Rietveld, H.M. Line profiles of neutron powder-diffraction peaks for structure refinement. *Acta Crystallogr.* **1967**, *22*, 151–152. [CrossRef]
23. Rietveld, H.M. A profile refinement method for nuclear and magnetic structures. *J. Appl. Crystallogr.* **1969**, *2*, 65–71. [CrossRef]
24. Rodríguez-Carvajal, J. *FullProf*; Institut Laue-Langevin: Grenoble, France, 2012.
25. *TOPAS*; Bruker AXS GmbH: Karlsruhe, Germany, 2014.
26. Önnerud, P.; Andersson, Y.; Tellgren, R.; Nordblad, P. The Magnetic Structure of Ordered Cubic Pd$_3$Mn. *J. Solid State Chem.* **1997**, *128*, 109–114. [CrossRef]
27. Worsham, J.E., Jr.; Wilkinson, M.K.; Shull, C.G. Neutron-diffraction observations on the palladium-hydrogen and palladium-deuterium systems. *J. Phys. Chem. Solids* **1957**, *3*, 303–310. [CrossRef]
28. Pickart, S.J.; Nathans, R. Alloys of the first transition series with Pd and Pt. *J. Appl. Phys.* **1962**, *33*, 1336–1338. [CrossRef]
29. Delley, B.; Jarlborg, T.; Freeman, A.J.; Ellis, D.E. All-electron local density theory of local magnetic moments in metals. *J. Magn. Magn. Mater.* **1983**, *31–34*, 549–550. [CrossRef]
30. Pathak, R.; Kashyap, A. Boron interstitials in ordered phases of Fe-Pd binary alloys: A first principle study. *J. Magn. Magn. Mater.* **2021**, *528*, 167766. [CrossRef]
31. Götze, A.; Urban, P.; Oeckler, O.; Kohlmann, H. Synthesis and Crystal Structure of Pd$_5$InSe. *Z. Nat. B Chem. Sci.* **2014**, *69*, 417–422. [CrossRef]
32. Wannek, C.; Harbrecht, B. Iodine-promoted synthesis of structurally ordered AlPd$_5$. *Z. Anorg. Allg. Chem.* **2007**, *633*, 1397–1402. [CrossRef]
33. Mohn, P.; Supanetz, E.; Schwarz, K. Electronic structure and spin fluctuations in the fcc Fe-Pd system. *Aust. J. Phys.* **1993**, *46*, 651. [CrossRef]
34. Kaul, S.N. Static critical phenomena in ferromagnets with quenched disorder. *J. Magn. Magn. Mater.* **1985**, *53*, 5–53. [CrossRef]
35. Ahlzén, P.-J.; Andersson, Y.; Tellgren, R.; Rodic, D.; Flanagan, T.B.; Sakamoto, Y. A Neutron Powder Diffraction Study of Pd$_3$MnD$_x$. *Z. Phys. Chem.* **1989**, *163*, 213–218. [CrossRef]
36. Sieverts, A.F. Absorption of gases by metals. *Z. Met.* **1929**, *21*, 37–46.
37. Önnerud, P.; Andersson, Y.; Tellgren, R.; Nordblad, P.; Bourée, F.; André, G. The crystal and magnetic structures of ordered cubic Pd$_3$MnD$_{0.7}$. *Solid State Commun.* **1997**, *101*, 433–437. [CrossRef]

38. Kohlmann, H.; Skripov, A.V.; Solonin, A.V.; Udovic, T.J. The anti-perovskite type hydride InPd$_3$H$_{0.89}$. *J. Solid State Chem.* **2010**, *183*, 2461–2465. [CrossRef]
39. Götze, A.; Kohlmann, H. *Palladium Hydride and Hydrides of Palladium-Rich Phases. Reference Module in Chemistry, Molecular Sciences and Chemical Engineering*; Elsevier: Amsterdam, The Netherlands, 2017.
40. Jaswal, S.S. Electronic structure and magnetism in VPd$_3$ and FePd$_3$. *Phys. Rev. B* **1982**, *53*, 8213–8214. [CrossRef]
41. Kuhnen, C.A.; da Silva, E.Z. Electronic structure of Pd$_3$Fe: Ordered phase. *Phys. Rev. B* **1987**, *35*, 370–376. [CrossRef] [PubMed]
42. Wu, D.-H.; Wang, H.-C.; Wei, L.-T.; Pan, R.-K.; Tang, B.-T. First-principles study of structural stability and elastic properties of MgPd$_3$ and its hydride. *J. Magnes. Alloys* **2014**, *2*, 165–174. [CrossRef]
43. Jaswal, S.S. Electronic structure and magnetism in transition metal compounds: VNi$_3$, MnPd$_3$ and MoPd$_3$. *Solid State Commun.* **1984**, *52*, 127–129. [CrossRef]
44. Nautiyal, T.; Auluck, S. The electronic structure and magnetism of MoPd$_3$ and MnPd$_3$. *Phys. Rev. B* **1989**, *1*, 2211–2215. [CrossRef]
45. Mousa, A.A.; Jaradat, R.; Abu-Jafar, M.; Mahmoud, N.T.; Al-Qaisi, S.; Khalifeh, J.M.; Abusaimeh, H. Theoretical investigation of the structural, electronic, and elastic properties of TM_3In (TM = Pd and Pt) intermetallic compounds. *AIP Adv.* **2020**, *10*, 65317. [CrossRef]
46. Kunkel, N.; Sander, J.; Louis, N.; Pang, Y.; Dejon, L.M.; Wagener, F.; Zang, Y.N.; Sayede, A.; Bauer, M.; Springborg, M.; et al. Theoretical investigation of the hydrogenation induced atomic rearrangements in palladium rich intermetallic compounds MPd$_3$ (M=Mg, In, Tl). *Eur. Phys. J. B* **2011**, *82*, 1–6. [CrossRef]
47. Marker, V.; Wolf, G.; Baranowski, B. Effect of long-range order and hydrogen content on the low-temperature heat capacity of Pd$_3$Fe. *Phys. Status Solidi A* **1974**, *26*, 167–173. [CrossRef]
48. Jung, C.G. *Collected Works of C.G. Jung, Volume 9 (Part 1): Archetypes and the Collective Unconscious*; Princeton University Press: Princeton, NJ, USA, 1969.

Article

Investigation of Magnetocaloric Properties in the TbCo$_2$-H System

Galina Politova [1,2,*], Irina Tereshina [3], Ioulia Ovchenkova [3], Abdu-Rahman Aleroev [3], Yurii Koshkid'ko [4], Jacek Ćwik [4] and Henryk Drulis [4]

[1] Baikov Institute of Metallurgy and Material Science RAS, Moscow 119991, Russia
[2] Institute of Electronics and Telecommunications, Peter the Great St. Petersburg Polytechnic University, Saint Petersburg 195251, Russia
[3] Faculty of Physics, Lomonosov Moscow State University, Moscow 119991, Russia
[4] Institute of Low Temperature and Structure Research PAS, 50-950 Wroclaw, Poland
* Correspondence: gpolitova@imet.ac.ru

Abstract: In this work the magnetocaloric effect in the TbCo$_2$-H system in the region of the Curie temperature was studied both by direct and indirect methods in external magnetic fields up to ~1.4 and 14 T, respectively. We have paid special attention to the magnetic and magnetothermal properties of the TbCo$_2$–H with high hydrogen content. The mechanisms responsible for the change in the Curie temperature were established, and the field and temperature dependences of the magnetocaloric effect were analyzed in detail. In addition, the magnetocaloric properties (including critical parameters) for various systems based on the TbCo$_2$ compound were compared. The main regularities of the change in the MCE value and the Curie temperature depending on the composition are discovered and discussed.

Keywords: rare-earth intermetallic compound; Laves phase structure; hydride; magnetic phase transition; magnetocaloric effect

1. Introduction

RCo$_2$ intermetallic compounds with a Laves phase structure demonstrate appreciable magnetocaloric effect (MCE) (quantified as the adiabatic temperature change ΔT_{ad} or isothermal entropy change ΔS when exposed to a magnetic field) near the magnetic phase transition, at the Curie temperature (T_C) [1–6]. It is well known [7], that the Curie temperatures of magnetically ordered rare-earth compounds cover a wide range from ~400 K (GdCo$_2$) to ~4 K (TmCo$_2$). The fact that LuCo$_2$ and YCo$_2$ compounds are merely the enhanced Pauli paramagnets emphasizes the special role of rare-earth ions in the magnetic properties of these compounds.

Among RCo$_2$ compounds, the highest MCE values have been found for compounds exhibiting first-order magnetic phase transitions: at T_C = 140 K for DyCo$_2$, at T_C = 75 K for HoCo$_2$ and at T_C = 32 K for ErCo$_2$ [7]. RCo$_2$-type compounds with heavy rare-earth elements such as Gd or Tb demonstrate second-order transitions (SOTs) at temperatures above 200 K. Although they have lower MCE values, their practical use in magnetic refrigerators is preferable due to the absence of magnetic hysteresis.

Doping of RCo$_2$-type compounds with suitable impurities (strongly or weakly magnetic [8–14]) can provide control of the Curie temperature and, consequently, the position of the peak in the $\Delta T_{ad}(T)$ and $\Delta S(T)$ dependences, and its magnitude. The patterns of change in the magnitude of the magnetocaloric effect in doped compounds have not been fully disclosed, despite the large number of experimental works available in the literature. An important contribution to the study of magnetothermal phenomena in pseudo-binary compounds of the (R,R')Co$_2$-type in the region of magnetic phase transitions was made by N.A. de Oliveira [15]. By using theoretical calculations, he confirmed the experimentally observed magnitude of the magnetocaloric effect, for compounds exhibiting phase transitions of both the second and/or first order.

It is known that the analysis of the field dependences of the MCE is of great importance in the study of the magnetocaloric properties of materials. An initial assumption about the field dependence of the MCE was made by Belov K.P. [16]. Based on Landau's theory of phase transitions, he showed that $\Delta T_{ad} \sim H^n$, where n = 2/3. An attempt to explain the field dependence of ΔS_{max} for magnetic materials with a second-order phase transition was made by H. Oesterreicher and F. T. Parker. They assumed that in the mean field approximation $\Delta S_{max} \sim H^n$, where n = 2/3 [17]. However, experimental results often deviated from this n value. An alternative approach was proposed by Romanov and Silin [18]. Their analysis of the MCE in inhomogeneous magnets was based on Landau's theory of second-order phase transitions [19]. As a result, rather complicated equations were obtained. A great deal of work comparing theoretical and experimental data on $\Delta S(H)$ for various compounds was done by Franco V. et al. [20–22].

Among RCo_2-type compounds, $TbCo_2$ attracts special attention, since its magnetic ordering temperature is in the ambient temperature range and is ~235 K (~−38 °C) (Curie temperatures vary from 231 to 238 K in various sources [7–9,23–29]). Moreover, such working temperature is important to design magnetic refrigerators for long-term storage of various biomaterials, vaccines and drugs. That is why the study of the magnetocaloric effect in $TbCo_2$-based compounds modified by interstitial or substitutional atoms will not only significantly expand the possibilities of practical use of these materials, but will also allow to collect new experimental data to test existing theoretical models [8,9,11,24–29].

Note that only a few works have been devoted to study the effect of interstitial atoms (such as hydrogen), on the magnetic and magnetocaloric properties of the $TbCo_2$ compound [30,31]. Thus, in the work of Mushnikov [30], the magnetic properties of $TbCo_2H_x$ samples with x = 0, 0.7, 2, 2.4, 3.9 were obtained and investigated. The magnetic moment at Co atoms and the Curie temperature were found to exhibit an increase at low hydrogen contents, whereas at high hydrogen contents, both magnetic characteristics decrease substantially. MCE was not studied in $TbCo_2$-H system.

It is known [32] that high MCE values for RCo_2 compounds with Laves phase structure are obtained only at temperatures below 200 K. That is why new compositions with Tc less than 200 K are of particular interest. The purpose of this work was, first of all, to study and analyze the effect of hydrogen on the Curie temperature and MCE, and to compare the magnetocaloric characteristics of $TbCo_2H_x$ hydrides with other substituted $TbCo_2$-based compositions to establish the main patterns of their changes depending on the composition.

2. Materials and Methods

We obtained $TbCo_2H_x$ compounds with x = 0, 0.5, 2.4 (the initial sample and samples with low and high hydrogen content). Details of the synthesis and certification of $TbCo_2$ compounds were described in our previous work [12]. The $TbCo_2$ sample weighing 0.6 g was placed in a 6 cm^3 reaction tube of fully automated stainless steel Sievert's-type volumetric apparatus. Before hydrogen absorption began, the sample was thermally activated under a high vacuum of 10^{-4} Torr, at 250 °C for 1 h. After cooling the sample to room temperature, the pure (99.999%) hydrogen gas was introduced into the reaction tube under a pressure of about 10 bars and left for 12 h at room temperature. The hydrogen contents were determined by monitoring pressure changes in a system with a known volume before and after the reaction. Pressures were monitored using a Honeywell ST3000 strain gauge. The accuracy of determination of absorbed hydrogen concentration is ±0.02 H atoms per formula unit (H/f.u.). The XRD patterns were recorded at scanning step of 0.02 (at the 2-s exposition) on a Rigaku Ultima IV powder diffractometer with a CuK_α radiation. The qualitative and quantitative phase analysis was performed using a program PDXL by Rigaku (Japan) integrated with the international database ICDD.

Field-dependent magnetization measurements (field range 0–14 T) were carried out using a vibration sample magnetometer (VSM) [33]. Magnetization isotherms were obtained in the temperature range 4.2–300 K. The temperature and type of magnetic phase transition were determined by the Belov–Arrott method [34]. The analysis of the magnetocaloric effect was carried out by calculating the change in entropy from magnetization isotherms (indirect method [1]).

Direct measurements of the magnetocaloric effect were performed using a special setup in fields up to 1.35 T in the temperature range 78–310 K. The measurements were carried out by recording the temperature change of the sample during the adiabatic increase in the magnetic field (ΔT_{ad}). Adiabaticity was achieved by good thermal insulation of the sample, by placing a copper–constantan thermocouple inside the sample, and by quickly turning on the magnetic field. The temperature changes of the sample were monitored with accuracy better than ±0.01 K. The study of the MCE by the direct method was possible only for samples with x = 0 and 0.5 obtained in the cast state. The sample with x = 2.4 was obtained in the form of a powder (indirect method of MCE estimation was used for it).

3. Results and Discussion

X-ray diffraction analysis (see Figure 1) showed that in all the obtained $TbCo_2H_x$ (x = 0, 0.5, 2.4) compounds, the content of the main phase, which has a cubic structure of the $MgCu_2$ type, is not less than 96%. The parameters of the $TbCo_2$ initial sample are in good agreement with the data in [30]. The increase of the relative unit cell volume $\Delta V/V$ varies depending on the hydrogen content, from 0.5% for the $TbCo_2H_{0.5}$ compound to 15% for the $TbCo_2H_{2.4}$.

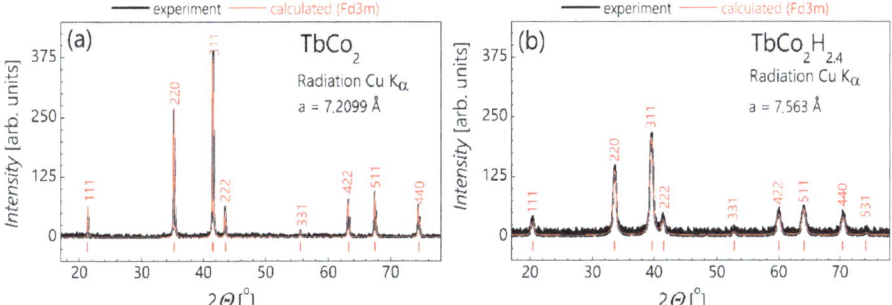

Figure 1. X-ray diffraction patterns for $TbCo_2$ (**a**) and its hydride $TbCo_2H_{2.4}$ (**b**).

Figure 2a shows the magnetization isotherms $M(B)$ at T = 4.2 K for the $TbCo_2$ and $TbCo_2H_{2.4}$ hydride in comparison with the known data for the initial composition and for the $TbCo_2H_2$ dihydride [30]. It can be seen that the data are in good agreement with each other. The type of magnetic phase transition for $TbCo_2H_{2.4}$ was analyzed in detail by Belov–Arrott plots (M^2 versus B/M) and the Banerjee criterion [35], as illustrated in Figure 2b. No negative slope or inflection was found as a characteristic of the first-order magnetic transition, which suggests that the phase transition in the $TbCo_2H_{2.4}$ is of a second-order type. It is known that the initial $TbCo_2$ compound exists on the instability boundary, and the type of its transition can be considered both as the first kind [36] or the second kind [8,9,37]. Therefore, the insertion of interstitial and/or substitution atoms in $TbCo_2$ can easily shift the boundary and make the transition type either first or second. This phenomenon is mainly associated with the instability of the magnetism of the cobalt sublattice, which demonstrates a strong dependence on the crystal lattice parameter a in RCo_2 compounds [37].

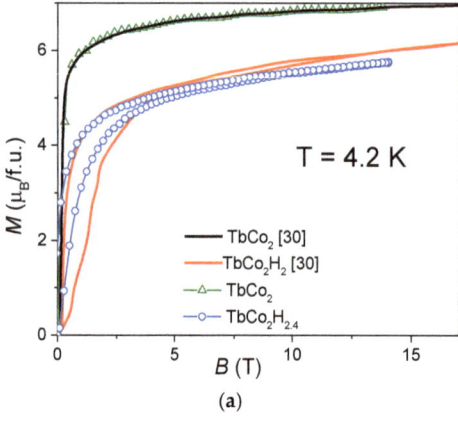 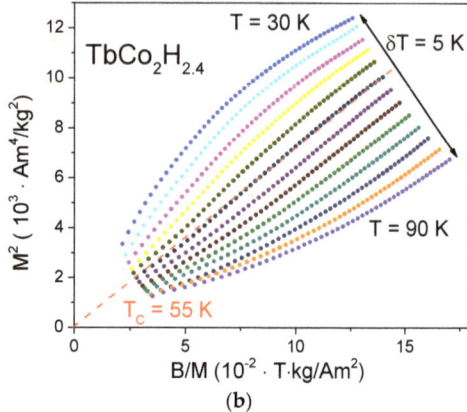

Figure 2. (a) Field dependence of the magnetization of the initial compound TbCo$_2$ and its hydrides TbCo$_2$H$_2$ and TbCo$_2$H$_{2.4}$; (b) the Curie temperature determination of the TbCo$_2$H$_{2.4}$ hydride by the Belov–Arrott method.

The Curie temperature of the TbCo$_2$H$_{2.4}$ hydride was determined to be equal to T_C = 55 K. In terms of lattice expansion, the insertion of hydrogen is equivalent to the application of a negative hydrostatic pressure. The change in the Curie temperature in the hydride we considered as a result of a change in the unit cell volume.

The effect of pressure on the Curie temperature (dT$_C$/dp = −9 K/GPa for TbCo$_2$) [38] and compressibility κ ≈ 10^{-2} GPa^{-1} [39]) determined from the literature data are shown in Figure 3 by a dashed line. The rate of decrease in Tc with an increase in the volume of the unit cell can be calculated by the formula

$$d\ln T_C/dp = -(\kappa/T_C)dT_C/d\ln V, \quad (1)$$

where κ = −(dV/V)/p. Hence dT$_C$/dlnV = dT$_C$/(dV/V) = 9 K per 1% change in unit cell volume. To determine the Curie temperatures for the TbCo$_2$-H system, we also used the inflection point technique based on analysis of the behavior of the temperature dependences of the magnetization in the magnetic field [30]. Figure 3 shows that the experimentally determined decrease in Tc is less than that expected in consequence of an increase in the unit cell volume when ΔV/V exceeds 20%. Herewith, when values of ΔV/V are close to 13–15%; the experimental and calculated data practically coincide. This means that the volume effect is the dominant mechanism in the latter case. However, with a further increase in volume, other factors also come into play, the most important of which is a change in the electronic structure of the compound due to the insertion of hydrogen atoms into the crystal lattice of the TbCo$_2$ compound [40–42].

The Curie temperature of TbCo$_2$H$_x$ increases at low hydrogen concentrations. It should be noted that hydrogen can occupy two types of tetrahedral interstices in the structure of the C15 Laves phase: positions AB3 (32e) and A2B2 (96g) [43]. According to neutron diffraction data for ErFe$_2$H$_x$, at low hydrogen concentrations, A2B2 interstices are predominantly filled, while at high hydrogen concentrations, interstices of both types are partially filled [44], which has an additional effect on the functional properties.

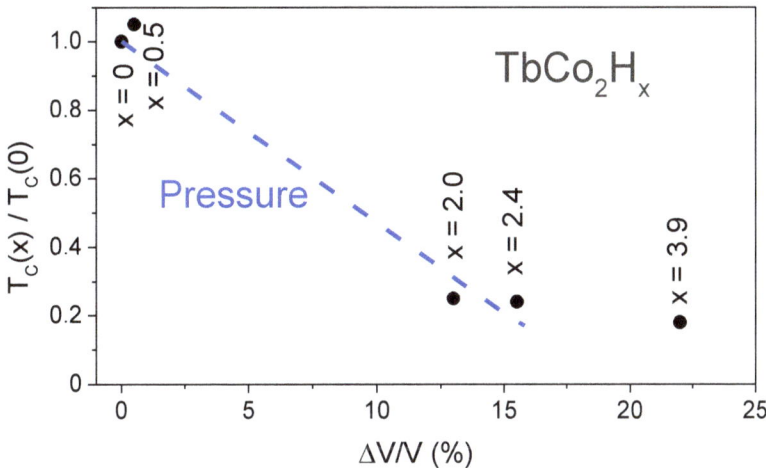

Figure 3. Curie temperature dependence on the relative increase in unit cell volume $\Delta V/V$ for the $TbCo_2$—H system and the expected change in T_C (dashed line), determined on the basis of literature data on the effect of hydrostatic pressure on the Curie temperature [30,38,39].

The magnitude of the magnetocaloric effect in $TbCo_2H_{2.4}$ was calculated from the experimentally obtained magnetization isotherms $M(B)$ by an indirect method [1]. The temperature dependences of the change in the magnetic part of the entropy in various magnetic fields (from 1 to 14 T) are shown in Figure 4a.

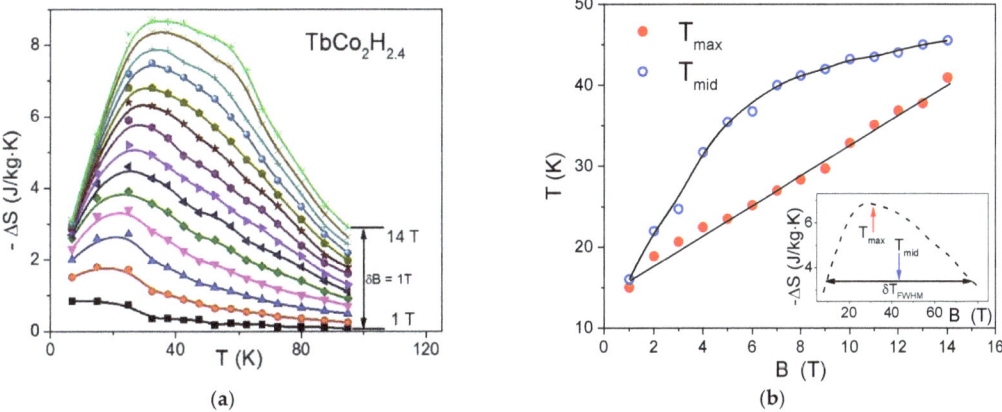

Figure 4. Temperature dependences of the change in the magnetic part of the entropy at various changes in the magnetic field of $TbCo_2H_{2.4}$ (**a**); field dependences of the maximum temperature $-\Delta S(T_{max})$ and the temperature of the middle of the working zone (T_{mid}) (**b**).

It is clearly seen that the temperature at which the maximum MCE (T_{max}) is observed increases with an increase in the external magnetic field. Figure 4b shows the field dependences of the T_{max} and the temperature (T_{mid}) in the middle of the working zone. The working zone is defined as the temperature range at which the values $\Delta S = 0.5 \cdot (\Delta S_{max})$ (see inset to Figure 4b). It can be seen that the dependence T_{max} (B) demonstrates a linear increase in applied fields, while the dependence T_{mid} (B) approximates to saturation.

Figure 5a shows a comparison of the MCE values for the $TbCo_2H_{2.4}$ hydride and the $TbCo_2$ sample at various magnetic field changes from 0 to 1, to 2, to 3, to 4, and up to

5 T (the most commonly used ranges of magnetic fields). The MCE values for the initial composition are in good agreement with the literature data [26]. It is clear that the value of the MCE of the hydrogenated sample decreases by a factor of 1.5, whereas the maximum temperature decreases by 200 K.

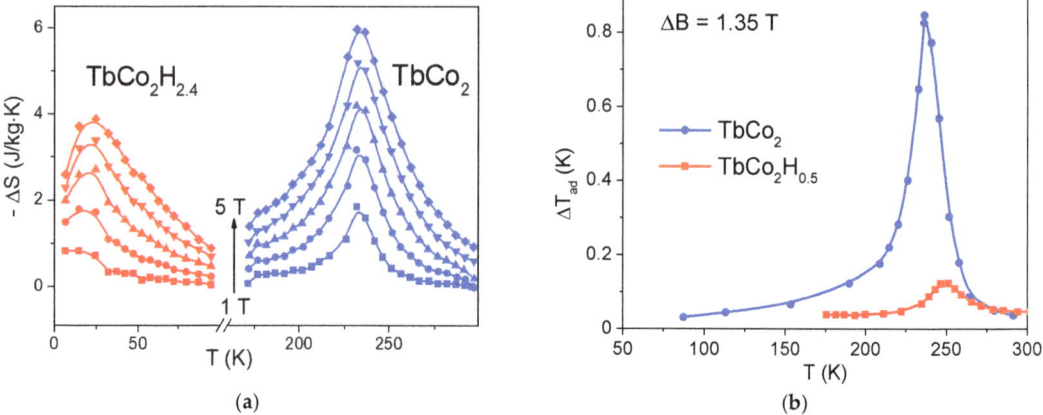

Figure 5. Temperature dependences of the change in the magnetic part of the entropy at various changes in the magnetic field (from 0 to 1–5 T) in the $TbCo_2H_{2.4}$ hydride and the initial composition $TbCo_2$ (**a**); Temperature dependences of the MCE at ΔB = 1.35 T measured by the direct method of $TbCo_2H_{0.5}$ and the initial composition $TbCo_2$ (**b**).

Figure 5b compares the MCE measured by the direct method for the initial $TbCo_2$ compound and for low-hydrogen content $TbCo_2H_{0.5}$ hydride within the magnetic field change ΔB = 1.35 T. It can be seen that the MCE of the hydrogenated sample is decreased by a factor of 10, while the Curie temperature (at which the maximum of the magnetocaloric effect is observed), on the contrary, is increased (by about 10 K). Such behavior of the Curie temperature and the MCE in hydrides with high ($TbCo_2H_{2.4}$) and low ($TbCo_2H_{0.5}$) hydrogen content may be related to the fact that hydrogen atoms occupy different positions in the $MgCu_2$-type cubic structure [30], but the type of the magnetic phase transition does not change in these cases.

For a detailed analysis of the nature of the magnetic phase transition in $TbCo_2H_{2.4}$ hydride we studied the critical exponents near the Curie temperature T_C. According to the scaling hypothesis [45,46] for a second-order phase transition in the T_C region, the critical exponents β (associated to spontaneous magnetization), γ (associated to the initial susceptibility), and δ (associated to the magnetization isotherm) are related by:

$$M_S(T) = M_S(-\varepsilon)^\beta, \varepsilon < 0 \quad (2)$$

$$\chi_0^{-1}(T) = (h_0/M_0)\varepsilon^\gamma, \varepsilon > 0 \quad (3)$$

where ε is the reduced temperature equal to $(T - T_C)/T_C$.

At Curie temperature, the exponent δ relates magnetization M and applied magnetic field B by

$$M(B, T_C) = A_0(B)^{1/\delta}, \varepsilon = 0 \quad (4)$$

where A_0 are the critical amplitudes (Kouvel–Fisher method [47]).

According to Equation (4), the value of δ can be obtained by a linear fit to the high field plots ln(M) vs. ln(B) near T_C, as shown in the insets in Figure 6a. The δ value obtained for $TbCo_2H_{2.4}$ was compared by us with the data for the $TbCo_2$, as well as for $Tb(Co,Fe)_2$ compositions (see Table 1).

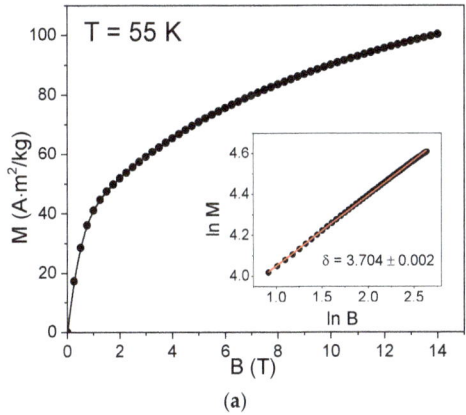
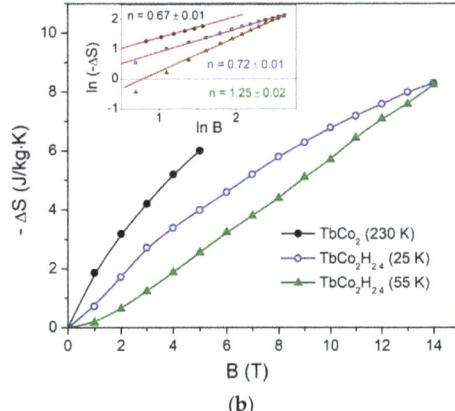

Figure 6. Critical exponent analysis of TbCo$_2$-H for: (**a**) Critical isotherm of M vs. B of TbCo$_2$H$_{2.4}$ close to the Curie temperature Tc = 55 K. Inset shows the same on log–log scale and the straight line is the linear fit following Equation (4); the critical exponent δ is obtained from the slope of the linear fit. (**b**) Critical isotherm of $-\Delta S$ vs. B of TbCo$_2$ close to the Curie temperature Tc = 230 K and TbCo$_2$H$_{2.4}$ close to the temperatures T = 25 and 55 K. Inset shows the same on log–log scale and the straight line is the linear fit following Equation (5); the critical exponent n is obtained from the slope of the linear fit.

Table 1. Critical exponents β, δ and n, Curie temperature T_C and maximum change of magnetic entropy $|-\Delta S|$ at 0–5 T, for the TbCo$_2$, substituted compounds Tb(Co,Fe)$_2$ and the hydride TbCo$_2$H$_{2.4}$.

Compounds	β	δ	n	Tc	$-\Delta S$ (0–5 T)	Reference
TbCo$_2$	0.386 (5)	4.83 (4)	0.67 (1)	234	6	This work
TbCo$_2$	-	-	0.67 (1)	235	6	[25]
TbCo$_2$	0.380 (4)	4.85 (3)	0.65 (1)	231	6.9	[26]
TbCo$_{1.94}$Fe$_{0.06}$	-	-	-	275	3.9	[25]
TbCo$_{1.9}$Fe$_{0.1}$	0.541 (1)	2.75 (4)	0.76 (2)	303	3.7	[25]
TbCo$_2$H$_{2.4}$	0.53 (3)	3.704 (2)	0.72 (1)	55	2.5	This work

It can be seen that both interstitial and substitutional atoms contribute to a decrease in the critical exponent δ. Moreover, the effect of interstitial atoms (hydrogen) on this indicator is less than the effect of substitution atoms.

According to the scaling hypothesis, the magnitude of the entropy change in the Curie temperature range is related to the magnitude of the external magnetic field by the following relation:

$$-\Delta S\ (T = T_C) \sim B^n \quad (5)$$

The exponent n is related to the critical exponents β and δ by the following relation [26]:

$$n = 1 + 1/\delta(1 - 1/\beta) \quad (6)$$

According to relation (6), we obtain n = 0.76, while a linear approximation of the dependences lnΔS vs. ln$B$ gives us the values n = 0.72 at T = 25 K and n = 1.25 at T_C = 55 K (Figure 5b). A similar analysis of the experimental results performed for several families of magnetically soft bulk alloys in the amorphous state [18] shows that the field dependence $-\Delta S(B)$ has the following features: at temperatures significantly below T_C, n = 1, while at temperatures significantly above T_C, n = 2. At a temperature corresponding to the maximum $|-\Delta S_{max}|$, the value of n is minimal and can approach 2/3. The high-temperature limit for n = 2 is a consequence of the Curie–Weiss law. Since the magnetization has a linear dependence on the field in the high-temperature region, the change in the magnetic part of the entropy also has a quadratic dependence on the field. The low temperature limit can also be explained by simple arguments: at temperatures well below the Curie temperature and in a moderate applied fields the magnetization does not show a strong field dependence [48]. The consequence of this fact is n = 1.

Table 1 contains data on the MCE for the substituted $Tb(Co,Fe)_2$ compositions. It can be seen that with an increase in the Fe content, the MCE value decreases. Note also that a decrease in the MCE in this system is accompanied by a significant increase in the Curie temperature: from 235 (4) K (for $TbCo_2$) to 303 K (for $TbCo_{1.9}Fe_{0.1}$). The insertion of hydrogen atoms into the crystal lattice of the $TbCo_2$ compound (as in the case of the $Tb(Co,Fe)_2$ system) leads to a significant decrease in the MCE, not only at low (in the $TbCo_2H_{0.5}$), but also at high (in the $TbCo_2H_{2.4}$) hydrogen concentrations. In the $TbCo_2$–H systems, at low concentrations of hydrogen, one can observe a slight increase in T_C, however, in $TbCo_2H_{2.4}$ the Curie temperature decreases from 234 K (in $TbCo_2$) to 55 K.

Neither interstitial (hydrogen) nor substitutional (iron) atoms change the type of magnetic phase transition in the $TbCo_2$ compound. This task, as well as the problem of a significant increase in the MCE in $TbCo_2$, can be solved due to substitutions in the rare earth sublattice, namely, the replacement of Tb atoms with Dy, Ho, Er atoms [12,13,32].

4. Conclusions

$TbCo_2H_x$ (x = 0, 0.5, and 2.4) with cubic $MgCu_2$-type structure have been successfully synthesized. It has been established that hydrogenation leads to a significant (almost 200 K) decrease in T_C at hydrogen concentration of 2.4 at./f.u when the increase in relative unit cell volume, $\Delta V/V$, is close to 15%, which agrees well with calculations made using compressibility and dT_C/dp. This means that the volume effect is dominant in its influence on the Curie temperature. As a consequence of the strong influence of the volume effect on the magnetic properties, the MCE value in hydrides, both with low and high hydrogen content (hydrogen fills different types of tetrahedral interstices in the structure of the C15 Laves phase: namely, positions AB3 (32e) and A2B2 (96g)), decreases through the increase in the distances between magnetically active ions. The type of magnetic phase transition from a magnetically ordered to a disordered state does not change upon hydrogenation.

An analysis of the field dependences of the MCE of the $TbCo_2H_{2.4}$ hydride showed that interstitial atoms contribute to a decrease in the critical index δ and an increase in the indices β and n, similarly to the partial replacement of cobalt atoms by iron atoms in $TbCo_2$ compound.

Author Contributions: Conceptualization, I.T. and J.Ć.; methodology, Y.K., H.D. and I.O.; formal analysis, I.T. and G.P.; investigation, Y.K., J.Ć. and I.O.; resources, G.P., I.T. and A.-R.A.; data curation, I.T.; writing—original draft preparation, G.P.; writing—review and editing, I.T.; visualization, G.P.; supervision, H.D.; project administration, I.T. and J.Ć. All authors have read and agreed to the published version of the manuscript.

Funding: The work is supported by the Russian Science Foundation, pr. No. 22-29-00773. The work of J. Ćwik was supported by the National Science Center, Poland, through the OPUS Program under Grant No. 2019/33/B/ST5/01853.

Data Availability Statement: The main data is contained within the article. The data presented in this study are available on request from the corresponding author.

Conflicts of Interest: The authors declare no conflict of interest.

References

1. Tishin, A.M.; Spichkin, Y.I. *The Magnetocaloric Effect and Its Applications*; Institute of Physics Publishing: Bristol, PA, USA, 2003; p. 476.
2. Kitanovski, A.; Tušek, J.; Tomc, U.; Plaznik, U.; Ožbolt, M.; Poredoš, A. *Magnetocaloric Energy Conversion from Theory to Applications*; Springer International Publishing: Cham, Switzerland, 2015; p. 456.

3. Duc, N.H.; Kim, A.D.T.; Brommer, P.E. Metamagnetism, giant magnetoresistance and magnetocaloric effects in RCo$_2$-based compounds in the vicinity of the Curie temperature. *Phys. B* **2002**, *319*, 1–8. [CrossRef]
4. Gschneidner, K.A.; Pecharsky, V.K., Jr. Magnetocaloric materials. *Annu. Rev. Mater. Sci.* **2000**, *30*, 387–429. [CrossRef]
5. Singh, N.K.; Suresh, K.G.; Nigam, A.K.; Malik, S.K.; Coelho, A.A.; Gama, S. Itinerant electron metamagnetism and magnetocaloric effect in RCo$_2$-based Laves phase compounds. *J. Magn. Magn. Mater.* **2007**, *317*, 68–79. [CrossRef]
6. Stein, F.; Leineweber, A. Laves phases: A review of their functional and structural applications and an improved fundamental understanding of stability and properties. *J. Mater. Sci.* **2021**, *56*, 5321–5427. [CrossRef]
7. Gratz, E.; Markosyan, A.S. Physical properties of RCo$_2$ Laves phases. *J. Phys. Condens. Matter.* **2001**, *13*, 385–413. [CrossRef]
8. Zeng, Y.; Tian, F.; Chang, T.; Chen, K.; Yang, S.; Cao, K.; Zhou, C.; Song, X. Large magnetocaloric effect and near-zero thermal hysteresis in the rare earth intermetallic Tb$_{1-x}$Dy$_x$Co$_2$ compounds. *J. Phys. Condens. Matter* **2017**, *29*, 055804. [CrossRef] [PubMed]
9. Balli, M.; Fruchart, D.; Gignoux, D. A study of magnetism and magnetocaloric effect in Ho$_{1-x}$Tb$_x$Co$_2$ compounds. *J. Magn. Magn. Mater.* **2007**, *314*, 16–20. [CrossRef]
10. Cwik, J.; Palewski, T.; Nenkov, K.; Gutfleisch, O.; Klamut, J. The influence of Er substitution on magnetic and magnetocaloric properties of Dy$_{1-x}$Er$_x$Co$_2$ solid solutions. *Intermetallics* **2011**, *19*, 1656–1660. [CrossRef]
11. Zhouv, K.W.; Zhuang, Y.H.; Li, J.Q.; Deng, J.Q.; Zhu, Q.M. Magnetocaloric effects in (Gd$_{1-x}$Tb$_x$)Co$_2$. *Solid State Comm.* **2006**, *137*, 275–277.
12. Tereshina, I.S.; Chzhan, V.B.; Tereshina, E.A.; Khmelevskyi, S.; Burkhanov, G.S.; Ilyushin, A.S.; Paukov, M.A.; Havela, L.; Karpenkov, A.Y.; Cwik, J.; et al. Magnetostructural phase transitions and magnetocaloric effect in Tb-Dy-Ho-Co-Al alloys with a Laves phase structure. *J. Appl. Phys.* **2016**, *120*, 013901. [CrossRef]
13. Politova, G.A.; Tereshina, I.S.; Cwik, J. Multifunctional phenomena in Tb-Dy-Gd(Ho)-Co(Al) compounds with a Laves phase structure: Magnetostriction and magnetocaloric effect. *J. Alloys Compd.* **2020**, *843*, 155887. [CrossRef]
14. Tereshina, I.; Politova, G.; Tereshina, E.; Nikitin, S.; Burkhanov, G.; Chistyakov, O.; Karpenkov, A. Magnetocaloric and magnetoelastic effects in (Tb$_{0.45}$Dy$_{0.55}$)$_{1-x}$Er$_x$Co$_2$ multicomponent compounds. *J. Phys. Conf. Ser.* **2010**, *200*, 092012. [CrossRef]
15. De Oliveira, N.A.; von Ranke, P.J. Theoretical aspects of the magnetocaloric effect. *Phys. Rep.* **2010**, *489*, 89–159. [CrossRef]
16. Belov, K.P. *Magnetic Transitions*; Consultants Bureau: New York, NY, USA, 1961.
17. Oesterreicher, H.; Parker, F.T. Magnetic cooling near Curie temperatures above 300 K. *J. Appl. Phys.* **1984**, *55*, 4334–4338. [CrossRef]
18. Romanov, A.Y.; Silin, V.P. On the magnetocaloric effect in inhomogeneous ferromagnets. *Phys. Met. Metallogr.* **1997**, *83*, 111–115.
19. Landau, L.D. Theory of Phase Transformations. II. *Zh. Eksp. Teor. Fiz.* **1937**, *7*, 627. (In Russian)
20. Franco, V.; Blázquez, J.; Conde, A. Field dependence of the magnetocaloric effect in materials with a second order phase transition: A master curve for the magnetic entropy change. *Appl. Phys. Lett.* **2006**, *89*, 222512. [CrossRef]
21. Franco, V.; Conde, A.; Pecharsky, V.K.; Gschneidner, K.A. Field dependence of the magnetocaloric effect in Gd and (Er$_{1-x}$Dy$_x$)Al$_2$: Does a universal curve exist? *EPL* **2007**, *79*, 47009. [CrossRef]
22. Franco, V.; Blázquez, J.S.; Ipus, J.J.; Law, J.Y.; Moreno-Ramírez, L.M.; Conde, A. Magnetocaloric effect: From materials research to refrigeration devices. *Prog. Mater. Sci.* **2018**, *93*, 112–232. [CrossRef]
23. Huang, D.; Gao, J.; Lapidus, S.H.; Brown, D.E.; Ren, Y. Exotic hysteresis of ferrimagnetic transition in Laves compound TbCo$_2$. *Mater. Res. Lett.* **2020**, *8*, 97–102. [CrossRef]
24. Nikitin, S.A.; Tskhadadze, G.A.; Ovtenkova, I.A.; Zhukova, D.A.; Ivanova, T.I. The Magnetic Phase Transitions and Magnetocaloric Effect in the Ho(Co$_{1-x}$Al$_x$)$_2$ and Tb(Co$_{1-x}$Al$_x$)$_2$ Compounds. *Solid State Phenom.* **2011**, *168–169*, 119–121.
25. Ovchenkova, I.A.; Tskhadadze, G.A.; Zhukova, D.A.; Ivanova, T.I.; Nikitin, S.A. Magnetocaloric effect in RCo$_2$ compounds. *Solid State Phenom.* **2012**, *190*, 339–342. [CrossRef]
26. Janatova, M.; Poltierova, J.; Vejpravova, J.; Javorsky, P.; Prokleska, J.; Svoboda, P.; Danis, S. Effect of Si Substitution and Annealing on Magnetocaloric Properties in TbCo$_2$. *Acta Phys. Pol. A* **2008**, *113*, 311–314. [CrossRef]
27. Gerasimov, E.G.; Inishev, A.A.; Terentev, P.B.; Kazantsev, V.A.; Mushnikov, N.V. Magnetostriction and thermal expansion of nonstoichiometric TbCo$_2$Mn$_x$ compounds. *J. Magn. Magn. Mater.* **2021**, *523*, 167628. [CrossRef]
28. Halder, M.; Yusuf, S.M.; Mukadam, M.D. Magnetocaloric effect and critical behavior near the paramagnetic to ferrimagnetic phase transition temperature in TbCo$_{2-x}$Fe$_x$. *Phys. Rev. B* **2010**, *81*, 174402. [CrossRef]
29. Fang, C.H. Tuning the magnetic and structural transitions in TbCo$_2$Mn$_x$ compounds. *Phys. Rev. B* **2017**, *96*, 064425. [CrossRef]
30. Mushnikov, N.V.; Gaviko, V.S.; Goto, T. Magnetic Properties of Hydrides RCo$_2$H$_x$ with R = Gd, Tb, Dy, Ho, and Er. *Phys. Met. Metallogr.* **2005**, *100*, 338–348.
31. Tereshina, I.S.; Kaminskaya, T.P.; Chzhan, V.B.; Ovchenkova, Y.A.; Trusheva, A.S.; Viryus, A.A. The Influence of Hydrogenation on the Structure, Magnetic and Magnetocaloric Properties of Tb–Dy–Co Alloys with a Laves Phase Structure. *Phys. Solid State* **2019**, *61*, 1169–1175. [CrossRef]
32. Tereshina, I.; Cwik, J.; Tereshina, E.; Politova, G.; Burkhanov, G.; Chzhan, V.; Ilyushin, A.; Miller, M.; Zaleski, A.; Schultz, L.; et al. Multifunctional Phenomena in Rare-Earth Intermetallic Compounds With a Laves Phase Structure: Giant Magnetostriction and Magnetocaloric Effect. *IEEE Trans. Magn.* **2014**, *50*, 2504604. [CrossRef]
33. Nizhankovskii, V.I.; Lugansky, L.B. Vibrating sample magnetometer with a step motor. *Meas. Sci. Technol.* **2007**, *18*, 1533–1537. [CrossRef]
34. Belov, K.P. *Magnetic Transformations*; FizMatGiz: Moscow, Russia, 1959; p. 260.
35. Banerjee, S.K. On a generalised approach to first and second order magnetic transitions. *Phys. Lett.* **1964**, *12*, 16–17. [CrossRef]

36. Zhou, C.; Chang, T.; Dai, Z.; Chen, Y.; Guo, C.; Matsushita, Y.; Ke, X.; Murtaza, A.; Zhang, Y.; Tian, F.; et al. Unified understanding of the first-order nature of the transition in TbCo$_2$. *Phys. Rev. B.* **2022**, *106*, 064409. [CrossRef]
37. Khmelevskyi, S.; Mohn, P. The order of the magnetic phase transitions in RCo$_2$ (R = rare earth) intermetallic compounds. *J. Phys. Condens. Matter* **2000**, *12*, 9453–9464. [CrossRef]
38. Burzo, E.; Vlaic, P.; Kozlenko, D.P.; Kichanov, S.E.; Dang, N.T.; Lukin, E.V.; Savenko, B.N. Magnetic properties of TbCo$_2$ compound at high pressures. *J. Alloys Compd.* **2013**, *551*, 702–710. [CrossRef]
39. Brouha, M.; Buschow, K.H.J. The pressure dependence of the Curie temperature of rare earth—Cobalt compounds. *J. Phys. F Met. Phys.* **1973**, *3*, 2218–2226. [CrossRef]
40. Tereshina, E.A.; Khmelevskyi, S.; Politova, G.; Kaminskaya, T.; Drulis, H.; Tereshina, I.S. Magnetic ordering temperature of nanocrystalline Gd: Enhancement of magnetic interactions via hydrogenation-induced "negative" pressure. *Sci. Rep.* **2016**, *6*, 22553. [CrossRef] [PubMed]
41. Tereshina, E.A.; Yoshida, H.; Andreev, A.V.; Tereshina, I.S.; Koyama, K.; Kanomata, T. Magnetism of a Lu$_2$Fe$_{17}$H Single Crystal under Pressure. *J. Phys. Soc. Jpn.* **2007**, *76* (Suppl. A), 82–83. [CrossRef]
42. Nikitin, S.A.; Tereshina, I.S.; Verbetsky, V.N.; Salamova, A.A.; Skokov, K.P.; Pankratov, N.Y.; Skourski, Y.V.; Tristan, N.V.; Zubenko, V.V.; Telegina, I.V. Magnetostriction and magnetic anisotropy in TbFe$_{11}$TiH$_x$ single crystal. *J. Alloys Compd.* **2001**, *322*, 42–44. [CrossRef]
43. Shoemaker, D.P.; Shoemaker, C.B. Concerning atomic sites and capacities for hydrogen absorption in the AB2 Friauf-Laves phases. *J. Less-Common Met.* **1979**, *68*, 43–58. [CrossRef]
44. DeSaxce, T.; Berthier, Y.; Fruchart, D. Magnetic and structural properties of the ternary hydrides of ErFe$_2$. *J. Less-Common Met.* **1985**, *107*, 35–43. [CrossRef]
45. Stanley, H.E. *Introduction to Phase Transitions and Critical Phenomena*; Oxford University Press: New York, NY, USA, 1971.
46. Stanley, H.E. Scaling, universality, and renormalization: Three pillars of modern critical phenomena. *Rev. Mod. Phys.* **1999**, *71*, 358–366. [CrossRef]
47. Khan, N.; Midya, A.; Mydeen, K.; Mandal, P.; Loidl, A.; Prabhakaran, D. Critical behavior in single-crystalline La$_{0.67}$Sr$_{0.33}$CoO$_3$. *Phys. Rev. B* **2010**, *82*, 064422. [CrossRef]
48. Sokolovskiy, V.V.; Miroshkina, O.N.; Buchelnikov, V.D.; Marchenkov, V.V. Magnetocaloric Effect in Metals and Alloys. *Phys. Metals Metallogr.* **2022**, *123*, 315–318. [CrossRef]

Article

The Presence of Charge Transfer Defect Complexes in Intermediate Band CuAl$_{1-p}$Fe$_p$S$_2$

Christopher Dickens, Adam O. J. Kinsella, Matt Watkins and Matthew Booth *

School of Mathematics and Physics, University of Lincoln, Lincoln LN6 7TS, UK
* Correspondence: mbooth@lincoln.ac.uk

Abstract: Despite chalcopyrite (CuFeS$_2$) being one of the oldest known copper ores, it exhibits various properties that are still the subject of debate. For example, the relative concentrations of the ionic states of Fe and Cu in CuFeS$_2$ can vary significantly between different studies. The presence of a plasmon-like resonance in the visible absorption spectrum of CuFeS$_2$ nanocrystals has driven a renewed interest in this material over recent years. The successful synthesis of CuAl$_{1-p}$Fe$_p$S$_2$ nanocrystals that exhibit a similar optical resonance has recently been demonstrated in the literature. In this study, we use density functional theory to investigate Fe substitution in CuAlS$_2$ and find that the formation energy of neutral [Fe$_{Cu}$]$^{2+}$ + [Cu$_{Al}$]$^{2-}$ defect complexes is comparable to [Fe$_{Al}$]^0 antisites when $p \geq 0.5$. Analysis of electron density and density of states reveals that charge transfer within these defect complexes leads to the formation of local Cu^{2+}/Fe^{2+} ionic states that have previously been associated with the optical resonance in the visible absorption of CuFeS$_2$. Finally, we comment on the nature of the optical resonance in CuAl$_{1-p}$Fe$_p$S$_2$ in light of our results and discuss the potential for tuning the optical properties of similar systems.

Keywords: semiconductors; DFT; chalcopyrite; defects; electronic structure; charge transfer; nanocrystals

1. Introduction

Semiconductor nanocrystals, known as quantum dots (QDs), have received significant attention in recent decades due to their potential relative to a multitude of applications including bioimaging [1–3], light-emitting diodes [4,5], and photovoltaic devices [6,7]. The essential property of these materials is the quantum size effect, whereby the band gap becomes strongly size-dependent when the radius of the nanocrystal is smaller than the excitonic Bohr radius. Early QD materials of interest included binary heavy metal chalcogenides such as CdS, CdSe, CdTe, PbS and PbSe. However, over the past decade, the improvement of solvothermal synthesis techniques has made it possible to produce high quality multinary chalcogenide nanocrystals which do not contain toxic elements [8]. By tuning the reactivity of the different precursors, a high degree of control over the stoichiometry can be achieved, enabling the QD properties to be controlled by varying the composition as well as by modifying the size. For example, CuInS$_2$ QDs have been developed as promising candidates for various applications. Their band gap can be tuned between 1.5 eV (i.e., the band gap of bulk CuInS$_2$) and approximately 2 eV, corresponding to photoluminescence emission in the near infrared and yellow regions of the spectrum, respectively [9]. The ability to emit within the so-called biological window makes CuInS$_2$ QDs ideal fluorescent probes for bio-imaging [10–12], and their strong, broad absorption within the solar spectrum makes them a viable photoactive component in photovoltaic devices [13]. Although CuInS$_2$ nanocrystals are less toxic than their heavy metal chalcogenide counterparts, there are some considerable toxicity concerns regarding indium, and furthermore, indium is a rare and expensive metal. Therefore, alternatives containing relatively earth-abundant constituent elements such as Zn, Sn and Fe have been investigated [14,15]. In this study we focus on the replacement of indium with two earth-abundant and inexpensive elements, namely Al and Fe.

Citation: Dickens, C.; Kinsella, A.O.J.; Watkins, M.; Booth, M. The Presence of Charge Transfer Defect Complexes in Intermediate Band CuAl$_{1-p}$Fe$_p$S$_2$. *Crystals* **2022**, *12*, 1823. https://doi.org/10.3390/cryst12121823

Academic Editor: Jacek Ćwik

Received: 1 November 2022
Accepted: 10 December 2022
Published: 14 December 2022

Copyright: © 2022 by the authors. Licensee MDPI, Basel, Switzerland. This article is an open access article distributed under the terms and conditions of the Creative Commons Attribution (CC BY) license (https://creativecommons.org/licenses/by/4.0/).

CuAlS$_2$ is similar to CuInS$_2$ in that it is a [A^{I}][B^{III}][C^{VI}]$_2$ compound that is stable in the chalcopyrite crystal structure but has a much wider band gap (approximately 3.5 eV). Early computational studies by Zunger et al. found that CuAlS$_2$ is further from the ideal zincblende structure than CuInS$_2$, and that the interactions between the Cu 3d and S 3p states are stronger in CuAlS$_2$, leading to a narrower upper valence band (VB) [16]. Relatively little work has been conducted on CuAlS$_2$ nanocrystalline materials in comparison to CuInS$_2$. Bhattacharyya et al. demonstrated that the type II band offset between CuAlS$_2$ and ZnS or CdS in CuAlS$_2$/ZnS [17,18] and CuAlS$_2$/CdS [19] core/shell nanocrystals can be engineered to enable the effective band gap of the heterostructures to span the visible spectrum.

Chalcopyrite, CuFeS$_2$, is perhaps the most earth-abundant copper ore and, despite having been known for thousands of years, remains to be fully understood, since transition metals with partially filled d-orbitals tend to introduce complex properties when included as a constituent in compounds such as chalcogenides. Neutron diffraction investigations of CuFeS$_2$ have, for example, indicated that the magnetic moment of the Fe ion is substantially lower than that expected for a high spin Fe^{3+} ion [20,21], an observation that is consistent with a model of CuFeS$_2$ being a mixture of two distinct ionic states [Cu]$^{+}$[Fe]$^{3+}$[S]$_2^{2-}$ and [Cu]$^{2+}$[Fe]$^{2+}$[S]$_2^{2-}$ [22]. Experimental results regarding the oxidation states of Cu and Fe in CuFeS$_2$ are inconclusive [23–27] and likely depend on the sample preparation. CuFeS$_2$ is suspected by some to be a charge-transport-type insulator, with very small [28,29] or even possibly negative [30] charge-transfer energy, due to the particularly strong pd hybridisation between the Fe 3d and S 3p orbitals. Various authors have observed an optical resonance at approximately 500 nm in the absorbance spectrum of CuFeS$_2$ nanomaterials and attributed it to the presence of an "intermediate band" (IB) between the VB and conduction band (CB) in CuFeS$_2$ [31–37]. Several authors have demonstrated that this IB (also referred to as an "upper valence band" or "additional conduction band") leads to excellent photothermal conversion efficiency [31,32] and the potential for improved photovoltaic efficiency [38] or spintronic applications [39]. Gabka et al. demonstrated that the spectral position of the resonance peak in CuFeS$_2$ could be red-shifted by increasing the [Cu]/[Fe] ratio (i.e., by decreasing Fe fraction) [40], and, more recently, Lee et al. demonstrated a stoichiometry-dependent transition between the dielectric resonance associated with the Fe 3d IB and a copper-vacancy-induced localised surface plasmon resonance (LSPR) in Fe-poor CuFeS$_2$[36]. Gaspari et al. used a linear optical model to demonstrate that when an IB with fixed width is shifted towards the VB (i.e., reducing the VB–IB gap but increasing the IB–CB gap) the resonant absorption associated with the so-called Fröhlich condition, which for all-dielectric materials typically occurs in the deep UV, can be red-shifted into the visible region of the spectrum [32]. Yao et al. investigated the influence of the Fe and Cu oxidation states on the resonance feature and found that the incorporation of Fe^{2+} in CuFeS$_2$ reduces the VB–IB gap, leading to a more prominent resonance feature [37]. In their work, they used the same model as Gaspari et al. but allowed the IB width to vary such that the VB–IB gap could be varied independently of the IB–CB gap. It was found that reducing the VB–IB gap in this manner blue-shifted the resonance feature in their simulated absorption spectra. These results illustrate that both the VB–IB gap and the IB width have a significant influence on the spectral position of the optical resonance.

Some work has been performed on transition-metal-doped CuAlS$_2$ [29,41–43]. Two absorption bands, located at 1.3 eV and 2.0 eV, were identified by Teranishi et al. in the optical absorption spectrum of Fe:CuAlS$_2$ (i.e., Fe-doped CuAlS$_2$) [41] that increased in intensity with increasing Fe content. Sato et al. investigated CuAl$_{0.9}$Fe$_{0.1}$S$_2$ and noted a small divalent component in the Cu 2p XPS peak in CuFeS$_2$ (again, consistent with the model of Pauling) but not in CuAl$_{0.9}$Fe$_{0.1}$S$_2$, suggesting hybridisation between the unoccupied Fe 3d orbitals with the valence Cu 3d orbitals mediated by the S 3p valence orbitals [29]. Wang et al. doped CuAlSe$_2$ with various transition metals and concluded that Ti:CuAlSe$_2$ is more promising for intermediate-band-based photovoltaics since it possesses a partially filled IB as opposed to the completely unoccupied IB in Fe:CuAlSe$_2$ [44].

Recently, Yadav et al. demonstrated the ability to synthesise quaternary earth-abundant $CuAl_xFe_{1-x}S_2$ nanocrystals with a band gap that is tunable across the entire UV/vis/NIR spectrum, whilst maintaining the chalcopyrite crystal structure [45]. As part of this work, they performed a limited DFT investigation into the electronic structure of the $CuAl_xFe_{1-x}S_2$ system showing the projected density of states (PDOS) calculated for the $CuAl_{0.75}Fe_{0.25}S_2$ system, the composition of which was obtained by substituting two Al^{3+} ions with Fe^{3+} ions, i.e., by introducing two $[Fe_{Al}]^0$ antisite defects. However, it is plausible that the site preference for Fe substitutions in $CuAlS_2$ is on Cu sites rather than on Al sites. First, since the effective ionic radius of Fe^{3+} in tetrahedral coordination is larger than that of Al^{3+} and smaller than that of Cu^+ (see Table 1), a $[Fe_{Al}]^0$ substitution is likely to introduce more local distortion in the lattice than a $[Fe_{Cu}]^{2+}$ substitution. Second, since the Cu-S and Fe-S bonds are significantly more covalent than the Al-S bond, which is mostly ionic in nature [46], a $[Fe_{Al}]^0$ substitution would result in a more covalent local environment. Navrátil et al. proposed that intrinsic point defects such as $[Fe_{Cu}]^{2+}$ antisites play an integral role alongside the charge transport phenomenon in explaining the low effective magnetic moment and weak ferromagnetism [47].

Table 1. Effective ionic radius of Cu^+, Fe^{3+} and Al^{3+} in tetrahedral coordination [48] (the Fe^{3+} ion is assumed to be in the high spin state).

	Effective Ionic Radius, Å
Cu^+	0.60
Fe^{3+}	0.49
Al^{3+}	0.39

If it were indeed the case that $[Fe_{Cu}]^{2+}$ antisites were more abundant than $[Fe_{Al}]^0$ antisites, the presence of interstitial Cu_i or $[Cu_{Al}]^{2-}$ antisite defects would be necessary to maintain the stoichiometric [Cu]/([Al]+[Fe]) ratio. Following the same reasoning regarding the site preference for Fe outlined above, the formation of $[Cu_{Al}]^{2-}$ antisite defects would likely introduce significant strain in the lattice. However, it may be that neutral $[Fe_{Cu}]^{2+} + [Cu_{Al}]^{2-}$ defect pairs have relatively low formation energies. It is well-known that the formation energies of intrinsic point defects, specifically cation vacancies and cation antisites, are comparatively low in ternary chalcogenides, such as $CuInS_2$ and $CuInSe_2$, with the chalcopyrite crystal structure, due to the lattice strain generated by the different cation–anion bond lengths. Furthermore, these charged defects can cluster to form neutral defect complexes such as $[In_{Cu}]^{2+} + 2[V_{Cu}]^-$ or $[In_{Cu}]^{2+} + [Cu_{In}]^{2-}$ [49–52]. Harvie et al. used electron energy-loss spectroscopy mapping to image the elemental distribution of copper and indium in single chalcopyrite $CuInS_2$ QDs and found that copper-rich and indium-rich domains can exist within individual nanocrystals [53]. Another consideration then is that the additional configuration entropy associated with such antisite defect complexes when compared to the same number of isolated $[Fe_{Al}]^0$ antisite defects may be a significant stabilising factor, especially in $CuAl_{1-p}Fe_pS_2$ nanocrystals such as those synthesised at elevated temperature by Yadav et al. This possibility has not been explored in detail here but may be investigated using the methods developed by Grau-Crespo et al. [54]. Overall, then, it is not obvious whether $[Fe_{Al}]^0$ antisite defects or $[Fe_{Cu}]^{2+} + [Cu_{Al}]^{2-}$ defect complexes would be the more prominent Fe substitution mode in $CuAlS_2$.

We expand on the work undertaken by Yadav et al. by examining these two distinct modes of Fe substitution in $CuAlS_2$. We first show that the site preference for Fe substitution is indeed on Cu sites as opposed to Al sites and that a $[Fe_{Cu}]^{2+}$ defect produces a local relaxation that reduces the formation energy of nearby $[Cu_{Al}]^{2-}$ defects. We then show that the lowest unoccupied Fe 3d orbitals exist inside the band gap of $CuAlS_2$ for both substitution modes, but that the $[Fe_{Cu}]^{2+} + [Cu_{Al}]^{2-}$ antisite complexes induce a charge transfer from the highest occupied Cu 3d orbital to the lowest unoccupied Fe 3d orbital, resulting in a $[Fe_{Cu}]^+ + [Cu_{Al}]^-$ defect complex and a significantly reduced band gap.

Finally, we calculate the per-defect formation energies and the PDOS of the $CuAl_{1-p}Fe_pS_2$ system for each Fe substitution mode, for several values of p between zero and one obtained by introducing uniformly distributed defects. We find that the per-defect formation energy for $[Fe_{Al}]^0$ antisites increases linearly with the Fe fraction across the whole range of compositions studied (from $CuAlS_2$ to $CuFeS_2$), whereas the per-defect formation energy of the $[Fe_{Cu}]^{2+} + [Cu_{Al}]^{2-}$ defect complexes is reduced as the Fe fraction increases. We comment on this result in the context of some of the experimental results presented in the literature.

2. Methods

DFT calculations for this work were performed within the open-source CP2K framework [55]. The majority of data displayed in this paper was obtained using a generalised gradient approximation (GGA) specifically the PBE functional [56]. The Gaussian and Plane Wave method was used with Gaussian basis sets [57] of triple-ζ quality, a plane wave cutoff of 600 Ry for geometry relaxations (750 Ry for cell optimizations) and corresponding GTH pseudopotentials [58,59] which include 11, 16, 3 and 6 valence electrons for Cu, Fe, Al and S, respectively. Self Consistent Field (SCF) convergence criteria was used with the orbital transform algorithm [60] with the convergence set to EPS_SCF = 1×10^{-7}, giving energetic convergence of below a micro Hartree in total energy. All systems were fully relaxed with respect to both internal atomic coordinates and cell parameters. The defective systems were constructed from a 64-atom supercell, $2 \times 2 \times 1$ of the default 16 atom chalcopyrite structure, thus giving 16 cation A sites (Cu for $CuAlS_2$), 16 cation B sites (Al for $CuAlS_2$) and 32 anion sites (S) allowing both flexibility in configurations and the possibility of reasonable sampling. Brillouin zone sampling was restricted to the Γ point.

For convenience, all systems were assumed to be fully ferro-magnetic: Whilst $CuFeS_2$ should be antiferromagnetic, and, doubtless, the heavily defective systems will have a complex spin ordering, the effect on overall energetics and positioning of single particle levels will be small and of limited importance for this study. We use the minority (beta) spin state channel when reporting band gaps.

When calculating the relative energies of defective systems, the chemical potentials of individual species were calculated using the now standard ab initio thermodynamics approach and derived from total energy calculations of the unary, binary and ternary compounds. In the few cases where charged defects were considered, Lany–Zunger [61] charge corrections were applied.

We refer to Fe substitution via $[Fe_{Al}]^0$ antisites as "mode α" and to Fe substitution via $[Fe_{Cu}]^{2+} + [Cu_{Al}]^{2-}$ defect complexes as "mode β" (Figure 1). The configuration containing a single Fe substitution via mode α is referred to as 'α_1', and the configuration containing a single Fe substitution via mode β is referred to as 'β_1'. In the case of two Fe substitutions, for which we have studied several configurations, each configuration is labelled, for example, '$\alpha_{2.1}$', '$\alpha_{2.2}$', '$\alpha_{2.3}$', etc. (see Figure 1). When referring to a system with more than two Fe substitutions, in general we will use the p notation (as in $CuAl_{1-p}Fe_pS_2$) or refer to specific configurations containing n defects per supercell as α_n or β_n. Since the $2 \times 2 \times 1$ supercell contains 16 Al sites, the value of p for a configuration containing n defects is $p = n/16$. We used the $p = 0.125$ ($n = 2$) and $p = 0.25$ ($n = 4$) configurations to generate what we call the "inverted" $p = 0.875$ ($n = 14$) and $p = 0.75$ ($n = 12$) systems, respectively, where the "inverted" configuration contains n defects located at the non-defective sites of the $(16 - n)$ configuration.

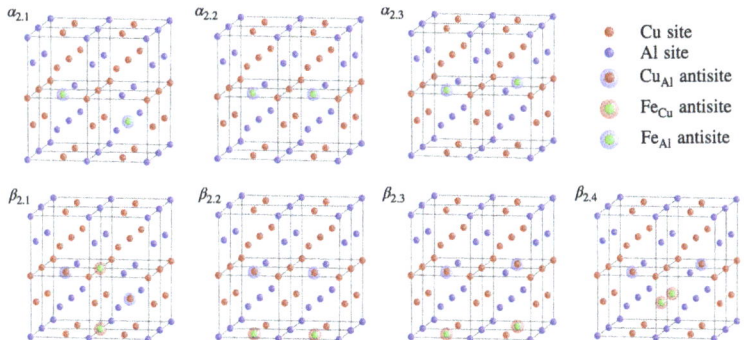

Figure 1. Illustrations showing the configurations $\alpha_{2.1}$, $\alpha_{2.2}$, $\alpha_{2.3}$, $\beta_{2.1}$, $\beta_{2.2}$, $\beta_{2.3}$ and $\beta_{2.4}$.

In order to determine the most uniformly distributed "α" configurations for a particular value of p, we utilised the ASE package [62] to generate all possible locations of n $[Fe_{Al}]^0$ defects. The average Fe-Fe distance in a given configuration was calculated using a minimum image convention (e.g., Fe_i to Fe_j would use the shortest of the distances between Fe_i and all periodic images of Fe_j). We find that the possible configurations split into multiple groups containing configurations with very similar mean Fe-Fe separation. From the group with the largest mean Fe-Fe separation, we selected the configuration with the lowest standard deviation about the mean as the most uniformly distributed configuration. To generate the β systems, the $[Fe_{Al}]^0$ were replaced with $[Cu_{Al}]^{2-}$, and the nearest Cu site in the relaxed geometry was replaced by an Fe to form a $[Fe_{Cu}]^{2+}$ defect.

Owing to the apparent importance of on-site electronic interactions and correlations for understanding the properties of $CuFeS_2$, DFT investigations based on LDA and GGA have struggled to provide reliable predictions of electronic structures of it and related materials. As a pragmatic approach, we have checked that the results are not qualitatively sensitive to the inclusion of exact exchange. After completion of PBE calculations, additional SCF calculations were performed using a generalised PBE0 type functional with 8, 16, or 24% exact exchange on the optimized PBE geometry.

3. Results and Discussion

3.1. Single ($p = 0.0625$) Fe Substitution

We calculate the band gap of $CuAlS_2$ to be approximately 1.65 eV (Figure 2), which is significantly smaller than the experimental value of 3.49 eV [63], as is typical for PBE [64]. The VB of $CuAlS_2$ consists of mostly Cu 3d and S 3p character, implying pd hybridisation, consistent with the strong covalent character of the Cu-S bonds.

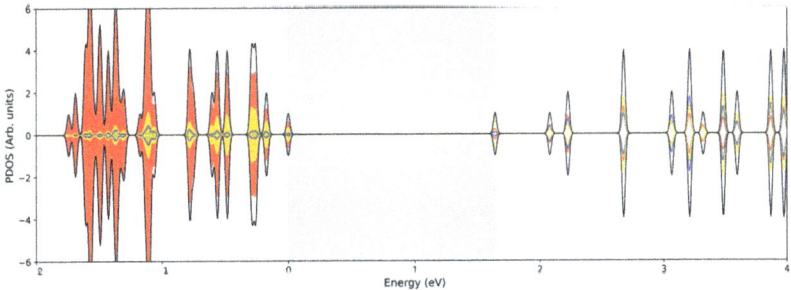

Figure 2. PDOS of $CuAlS_2$ (copper = red, sulfur = yellow, aluminium = blue).

For the case of a single Fe substitution, we compared the formation energy of a $[Fe_{Al}]^0$ defect (configuration α_1) with that of a single neutral $[Fe_{Cu}]^{2+} + [Cu_{Al}]^{2-}$ defect pair (configuration β_1). Because of the symmetry of the CuAlS$_2$ lattice, the site selection has a negligible effect on the calculated defect formation energy of the α substitution mode. In the case of the β substitution mode, the Cu site closest to the $[Cu_{Al}]^{2-}$ defect was chosen for the Fe substitution. We also calculated the formation energy of single isolated $[Fe_{Cu}]^{2+}$ and $[Cu_{Al}]^{2-}$ defects. The calculated values for all defect types are shown in Table 2.

Table 2. Calculated defect formation energies in CuAlS$_2$.

	Defect Formation Energy, eV
$[Fe_{Al}]^0$	2.789
$[Fe_{Cu}]^{2+}$	0.471
$[Cu_{Al}]^{2-}$	4.112
$[Fe_{Cu}]^{2+} + [Cu_{Al}]^{2-}$	3.559

We find that the formation energy of a $[Fe_{Cu}]^{2+}$ antisite (0.471 eV) is significantly lower than that of a neutral $[Fe_{Al}]^0$ antisite (2.789 eV), as expected. To reiterate, we primarily attribute this to the difference in effective ionic radii and cation–anion bond lengths of the relevant ions in tetrahedral coordination. We also find that the combined formation energies of the isolated $[Fe_{Cu}]^{2+}$ and $[Cu_{Al}]^{2-}$ defects (4.583 eV) is almost 1 eV larger than the formation energy of a neutral $[Fe_{Cu}]^{2+} + [Cu_{Al}]^{2-}$ defect pair (3.599 eV). We suggest, therefore, that when a relatively weak Cu-S (BDE$_{Cu-S}$ = 2.86 eV [65]) bond is replaced with a relatively strong Fe-S bond (BDE$_{Fe-S}$ = 3.34 eV [65]), the surrounding Cu-S and Al-S bonds weaken in response and that it is this local weakening of the Al-S bonds that reduces the formation energy of $[Cu_{Al}]^{2-}$ antisites close to the $[Fe_{Cu}]^{2+}$ defect. Indeed, when analysing the distribution of Al-S bond lengths in the system containing a single $[Fe_{Cu}]^{2+}$ antisite, we find that for Al sites far away from (i.e., more than 4 Å) the Fe substitution, the Al-S bond lengths at each site are uniform (between 2.26 Å and 2.27 Å with a standard deviation of less than 0.005 Å). However, close to (i.e., within 4 Å of) the Fe substitution, the four Al-S bond lengths at each Al site vary to a much greater degree. The mean values remain consistent between 2.26 Å and 2.27 Å, but the Al-S bond involving the sulfur that is also bonded to the substituted Fe weakens significantly (this Al-S bond length increases to more than 2.29 Å), and the other three strengthen slightly (see Figures A1–A3).

When a single Fe substitution is made, the atomic PDOS indicates that the fully occupied Fe $3d$ orbitals are quite broad and exist throughout the VB suggesting hybridisation with the S $3p$ orbitals (Figure 3), consistent with the strong covalency of the Fe-S bond in chalcopyrite [26,29]. The unoccupied Fe $3d$ orbitals exist inside the band gap of the CuAlS$_2$ host and display some S character, suggesting weak pd hybridisation. In the case of configuration α_1, the five unoccupied $3d$ orbitals resemble the canonical localised d-electron states Fe $3d_{z^2}$, Fe $3d_{x^2-y^2}$, Fe $3d_{xy}$, Fe $3d_{xz}$ and Fe $3d_{yz}$ (see Figure A4) and appear in two clearly separated regions within the band gap (Figure 3a), corresponding to the minority spin e and t_2 orbitals separated by a crystal field splitting of approximately 0.5 eV, which is smaller than that observed in Fe:ZnS [66], for example. We observe some internal structure in each of these $3d$ bands, which implies that the degeneracy of the two e orbitals (d_{z^2} and $d_{x^2-y^2}$) and three t_2 orbitals (d_{xy}, d_{xz} and d_{yz}) has been lifted, presumably because the tetrahedral coordination is not ideal due to the varying cation–anion bond lengths.

In the case of configuration β_1, the asymmetric crystal field is even more evident, with the five $3d$ orbitals appearing to be well-separated (Figure 3b)). This asymmetry is reflected in the Fe-S bond lengths: the four Fe-S bond lengths in the α_1 configuration are all approximately 2.297 Å (with a standard deviation of <0.001 Å), whereas the four Fe-S bond lengths in the β_1 configuration are 2.380 Å, 2.367 Å, 2.354 Å and 2.268 Å (corresponding to a standard deviation of 0.050 Å). By comparing the density of states of configuration β_1 (Figure 3b)) with that of configuration α_1, we observe that the Fe $3d_{x^2-y^2}$ orbital, which is the lowest unoccupied state in configuration α_1, is the highest occupied state in configuration

β_1, and the Cu $3d_{xy}$ orbital, which is occupied in configuration α_1, is shifted up in energy to about 100 meV above the VB edge. This is observed in the system containing only the isolated $[Fe_{Cu}]^{2+}$ antisite (see Figure A5), but the occupied Fe $3d$ and unoccupied Cu $3d$ are essentially degenerate (25 meV separation). We therefore attribute it to a charge transfer from Cu $3d_{xy}$ to Fe $3d_{x^2-y^2}$, that is significantly enhanced by the presence of the $[Cu_{Al}]^{2-}$ antisite and which results in a local $[Cu]^{2+}[Fe]^{2+}[S]_2^{2-}$ ionic state. If we include the oxidation state explicitly in our defect notation, this corresponds to

$$[Fe_{Cu^I}^{III}]^{2+} + [Cu_{Al^{III}}^{I}]^{2-} \to [Fe_{Cu^I}^{II}]^{+} + [Cu_{Al^{III}}^{II}]^{-}. \tag{1}$$

Various studies have highlighted the importance of the Fe^{2+} content on the properties of $CuFeS_2$ [37,47]. For example, Yao et al. have shown that the intensity of the plasmon-like feature in $CuFeS_2$ grows with increasing Fe^{2+} content [37]. It was suggested by the authors that the Fe^{2+} state corresponds to an additional electron in the lowest unoccupied Fe $3d$ orbital, which effectively narrows the VB–IB gap, as we also suggest here. We therefore expect that this possible route to forming Fe^{2+} in Fe-substituted $CuAlS_2$ may be relevant to understanding the nature of the optical resonance observed by Yadav et al. in the absorption spectrum of $CuAl_{1-p}Fe_pS_2$ [45].

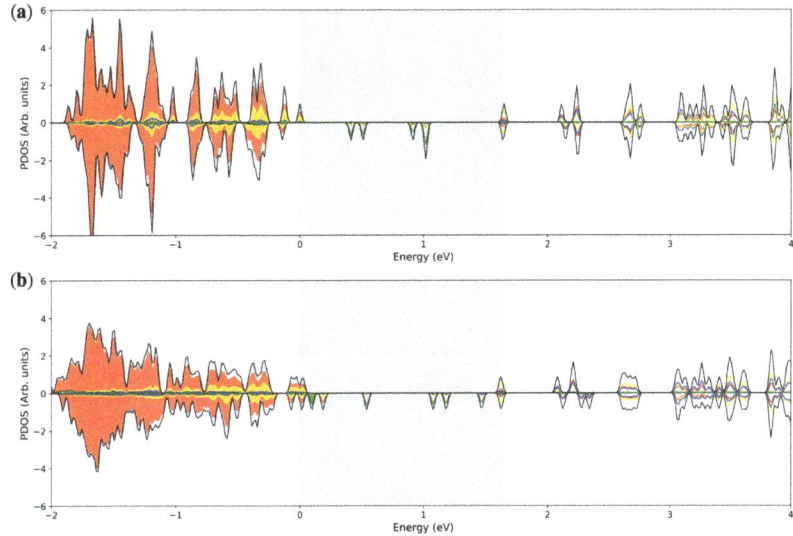

Figure 3. PDOS for configurations (**a**) α_1 and (**b**) β_1 (copper = red, sulfur = yellow, aluminium = blue and iron = green).

3.2. Double ($p = 0.125$) Fe Substitution

In the case of two Fe substitutions, we have investigated the effect of the precise configuration on the defect formation energies and PDOS. We find that the per-defect formation energy for the three configurations containing two $[Fe_{Al}]^0$ defects, labelled $\alpha_{2.1}$, $\alpha_{2.2}$ and $\alpha_{2.3}$ in Figure 1, is similar in all three cases (2.811 eV, 2.813 eV and 2.820 eV, respectively) and only slightly higher than the single defect case (2.789 eV). However, we do see some small variation in the PDOS (Figure 4), specifically that the variations in the local distortion already are sufficient for defect levels to span the range of energies associated with an IB such as that observed in $CuFeS_2$.

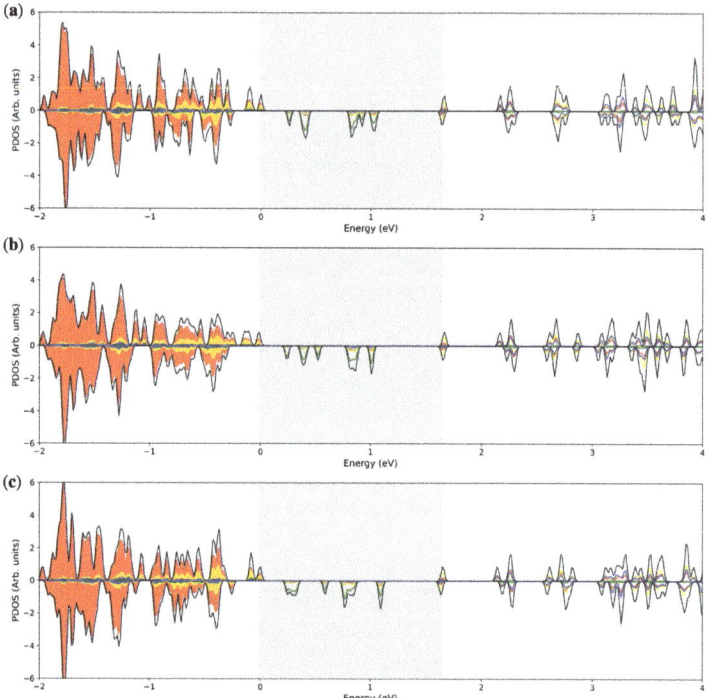

Figure 4. PDOS for configurations (**a**) $\alpha_{2.1}$, (**b**) $\alpha_{2.2}$ and (**c**) $\alpha_{2.3}$ (copper = red, sulfur = yellow, aluminium = blue and iron = green).

In the case of $[Fe_{Cu}]^{2+} + [Cu_{Al}]^{2-}$ defect complexes we examined four configurations ($\beta_{2.1}$, $\beta_{2.2}$, $\beta_{2.3}$ and $\beta_{2.4}$ shown in Figure 1). We find that the per-defect formation energies in the β_2 configurations are more variable than for the α_2 configurations: for configurations $\beta_{2.1}$, $\beta_{2.2}$, $\beta_{2.3}$ and $\beta_{2.4}$, we calculated it to be 3.455 eV, 3.789 eV, 3.732 eV and 3.541 eV, respectively. We note that the two configurations with lower formation energies ($\beta_{2.1}$ and $\beta_{2.4}$) are those in which the $[Fe_{Cu}]^{2+}$ defects exist 'in between' the $[Cu_{Al}]^{2-}$, which will become increasingly likely as the Fe fraction is increased via this Fe substitution mode. The PDOS for configurations $\beta_{2.1}$, $\beta_{2.2}$, $\beta_{2.3}$ and $\beta_{2.4}$ are shown in Figure 5. It is evident that the presence of the charge transfer is insensitive to small variations in the configuration.

3.3. Multiple ($p > 0.125$) Fe Substitutions

The calculated per-defect formation energies for both substitution modes across the entire composition space between $CuAlS_2$ and $CuFeS_2$ are shown in Figure 6. The per-defect formation energy of the $[Fe_{Al}]^0$ antisite increases linearly with the Fe fraction. In contrast, the formation energy per defect of the $[Fe_{Cu}]^{2+} + [Cu_{Al}]^{2-}$ complex decreases relatively quickly as p increases from 0 to 0.5, such that the two modes have almost indistinguishable defect formation energies for the ideal mixed system $CuAl_{0.5}Fe_{0.5}S_2$.

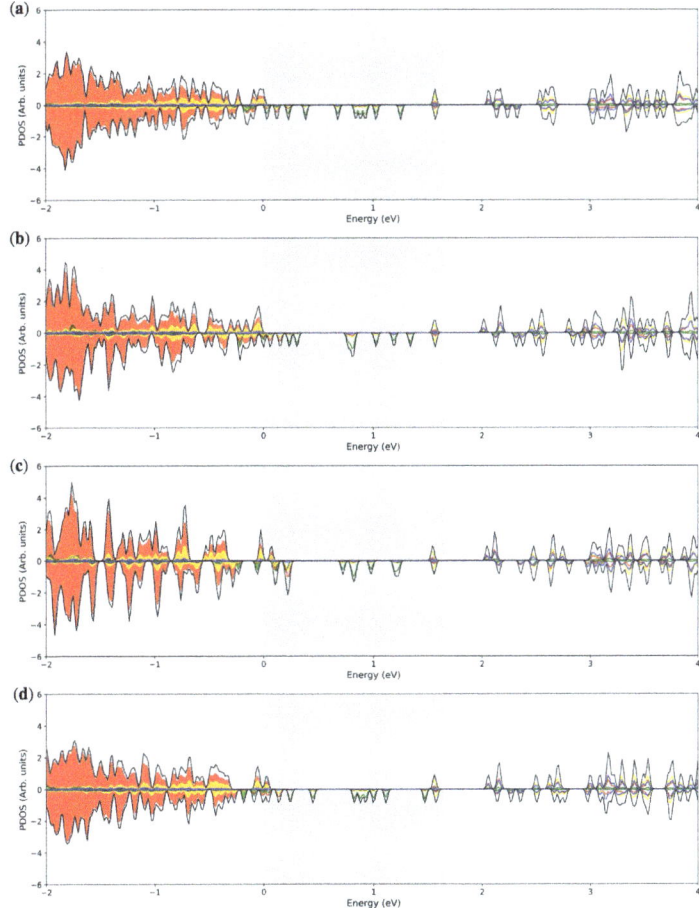

Figure 5. PDOS for configurations (**a**) $\beta_{2.1}$, (**b**) $\beta_{2.2}$, (**c**) $\beta_{2.3}$ and (**d**) $\beta_{2.4}$ (copper = red, sulfur = yellow, aluminium = blue and iron = green).

In general, to generate the β_n configurations from the α_n configurations, we replaced the n $[Fe_{Al}]^0$ defects with $[Cu_{Al}]^{2-}$ defects and chose to substitute the closest Cu to each $[Cu_{Al}]^{2-}$ site for Fe to form n $[Fe_{Cu}]^{2+} + [Cu_{Al}]^{2-}$ defect pairs. We have investigated how this choice affects the formation energies at either very low ($p < 0.2$) or very high ($p > 0.8$) Fe content, where the configuration space is reasonably accessible. The four different values of the per-defect formation energy at $p = 0.125$ for the β configurations (red curve in Figure 6) correspond to the $\beta_{2.1}$, $\beta_{2.2}$, $\beta_{2.3}$ and $\beta_{2.4}$ systems (see Figure 1). It is clear that slight variations in the β_2 configurations can lead to significantly different defect formation energies. The reason for this fluctuation is likely the additional degree of freedom associated with the $[Fe_{Cu}]^{2+} + [Cu_{Al}]^{2-}$ defect pair relative to the isolated $[Fe_{Al}]^0$ defect. The order of the β_2 configurations in terms of defect formation energy is as follows: $\beta_{2.2} > \beta_{2.3} > \beta_{2.4} > \beta_{2.1}$. It appears, therefore, that when the two $[Fe_{Cu}]^{2+}$ defects are located in between the two $[Cu_{Al}]^{2-}$ defects (see Figure 1), the formation energy of the cluster as a whole is decreased. This is consistent with our reasoning in Section 3.1 regarding the lower formation energy of the $[Fe_{Cu}]^{2+} + [Cu_{Al}]^{2-}$ defect pairs relative to the sum of the formation energies of the isolated $[Fe_{Cu}]^{2+}$ and $[Cu_{Al}]^{2-}$ defects and is also consistent with the observation of compositional domains in similar systems [53]. What is not clearly visible in Figure 6 is that we have also performed similar calculations for the three α_2 configurations $\alpha_{2.1}$, $\alpha_{2.2}$ and $\alpha_{2.3}$ (blue curve at $p = 0.125$), but the variation in per-defect formation energies is almost imperceptible. As the Fe fraction increases, we would expect this additional configurational freedom in the $\beta_{n>1}$ systems and, therefore, the variation in the calculated defect formation energies, to also decrease. Indeed, we also calculated the variation in the per-defect formation energy of the four β_{14} systems (at $p = 0.875$) generated by 'inverting' the β_2 systems and found that the fluctuation is significantly smaller than at $p = 0.125$. However, the full exploration of this hypothesis is beyond the scope of this study.

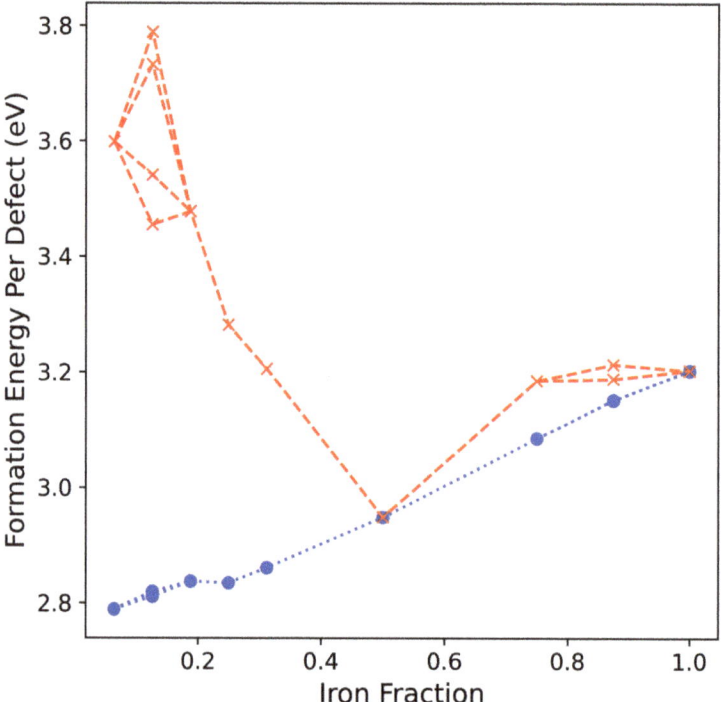

Figure 6. Defect formation energies (per-defect) for $[Fe_{Al}]^0$ antisites (blue circles) and $[Fe_{Cu}]^{2+} + [Cu_{Al}]^{2-}$ defect complexes (red crosses). The multiple values at $p = 0.125$ and $p = 0.875$ correspond to the $\alpha_{2.1}$, $\alpha_{2.2}$, $\alpha_{2.3}$, $\beta_{2.1}$, $\beta_{2.2}$, $\beta_{2.3}$ and $\beta_{2.4}$ configurations and their inverted systems, respectively.

To explain the observed trends in the per-defect formation energies of the two Fe substitution modes, we examine the composition-induced structural changes in both systems. $[A^I][B^{III}][C^{VI}]_2$ compounds in the chalcopyrite crystal structure are derived from the zinc-blende structure by replacing the group II element of a binary chalcogenide compound with group I and group III elements in an ordered manner such that the average number of valence electrons per atom is four (i.e., according to the extension of the Grimm–Sommerfeld rules to ternary compounds by Goryunova [67]). In such ternary systems, where the cationic sub-lattice contains two different cation species, the two cation–anion bond lengths can differ ($d_{A-C} \neq d_{B-C}$), potentially causing significant strain in the crystal lattice. The tetragonal distortion parameter $\eta = c/2a$, where c is the lattice parameter in the z direction and a is the lattice parameter in the x direction, characterises the compression ($\eta < 1$) or tension ($\eta > 1$) along the z direction. Figure 7a) shows the tetragonal distortion parameter η for the range of quaternary compositions between CuAlS$_2$ and CuFeS$_2$. It can be seen that η approaches a value of approximately 0.997 when $p \geq 0.75$, which is significantly closer to the ideal chalcopyrite structure ($\eta = 1$) than the value of 0.983 observed when $p = 0$. This indicates an expansion of the primitive cell along the z direction, which is consistent with the observed increase in primitive cell volume of about 1% (Figure 7b)).

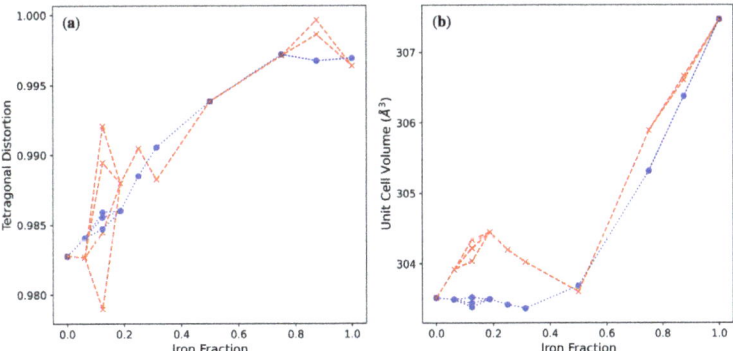

Figure 7. (a) Tetragonal distortion parameter η for both modes of Fe-substitution over the range $0 < p < 1$; and (b) Unit cell volume for both modes of Fe-substitution over the range $0 < p < 1$ (blue circles are α configurations and red crosses are β configurations. The multiple values at $p = 0.125$ and $p = 0.875$ correspond to the $\alpha_{2.1}$, $\alpha_{2.2}$, $\alpha_{2.3}$, $\beta_{2.1}$, $\beta_{2.2}$, $\beta_{2.3}$ and $\beta_{2.4}$ configurations and their inverted systems, respectively.

In CuAl$_{1-p}$Fe$_p$S$_2$, the variation in the cation–anion bond lengths increases substantially as p increases. As can be seen in Figure A6, the variation about the mean Fe-S, Al-S and Cu-S bond lengths is greater for systems containing the defect complexes than it is for systems containing the isolated antisite defects, reflecting the additional degree of freedom associated with the precise [Fe$_{Cu}$]$^{2+}$ + [Cu$_{Al}$]$^{2-}$ arrangement. As we mentioned in the introduction, the large configuration entropy in materials with multiple components in a disordered configuration can be a significant stabilising factor [54,68]. We therefore anticipate that the [Fe$_{Cu}$]$^{2+}$ + [Cu$_{Al}$]$^{2-}$ complexes become even more competitive with increasing Fe fraction than is suggested by the data presented in Figure 7.

When the Fe substitutions occur in the form of a [Fe$_{Al}$]^0 antisite defect, $d^p_{\text{Fe-S}}$ increases gradually with the Fe fraction, only broadening slightly for intermediate values of p (Figure A6c)). When analysing the distributions of Cu-S and Al-S bond lengths in the α configurations, we see that as soon as a single [Fe$_{Al}$]^0 defect is introduced into the supercell ($p = 0.0625$), the $d^p_{\text{Cu-S}}$ values appear to split into two clusters. We observe a similar splitting, albeit less prominent, in the $d^p_{\text{Al-S}}$ values for $0 < p < 1$ (Figure A6b)), which may be rationalised as follows. The secondary cluster of shorter bond lengths is likely to be made up of those Al-S bonds close to the [Fe$_{Al}$]^0 antisite, which are strengthened due to the replacement of the Al-S bond, which has a bond dissociation energy BDE$_{\text{Al-S}}$ = 3.88 eV [65], with a slightly weaker bond (BDE$_{\text{Fe-S}}$ = 3.34 eV [65]). This local strengthening of the Al-S bonds close to the existing [Fe$_{Al}$]^0 antisites hinders the formation of additional [Fe$_{Al}$]^0 antisites.

In the case of Fe substitutions via the introduction of [Fe$_{Cu}$]$^{2+}$ + [Cu$_{Al}$]$^{2-}$ defect complexes, we again see that as soon as a single defect is introduced, the $d^p_{\text{Cu-S}}$ values appear to split into two distinct clusters (Figure A6d)). Since the bond dissociation energy of Cu-S is significantly lower than that of Fe-S, the [Fe$_{Cu}$]$^{2+}$ defects substitute a relatively weak bond with a relatively strong bond, and the surrounding Cu-S and Al-S bonds weaken. It is likely that this local weakening of the Al-S bonds reduces the formation energy of

$[Cu_{Al}]^{2-}$ antisites close to existing $[Fe_{Cu}]^{2+}$ defects. This is consistent with our earlier observation (Section 3.1) that the combined formation energies of the isolated $[Fe_{Cu}]^{2+}$ and $[Cu_{Al}]^{2-}$ defects is larger than that of the neutral $[Fe_{Cu}]^{2+} + [Cu_{Al}]^{2-}$ defect complex. It is also possible that the local weakening of Cu-S bonds reduces the formation energy for additional $[Fe_{Cu}]^{2+}$ antistites. This would be consistent with the observation (Figure 6) that the per-defect formation energy of a $[Fe_{Cu}]^{2+} + [Cu_{Al}]^{2-}$ defect complex initially decreases with increasing defect density.

Figure 8a,b show the PDOS of $CuAl_{0.75}Fe_{0.25}S_2$ obtained via uniformly distributed $[Fe_{Al}]^0$ antisite defects (configuration α_4) and $[Fe_{Cu}]^{2+} + [Cu_{Al}]^{2-}$ defect complexes (configuration β_4), respectively. We still see quite a large difference between the α and β configurations. In the β_4 configuration, the highest occupied minority spin states have significantly more Fe $3d$ character than in the α_4 configuration, and the lowest unoccupied minority spin states have significantly more Cu $3d$ character.

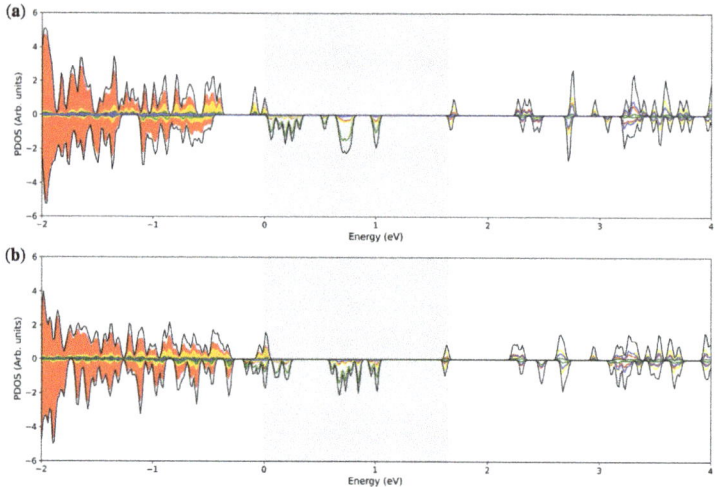

Figure 8. PDOS for (a) α_4 and (b) β_4 systems (i.e., $p = 0.25$) (copper = red, sulfur = yellow, aluminium = blue and iron = green).

Figure 9a,b show the PDOS of $CuAl_{0.25}Fe_{0.75}S_2$ obtained via uniformly distributed $[Fe_{Al}]^0$ antisite defects (configuration α_{12}) and $[Fe_{Cu}]^{2+} + [Cu_{Al}]^{2-}$ defect complexes (configuration β_{12}), respectively. It is clear that once the Fe content is this high, both systems begin to resemble $CuFeS_2$. However, there is still a major difference between the α_{12} and β_{12} configurations, in that the latter has a significantly narrower VB–IB gap.

Figure 10 shows the calculated band gap of $CuAl_{1-p}Fe_pS_2$ obtained via uniformly distributed $[Fe_{Al}]^0$ antisite defects and $[Fe_{Cu}]^{2+} + [Cu_{Al}]^{2-}$ defect complexes. In their 2021 article, Yadav et al. calculated a band gap of 1.78 eV for the $CuAlS_2$ system, a band gap of 0.21 eV for the $CuFeS_2$ system and a band gap of 0.38 eV for the $CuAl_{0.75}Fe_{0.25}S_2$ system. It can be seen from Figure 10 that our calculations are in approximate agreement. We see that a single Fe substitution in the form of a $[Fe_{Cu}]^{2+} + [Cu_{Al}]^{2-}$ defect pair reduces the band gap from 1.6 eV to approximately 0.1 eV, whereas for a single $[Fe_{Al}]^0$ defect the band gap is reduced to 0.6 eV. By the time $p = 0.5$, the band gap of both the α and β configurations is approximately 0.2 eV. When $p > 0.5$, we see that the band gap of the α configuration increases again to about 0.4 eV for $p = 0.75$ before dropping to about 0.2 eV for $p = 1.0$, whereas the band gap of the β configuration remains at about 0.2 eV.

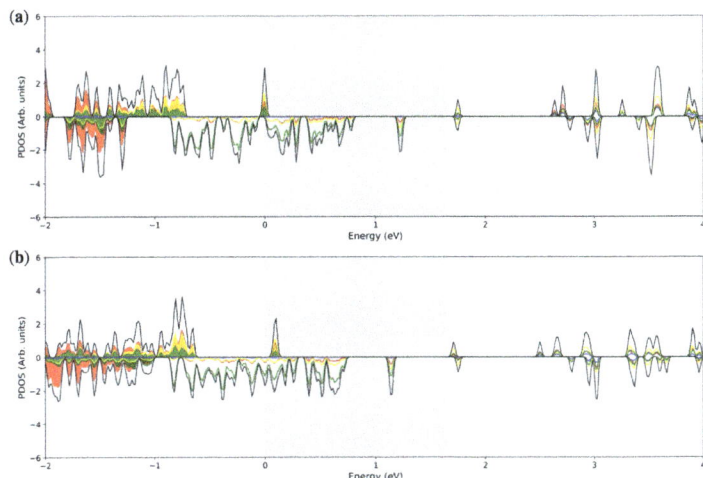

Figure 9. PDOS (**a**) α_{12} and (**b**) β_{12} systems (i.e., $p = 0.75$) (copper = red, sulfur = yellow, aluminium = blue and iron = green).

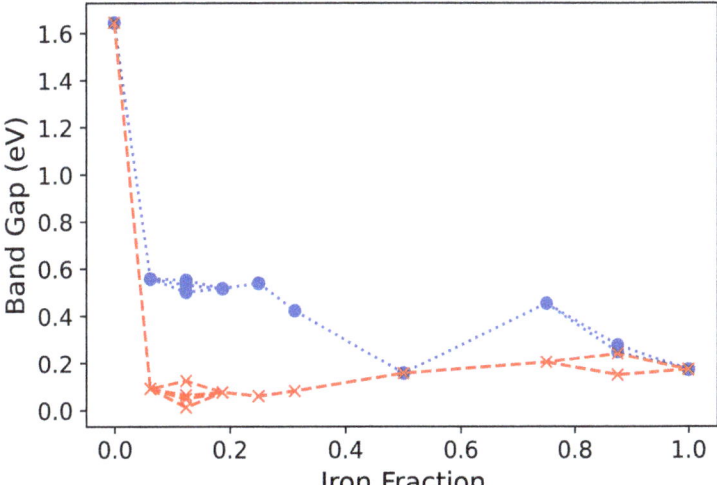

Figure 10. Band gap of the $CuAl_{1-p}Fe_pS_2$ obtained via uniformly distributed $[Fe_{Al}]^0$ antisite defects (blue circles) and $[Fe_{Cu}]^{2+} + [Fe_{Cu}]^{2+}$ defect complexes (red crosses). The multiple values at $p = 0.125$ and $p = 0.875$ correspond to the $\alpha_{2.1}$, $\alpha_{2.2}$, $\alpha_{2.3}$, $\beta_{2.1}$, $\beta_{2.2}$, $\beta_{2.3}$ and $\beta_{2.4}$ configurations and their inverted systems, respectively.

As was mentioned previously, Yao et al. observed a plasmon-like resonance at 500 nm in the absorbance spectrum of $CuFeS_2$ and attributed it to the presence of Fe^{2+} [37]. We notice that in the data presented by Yadav et al. a similar resonance, also centred at 500 nm, appears in the absorbance spectrum of $CuAl_xFe_{1-x}S_2$ nanocrystals only when $x < 0.5$ (in our notation this corresponds to $p > 0.5$ in $CuAl_{1-p}Fe_pS_2$) rather than increasing gradually in amplitude as soon as Fe is introduced [45]. We therefore speculate that for systems with $p > 0.5$ in which the $[Fe_{Cu}]^{2+} + [Cu_{Al}]^{2-}$ complex becomes relatively abundant, the charge transfer that we observe for the β configurations may be responsible for the experimentally observed plasmon-like absorption feature. This also is consistent with the fact that Sato et al. identified a small divalent component in the Cu $2p$ XPS peak in $CuFeS_2$ but not in $CuAl_{0.9}Fe_{0.1}S_2$ [29]. We therefore expect that a comprehensive XPS study of $CuAl_{1-p}Fe_pS_2$ nanocrystals will enable the determination of the precise Fe fraction at which the Cu^{2+} component emerges. We also expect that a detailed XRD study, similar to the one performed by Yao et al., who demonstrated the site preference for Mn on the Cu site in $CuInSe_2$ by performing Rietveld refinements of X-ray diffraction data [69], will shed light on the prevalence of the $[Fe_{Cu}]^{2+} + [Cu_{Al}]^{2-}$ defect complexes we propose here.

4. Conclusions

In summary, we have used density functional theory to investigate Fe substitution in $CuAlS_2$ via two modes, namely $[Fe_{Al}]^0$ antisites and $[Fe_{Cu}]^{2+} + [Cu_{Al}]^{2-}$ defect complexes. We find that the formation energy of $[Fe_{Cu}]^{2+}$ antisites is significantly lower than that of $[Fe_{Al}]^0$ antisites. Furthermore, we find that the presence of $[Fe_{Cu}]^{2+}$ promotes the formation of nearby $[Cu_{Al}]^{2-}$ antisites and that neutral $[Fe_{Cu}]^{2+} + [Cu_{Al}]^{2-}$ defect complexes are formed competitively when $p \geq 0.5$. Analysis of electron density and density of states reveals that charge transfer within these defect complexes leads to the formation of local Cu^{2+}/Fe^{2+} ionic states, whereas the dominant ionic state in the $[Fe_{Al}]^0$ substituted systems is Cu^+/Fe^{3+}. We speculate that charge transfer processes such as the one discussed in this article can lead to the formation of broad IBs with narrow VB–IB gaps in similar transition metal-containing quaternary and penternary chalcogenides. A clearer understanding of these processes and how they can be utilised to tune the electronic structure of materials in order to generate desirable optoelectronic properties may pave the way for their application in photovoltaics and spintronics.

Author Contributions: Conceptualization, M.B. and C.D.; data curation, C.D.; formal analysis, C.D. and A.O.J.K.; funding acquisition, M.W. and M.B.; investigation, C.D. and A.O.J.K.; methodology, M.W. and C.D.; project administration, M.B.; resources, M.W.; software, M.W.; supervision, M.B. and M.W.; validation, M.W.; visualization, C.D. and A.O.J.K.; writing—original draft, C.D. and M.B.; writing—review and editing, M.B., C.D. and M.W. All authors have read and agreed to the published version of the manuscript.

Funding: This research was supported via our membership of the UK's HEC Materials Chemistry Consortium, which is funded by EPSRC (EP/R029431), this work used the ARCHER2 UK National Supercomputing Service and the UK Materials and Molecular Modelling Hub for computational resources, MMM Hub, which is partially funded by EPSRC (EP/T022213) for YOUNG.

Data Availability Statement: Data available on request due to storage size of data set.

Acknowledgments: M.B. would like to acknowledge Richard Robinson for clarifying the band structure used to generate their simulated absorption spectra in reference [37]. A.K. would like to thank the University of Lincoln Undergraduate Research Opportunities Scheme for funding his summer research project.

Conflicts of Interest: The authors declare no conflict of interest.

Abbreviations

The following abbreviations are used in this manuscript:

CB	Conduction Band
DFT	Density Functional Theory
IB	Intermediate Band
PDOS	Projected Density of States
QD	Quantum Dot
SCF	Self Consistent Field
VB	Valence Band

Appendix A

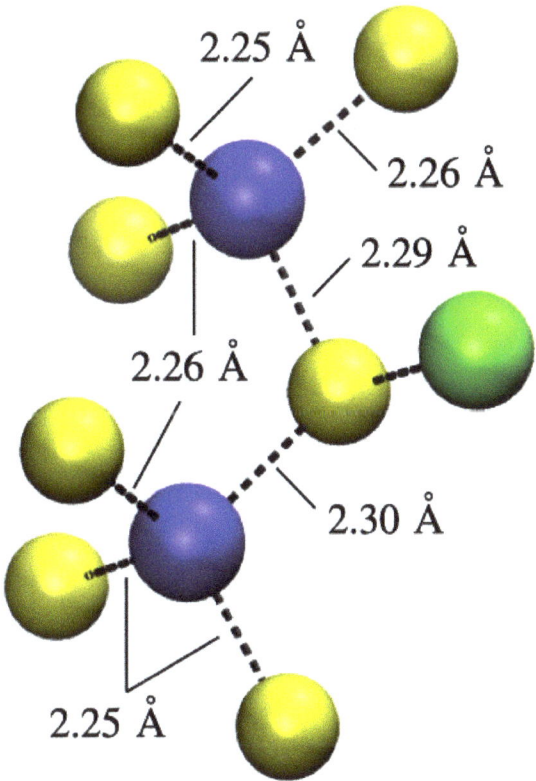

Figure A1. Al-S bond lengths for Al close to $[Fe_{Cu}]^{2+}$ antisite defect (only half of the nearest neighbour Al sites are shown for purposes of clarity).

Al-Fe separation	Al-S (1)	Al-S (2)	Al-S (3)	Al-S (4)	Mean Al-S bond length	Standard deviation
3.735207137	2.297828	2.255122	2.250755	2.258143	2.265461696	0.021789587
3.737644567	2.250836	2.257803	2.297611	2.255092	2.265335457	0.021707022
3.778091106	2.25442	2.301879	2.255023	2.256722	2.267010912	0.023265957
3.780861098	2.255315	2.256903	2.254215	2.300908	2.266835333	0.0227418
3.783227874	2.297335	2.253125	2.253837	2.260137	2.266108454	0.02105501
3.788007754	2.296802	2.252368	2.254556	2.26058	2.266076472	0.020775845
3.797758907	2.25575	2.291614	2.252157	2.256326	2.263961808	0.018526723
3.800521388	2.256574	2.292255	2.252669	2.255715	2.264303357	0.018709902
5.178851846	2.262945	2.26151	2.262503	2.260641	2.26190011	0.001031757
6.47086701	2.263136	2.267112	2.265553	2.259317	2.26377949	0.003394632
6.471146075	2.265671	2.260256	2.262931	2.267235	2.26402313	0.003077826
6.498807275	2.259852	2.266985	2.261333	2.26687	2.263760123	0.003707239
6.504763273	2.262556	2.266535	2.259995	2.266885	2.263992688	0.003310383
7.420906941	2.267574	2.265295	2.267612	2.264398	2.266219836	0.001627243
7.432537921	2.265021	2.26852	2.264399	2.268078	2.266504757	0.002095482
9.12358328	2.263981	2.265024	2.264333	2.264369	2.264426684	0.000435051

Figure A2. Al-Fe separation and Al-S bond lengths for all Al sites in $CuAlS_2$ containing an isolated $[Fe_{Cu}]^{2+}$ antisite defect (blue cells indicate a larger value relative to the mean and red cells indicate a smaller value relative to the mean).

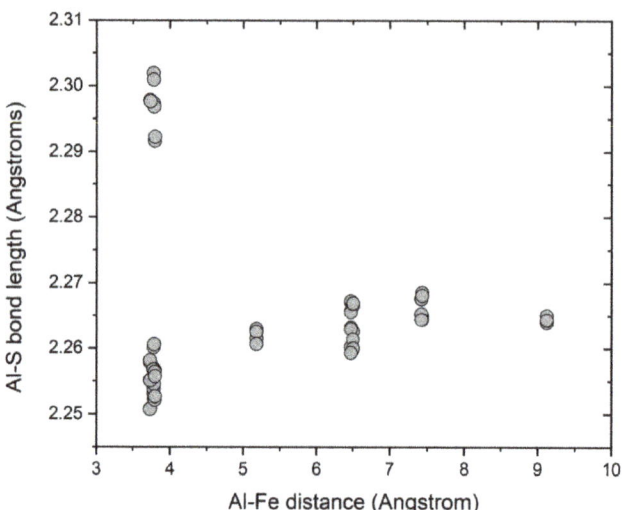

Figure A3. Al-S bond lengths plotted against Al-Fe separation for CuAlS$_2$ containing an isolated [Fe$_{Cu}$]$^{2+}$ antisite defect.

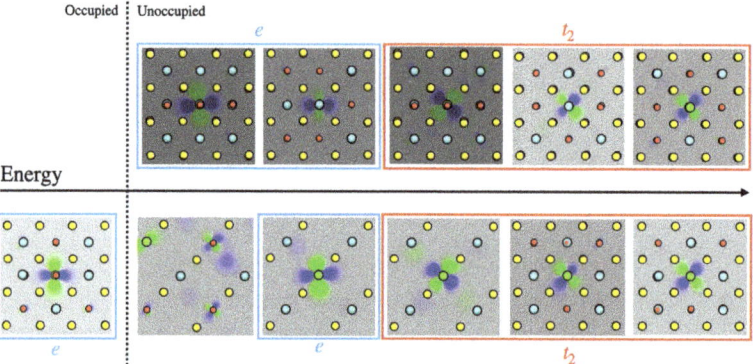

Figure A4. 3d orbitals in configuration α_1 (top row) and configuration β_1 (bottom row). Yellow circles = sulfur, red circles = copper, blue circles = aluminium and green circles = iron.

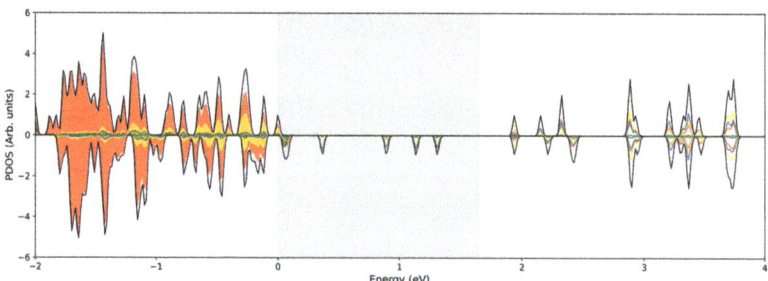

Figure A5. PDOS of CuAlS$_2$ containing an isolated [Fe$_{Cu}$]$^{2+}$ antisite defect (copper = red, sulfur = yellow, aluminium = blue and iron = green).

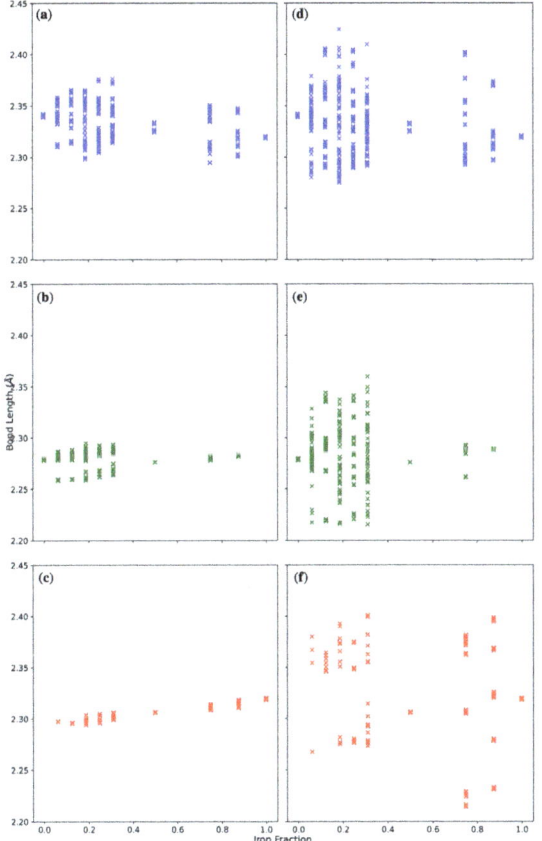

Figure A6. (a) Cu-S bond lengths in $[Fe_{Al}]^0$ substituted $CuAl_{1-p}Fe_pS_2$; (b) Al-S bond lengths in $[Fe_{Al}]^0$ substituted $CuAl_{1-p}Fe_pS_2$; (c) Fe-S bond lengths in $[Fe_{Al}]^0$ substituted $CuAl_{1-p}Fe_pS_2$; (d) Cu-S bond lengths in $[Fe_{Cu}]^{2+} + [Cu_{Al}]^{2-}$ substituted $CuAl_{1-p}Fe_pS_2$; (e) Al-S bond lengths in $[Fe_{Cu}]^{2+} + [Cu_{Al}]^{2-}$ substituted $CuAl_{1-p}Fe_pS_2$; (f) Fe-S bond lengths in $[Fe_{Cu}]^{2+} + [Cu_{Al}]^{2-}$ substituted $CuAl_{1-p}Fe_pS_2$.

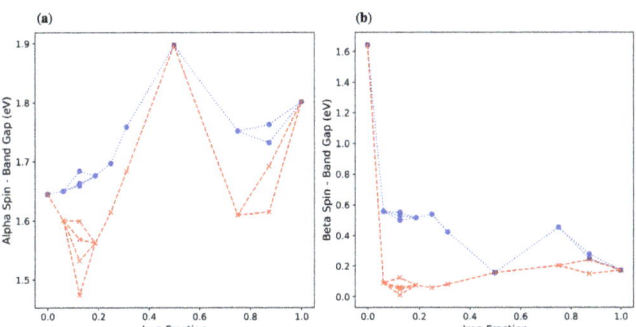

Figure A7. Band gaps for the α (blue circles) and β (red crosses) configurations in (a) major; and (b) minor spin channels.

References

1. Kairdolf, B.; Smith, A.; Stokes, T.; Wang, M.; Young, A.; Nie, S. Semiconductor Quantum Dots for Bioimaging and Biodiagnostic Applications. *Annu. Rev. Anal. Chem. (Palo Alto Calif.)* **2013**, *6*, 143–162. [CrossRef] [PubMed]
2. Bilan, R.; Nabiev, I.; Sukhanova, A. Quantum Dot-Based Nanotools for Bioimaging, Diagnostics, and Drug Delivery. *ChemBioChem* **2016**, *17*, 2103–2114. [CrossRef] [PubMed]
3. Pandey, S.; Bodas, D. High-quality quantum dots for multiplexed bioimaging: A critical review. *Adv. Colloid Interface Sci.* **2020**, *278*, 102137. [CrossRef] [PubMed]
4. Wood, V.; Bulović, V. Colloidal quantum dot light-emitting devices. *Nano Rev.* **2010**, *1*, 5202. [CrossRef] [PubMed]
5. Lu, M.; Guo, J.; Sun, S.; Lu, P.; Wu, J.; Wang, Y.; Kershaw, S.V.; Yu, W.W.; Rogach, A.L.; Zhang, Y. Bright $CsPbI_3$ perovskite quantum dot light-emitting diodes with top-emitting structure and a low efficiency roll-off realized by applying zirconium acetylacetonate surface modification. *Nano Lett.* **2020**, *20*, 2829–2836. [CrossRef]
6. Carey, G.H.; Abdelhady, A.L.; Ning, Z.; Thon, S.M.; Bakr, O.M.; Sargent, E.H. Colloidal quantum dot solar cells. *Chem. Rev.* **2015**, *115*, 12732–12763. [CrossRef]
7. Liu, Z.; Lin, C.H.; Hyun, B.R.; Sher, C.W.; Lv, Z.; Luo, B.; Jiang, F.; Wu, T.; Ho, C.H.; Kuo, H.C.; et al. Micro-light-emitting diodes with quantum dots in display technology. *Light. Sci. Appl.* **2020**, *9*, 1–23. [CrossRef]
8. Reiss, P.; Carriere, M.; Lincheneau, C.; Vaure, L.; Tamang, S. Synthesis of semiconductor nanocrystals, focusing on nontoxic and earth-abundant materials. *Chem. Rev.* **2016**, *116*, 10731–10819. [CrossRef]
9. Chuang, Y.J.; Zhen, Z.; Zhang, F.; Liu, F.; Mishra, J.P.; Tang, W.; Chen, H.; Huang, X.; Wang, L.; Chen, X.; et al. Photostimulable near-infrared persistent luminescent nanoprobes for ultrasensitive and longitudinal deep-tissue bio-imaging. *Theranostics* **2014**, *4*, 1112. [CrossRef]
10. Li, P.; Duan, X.; Chen, Z.; Liu, Y.; Xie, T.; Fang, L.; Li, X.; Yin, M.; Tang, B. A near-infrared fluorescent probe for detecting copper (II) with high selectivity and sensitivity and its biological imaging applications. *Chem. Commun.* **2011**, *47*, 7755–7757. [CrossRef]
11. Yu, K.; Ng, P.; Ouyang, J.; Zaman, M.B.; Abulrob, A.; Baral, T.N.; Fatehi, D.; Jakubek, Z.J.; Kingston, D.; Wu, X.; et al. Low-temperature approach to highly emissive copper indium sulfide colloidal nanocrystals and their bioimaging applications. *ACS Appl. Mater. Interfaces* **2013**, *5*, 2870–2880. [CrossRef] [PubMed]
12. Choi, J.Y.; Kim, G.H.; Guo, Z.; Lee, H.Y.; Swamy, K.; Pai, J.; Shin, S.; Shin, I.; Yoon, J. Highly selective ratiometric fluorescent probe for Au^{3+} and its application to bioimaging. *Biosens. Bioelectron.* **2013**, *49*, 438–441. [CrossRef] [PubMed]
13. Jara, D.H.; Yoon, S.J.; Stamplecoskie, K.G.; Kamat, P.V. Size-dependent photovoltaic performance of $CuInS_2$ quantum dot-sensitized solar cells. *Chem. Mater.* **2014**, *26*, 7221–7228. [CrossRef]
14. Zhou, H.; Hsu, W.C.; Duan, H.S.; Bob, B.; Yang, W.; Song, T.B.; Hsu, C.J.; Yang, Y. CZTS nanocrystals: A promising approach for next generation thin film photovoltaics. *Energy Environ. Sci.* **2013**, *6*, 2822–2838. [CrossRef]
15. Dalui, A.; Khan, A.H.; Pradhan, B.; Pradhan, J.; Satpati, B.; Acharya, S. Facile synthesis of composition and morphology modulated quaternary CuZnFeS colloidal nanocrystals for photovoltaic application. *RSC Adv.* **2015**, *5*, 97485–97494. [CrossRef]
16. Jaffe, J.; Zunger, A. Electronic structure of the ternary chalcopyrite semiconductors $CuAlS_2$, $CuGaS_2$, $CuInS_2$, $CuAlSe_2$, $CuGaSe_2$, and $CuInSe_2$. *Phys. Rev. B* **1983**, *28*, 5822. [CrossRef]
17. Bhattacharyya, B.; Simlandy, A.K.; Chakraborty, A.; Rajasekar, G.P.; Aetukuri, N.B.; Mukherjee, S.; Pandey, A. Efficient photosynthesis of organics from aqueous bicarbonate ions by quantum dots using visible light. *ACS Energy Lett.* **2018**, *3*, 1508–1514. [CrossRef]
18. Mukherjee, A.; Dutta, P.; Bhattacharyya, B.; Rajasekar, G.P.; Simlandy, A.K.; Pandey, A. Ultrafast spectroscopic investigation of the artificial photosynthetic activity of $CuAlS_2$/ZnS quantum dots. *Nano Sel.* **2021**, *2*, 958–966. [CrossRef]
19. Bhattacharyya, B.; Pandit, T.; Rajasekar, G.P.; Pandey, A. Optical transparency enabled by anomalous Stokes shift in visible light-emitting $CuAlS_2$-based quantum dots. *J. Phys. Chem. Lett.* **2018**, *9*, 4451–4456. [CrossRef]
20. Donnay, G.; Corliss, L.; Donnay, J.; Elliott, N.; Hastings, J. Symmetry of magnetic structures: Magnetic structure of chalcopyrite. *Phys. Rev.* **1958**, *112*, 1917. [CrossRef]
21. Woolley, J.; Lamarche, A.M.; Lamarche, G.; Quintero, M.; Swainson, I.; Holden, T. Low temperature magnetic behaviour of $CuFeS_2$ from neutron diffraction data. *J. Magn. Magn. Mater.* **1996**, *162*, 347–354. [CrossRef]
22. Pauling, L.; Brockway, L. The crystal structure of chalcopyrite $CuFeS_2$. *Z. Krist.-Cryst. Mater.* **1932**, *82*, 188–194. [CrossRef]
23. Hu, J.; Lu, Q.; Deng, B.; Tang, K.; Qian, Y.; Li, Y.; Zhou, G.; Liu, X. A hydrothermal reaction to synthesize $CuFeS_2$ nanorods. *Inorg. Chem. Commun.* **1999**, *2*, 569–571. [CrossRef]
24. Todd, E.; Sherman, D.; Purton, J. Surface oxidation of chalcopyrite ($CuFeS_2$) under ambient atmospheric and aqueous (pH 2–10) conditions: Cu, Fe L-and O K-edge X-ray spectroscopy. *Geochim. Cosmochim. Acta* **2003**, *67*, 2137–2146. [CrossRef]
25. Mikhlin, Y.; Tomashevich, Y.; Tauson, V.; Vyalikh, D.; Molodtsov, S.; Szargan, R. A comparative X-ray absorption near-edge structure study of bornite, Cu_5FeS_4, and chalcopyrite, $CuFeS_2$. *J. Electron Spectrosc. Relat. Phenom.* **2005**, *142*, 83–88. [CrossRef]
26. Boekema, C.; Krupski, A.; Varasteh, M.; Parvin, K.; Van Til, F.; Van Der Woude, F.; Sawatzky, G. Cu and Fe valence states in $CuFeS_2$. *J. Magn. Magn. Mater.* **2004**, *272*, 559–561. [CrossRef]
27. Pearce, C.; Pattrick, R.; Vaughan, D.; Henderson, C.; Van der Laan, G. Copper oxidation state in chalcopyrite: Mixed Cu d9 and d10 characteristics. *Geochim. Cosmochim. Acta* **2006**, *70*, 4635–4642. [CrossRef]
28. Zaanen, J.; Sawatzky, G. The electronic structure and superexchange interactions in transition-metal compounds. *Can. J. Phys.* **1987**, *65*, 1262–1271. [CrossRef]

29. Fujisawa, M.; Suga, S.; Mizokawa, T.; Fujimori, A.; Sato, K. Electronic structures of CuFeS$_2$ and CuAl$_{0.9}$Fe$_{0.1}$S$_2$ studied by electron and optical spectroscopies. *Phys. Rev. B* **1994**, *49*, 7155. [CrossRef]
30. Sato, K.; Harada, Y.; Taguchi, M.; Shin, S.; Fujimori, A. Characterization of Fe 3d states in CuFeS$_2$ by resonant X-ray emission spectroscopy. *Phys. Status Solidi (A)* **2009**, *206*, 1096–1100. [CrossRef]
31. Ghosh, S.; Avellini, T.; Petrelli, A.; Kriegel, I.; Gaspari, R.; Almeida, G.; Bertoni, G.; Cavalli, A.; Scotognella, F.; Pellegrino, T.; et al. Colloidal CuFeS$_2$ nanocrystals: Intermediate Fe d-band leads to high photothermal conversion efficiency. *Chem. Mater.* **2016**, *28*, 4848–4858. [CrossRef] [PubMed]
32. Gaspari, R.; Della Valle, G.; Ghosh, S.; Kriegel, I.; Scotognella, F.; Cavalli, A.; Manna, L. Quasi-static resonances in the visible spectrum from all-dielectric intermediate band semiconductor nanocrystals. *Nano Lett.* **2017**, *17*, 7691–7695. [CrossRef] [PubMed]
33. Bastola, E.; Bhandari, K.P.; Subedi, I.; Podraza, N.J.; Ellingson, R.J. Structural, optical, and hole transport properties of earth-abundant chalcopyrite (CuFeS$_2$) nanocrystals. *MRS Commun.* **2018**, *8*, 970–978. [CrossRef]
34. Kowalik, P.; Bujak, P.; Penkala, M.; Kotwica, K.; Kmita, M.; Gajewska, M.; Ostrowski, A.; Pron, A. Synthesis of CuFeS$_{2-x}$Se$_x$-alloyed nanocrystals with localized surface plasmon resonance in the visible spectral range. *J. Mater. Chem. C* **2019**, *7*, 6246–6250. [CrossRef]
35. Sugathan, A.; Bhattacharyya, B.; Kishore, V.; Kumar, A.; Rajasekar, G.P.; Sarma, D.; Pandey, A. Why does CuFeS$_2$ resemble gold? *J. Phys. Chem. Lett.* **2018**, *9*, 696–701. [CrossRef] [PubMed]
36. Lee, S.; Ghosh, S.; Hoyer, C.E.; Liu, H.; Li, X.; Holmberg, V.C. Iron-content-dependent, quasi-static dielectric resonances and oxidative transitions in bornite and chalcopyrite copper iron sulfide nanocrystals. *Chem. Mater.* **2021**, *33*, 1821–1831. [CrossRef]
37. Yao, Y.; Bhargava, A.; Robinson, R.D. Fe cations control the plasmon evolution in CuFeS$_2$ nanocrystals. *Chem. Mater.* **2021**, *33*, 608–615. [CrossRef]
38. Chen, P.; Qin, M.; Chen, H.; Yang, C.; Wang, Y.; Huang, F. Cr incorporation in CuGaS$_2$ chalcopyrite: A new intermediate-band photovoltaic material with wide-spectrum solar absorption. *Phys. Status Solidi (A)* **2013**, *210*, 1098–1102. [CrossRef]
39. Baltz, V.; Manchon, A.; Tsoi, M.; Moriyama, T.; Ono, T.; Tserkovnyak, Y. Antiferromagnetic spintronics. *Rev. Mod. Phys.* **2018**, *90*, 015005. [CrossRef]
40. Gabka, G.; Zybała, R.; Bujak, P.; Ostrowski, A.; Chmielewski, M.; Lisowski, W.; Sobczak, J.W.; Pron, A. Facile Gram-Scale Synthesis of the First n-Type CuFeS$_2$ Nanocrystals for Thermoelectric Applications. *Eur. J. Inorg. Chem.* **2017**, *2017*, 3150–3153. [CrossRef]
41. Teranishi, T.; Sato, K.; Kondo, K. Optical properties of a magnetic semiconductor: Chalcopyrite CuFeS$_2$: I. Absorption spectra of CuFeS$_2$ and Fe-Doped CuAlS$_2$ and CuGaS$_2$. *J. Phys. Soc. Jpn.* **1974**, *36*, 1618–1624. [CrossRef]
42. Liu, M.L.; Huang, F.Q.; Chen, L.D.; Wang, Y.M.; Wang, Y.H.; Li, G.F.; Zhang, Q. p-type transparent conductor: Zn-doped CuAlS$_2$. *Appl. Phys. Lett.* **2007**, *90*, 072109. [CrossRef]
43. Liu, M.L.; Huang, F.Q.; Chen, L.D. p-Type electrical conduction and wide optical band gap in Mg-doped CuAlS$_2$. *Scr. Mater.* **2008**, *58*, 1002–1005. [CrossRef]
44. Wang, T.; Li, X.; Li, W.; Huang, L.; Ma, C.; Cheng, Y.; Cui, J.; Luo, H.; Zhong, G.; Yang, C. Transition metals doped CuAlSe$_2$ for promising intermediate band materials. *Mater. Res. Express* **2016**, *3*, 045905. [CrossRef]
45. Yadav, R.; Bhattacharyya, B.; Saha, S.K.; Dutta, P.; Roy, P.; Rajasekar, G.P.; Narayan, A.; Pandey, A. Electronic Structure Insights into the Tunable Luminescence of CuAl$_x$Fe$_{1-x}$S$_2$/ZnS Nanocrystals. *J. Phys. Chem. C* **2021**, *125*, 2511–2518. [CrossRef]
46. Sato, K. *Chalcopyrite Crystals Doped with Transition Elements, Rare Earths, Ternary and Multinary Compounds in the 21st Century*; IPAP: Tokyo, Japan, 2001; pp. 228–243.
47. Navrátil, J.; Levinský, P.; Hejtmánek, J.; Pashchenko, M.; Knížek, K.; Kubíčková, L.; Kmječ, T.; Drašar, C. Peculiar magnetic and transport properties of CuFeS$_2$: Defects play a key role. *J. Phys. Chem. C* **2020**, *124*, 20773–20783. [CrossRef]
48. Shannon, R.D. Revised effective ionic radii and systematic studies of interatomic distances in halides and chalcogenides. *Acta Crystallogr. Sect. A Cryst. Phys. Diffr. Theor. Gen. Crystallogr.* **1976**, *32*, 751–767. [CrossRef]
49. Ueng, H.; Hwang, H. The defect structure of CuInS$_2$. part I: Intrinsic defects. *J. Phys. Chem. Solids* **1989**, *50*, 1297–1305. [CrossRef]
50. Zhang, S.; Wei, S.H.; Zunger, A.; Katayama-Yoshida, H. Defect physics of the CuInSe$_2$ chalcopyrite semiconductor. *Phys. Rev. B* **1998**, *57*, 9642. [CrossRef]
51. Zhang, S.; Wei, S.H.; Zunger, A. Stabilization of ternary compounds via ordered arrays of defect pairs. *Phys. Rev. Lett.* **1997**, *78*, 4059. [CrossRef]
52. Huang, D.; Persson, C. Stability of the bandgap in Cu-poor CuInSe$_2$. *J. Phys. Condens. Matter* **2012**, *24*, 455503. [CrossRef] [PubMed]
53. Harvie, A.J.; Booth, M.; Chantry, R.L.; Hondow, N.; Kepaptsoglou, D.M.; Ramasse, Q.M.; Evans, S.D.; Critchley, K. Observation of compositional domains within individual copper indium sulfide quantum dots. *Nanoscale* **2016**, *8*, 16157–16161. [CrossRef] [PubMed]
54. Grau-Crespo, R.; Hamad, S.; Catlow, C.R.A.; De Leeuw, N. Symmetry-adapted configurational modelling of fractional site occupancy in solids. *J. Physics: Condens. Matter* **2007**, *19*, 256201. [CrossRef]
55. Kühne, T.D.; Iannuzzi, M.; Del Ben, M.; Rybkin, V.V.; Seewald, P.; Stein, F.; Laino, T.; Khaliullin, R.Z.; Schütt, O.; Schiffmann, F.; et al. CP2K: An electronic structure and molecular dynamics software package—Quickstep: Efficient and accurate electronic structure calculations. *J. Chem. Phys.* **2020**, *152*, 194103,

56. Perdew, J.P.; Burke, K.; Ernzerhof, M. Generalized Gradient Approximation Made Simple. *Phys. Rev. Lett.* **1996**, *77*, 3865–3868. [CrossRef]
57. VandeVondele, J.; Hutter, J. Gaussian basis sets for accurate calculations on molecular systems in gas and condensed phases. *J. Chem. Phys.* **2007**, *127*, 114105.
58. Hartwigsen, C.; Goedecker, S.; Hutter, J. Relativistic separable dual-space Gaussian pseudopotentials from H to Rn. *Phys. Rev. B* **1998**, *58*, 3641–3662. [CrossRef]
59. Krack, M. Pseudopotentials for H to Kr optimized for gradient-corrected exchange-correlation functionals. *Theor. Chem. Accounts* **2005**, *114*, 145–152. [CrossRef]
60. VandeVondele, J.; Hutter, J. An efficient orbital transformation method for electronic structure calculations. *J. Chem. Phys.* **2003**, *118*, 4365–4369. [CrossRef]
61. Lany, S.; Zunger, A. Assessment of correction methods for the band-gap problem and for finite-size effects in supercell defect calculations: Case studies for ZnO and GaAs. *Phys. Rev. B* **2008**, *78*, 235104. [CrossRef]
62. Larsen, A.H.; Mortensen, J.J.; Blomqvist, J.; Castelli, I.E.; Christensen, R.; Dułak, M.; Friis, J.; Groves, M.N.; Hammer, B.; Hargus, C.; et al. The atomic simulation environment—A Python library for working with atoms. *J. Phys. Condens. Matter* **2017**, *29*, 273002. [CrossRef] [PubMed]
63. Sato, K.; Ishii, K.; Watanabe, K.W.K.; Ohe, K.O.K. Time-Resolved Photoluminescence Spectra in Single Crystals of $CuAlS_2$:Mn. *Jpn. J. Appl. Phys.* **1991**, *30*, 307. [CrossRef]
64. Perdew, J.P. Density Functional Theory and the Band Gap Problem. *Int. J. Quantum Chem.* **1985**, *28*, 497–523. [CrossRef]
65. Lide, D.R. *Handbook of Chemistry and Physics*, 80th ed.; CRC Press: Boca Raton, FL, USA, 2002.
66. Mahmood, Q.; Hassan, M.; Noor, N. Theoretical Study of Electronic, Magnetic, and Optical Response of Fe-doped ZnS: First-Principle Approach. *J. Supercond. Nov. Magn.* **2017**, *30*, 1463–1471. [CrossRef]
67. Adachi, S. *Properties of Semiconductor Alloys: Group-IV, III-V and II-VI Semiconductors*; John Wiley & Sons: Hoboken, NJ, USA, 2009.
68. Ahn, K.; Kim, M.G.; Park, S.; Ryu, B. Entropy stabilized off-stoichiometric cubic γ-Cu1-xIx phase containing high-density Cu vacancies. *AIP Adv.* **2021**, *11*, 095018. [CrossRef]
69. Yao, J.; Wang, Z.; van Tol, J.; Dalal, N.S.; Aitken, J.A. Site preference of manganese on the copper site in Mn-substituted $CuInSe_2$ chalcopyrites revealed by a combined neutron and X-ray powder diffraction study. *Chem. Mater.* **2010**, *22*, 1647–1655. [CrossRef]

Article

Dewetting Process in Ni-Mn-Ga Shape-Memory Heusler: Effects on Morphology, Stoichiometry and Magnetic Properties

Milad Takhsha Ghahfarokhi [1,*], Federica Celegato [2], Gabriele Barrera [2], Francesca Casoli [1,*], Paola Tiberto [2] and Franca Albertini [1]

[1] Institute of Materials for Electronics and Magnetism, National Research Council (IMEM-CNR), Parco Area delle Scienze 37/A, 43124 Parma, Italy
[2] Advanced Materials Metrology and Life Sciences (INRiM), Strada delle Cacce 91, 10135 Turin, Italy
* Correspondence: milad.takhsha@imem.cnr.it (M.T.G.); francesca.casoli@imem.cnr.it (F.C.)

Abstract: In this work, dewetting process has been investigated in shape-memory Heuslers. To this aim, series of high-temperature annealing (1100–1150 K) have been performed at high vacuum (time is varied in the range of 55–165 min) in Ni-Mn-Ga epitaxial thin films grown on MgO(001). The process kinetics have been followed by studying the evolution of morphology and composition. In particular, we report the initiation of the dewetting process by the formation of symmetric holes in the films. The holes propagate and integrate, leaving micrometric and submicron islands of the material, increasing the average roughness of the films by a factor of up to around 30. The dewetting process is accompanied by severe Ga and Mn sublimation, and Ni-Ga segregation, which significantly modify the magnetic properties of the films measured at each stage. The annealed samples show a relatively weak magnetic signal at room temperature with respect to the pristine sample.

Keywords: magnetic shape memory alloys; multifunctional Heusler compounds; solid-state dewetting; annealing; morphology; stoichiometry; magnetic properties

1. Introduction

Small-scale magnetic-shape-memory (MSM) Heuslers are promising for a vast variety of applications in automotive, aerospace, biology and robotics, serving as smart moving components, sensors, and energy harvesters [1–3], as well as being the active materials in multicaloric cooling systems [4,5]. The material down-scaling is specifically important, as it can improve the integration capability into small-scale devices [2,5,6]. It can also enhance the mechanical properties and enable manufacturing complex shapes, giving rise to the idea of "crushing down the material and building back better". For instance, there have recently been a few pioneering works reporting the use of MSM Heusler powder/polymer composites with excellent shaping capability [7] and improved mechanical properties [8]. In addition, additive manufacturing has been recently employed to fabricate Ni-Mn-Ga complex shapes starting from powder samples [9–11]. Finally, by down-scaling, one can reach the properties of the material that do not necessarily appear when the material is in the bulk format [12–15]. Different pioneering works have been dedicated in down-scaling MSM Heuslers using several bottom-up and top-down approaches [16–18]. Specifically, starting from thin films of MSM Heuslers, the down-scaling approaches have been limited to nanosphere-lithography [15], photolithography [13,19,20], electron-beam lithography [6,21–26] and focused ion-beam milling techniques [27,28].

Searching for cost-effective down-scaling techniques starting from the continuous thin films, the solid-state dewetting process is known as an appropriate alternative. This process is a thermally activated top-down method based on the agglomeration of the material on the substrate, forming arrays of islands below the melting point of the material [29–31]. The driving force leading to such an effect is the free-energy minimization on the interfaces between the film and the substrate at high-temperatures. The process initiates with the

formation of holes, proceeds with the propagation and integration of the holes, and ends with the 3D islands of the material on the substrate [29,30]. Various parameters influence the size and density of the islands, among which are the initial thickness of the film, the temperature and time of heat treatment, the crystallography of the film, substrate, defects, the morphology of the film, etc. [30,32,33]. Dewetting processes have been so far reported for successfully down-scaling a number of single-element and binary alloys and compounds [33–40]. The first approach exploiting dewetting during deposition to induce arrays of holes in Ni-Mn-Ga-Co epitaxial films has been recently reported by Lünser et al. [41]. They took advantage of the initial stage of the dewetting process by growing thin films of various thicknesses (80–800 nm) at elevated temperatures (673–873 K). By adjusting the thickness and the growth temperature, they succeeded in controlling the size and distribution of the holes. Nevertheless, they have reported a significant composition variation of the material throughout the process, which inevitably influences the martensitic transformation and the magnetic properties of the films.

In this study, a different approach has been followed: performing post-deposition annealing at high-temperatures (1100–1150 K) in epitaxial Ni-Mn-Ga thin films grown on MgO(001) with thickness of 75 nm. The aim of the work is to complete the dewetting process by the formation of Ni-Mn-Ga islands. Morphology, composition, phase transformation and magnetic properties of the material have been investigated at each stage of the process.

2. Materials and Methods

Experimental

An epitaxial Ni-Mn-Ga film was deposited on MgO(001) substrate at elevated temperature. The epitaxial crystal relation is Ni-Mn-Ga(001)[100]/MgO(001)[110]. The growth condition and the composition of the film are summarized in Table 1.

Table 1. Pristine sample: thickness, deposition rate, deposition temperature, composition and Curie temperature.

Thickness (nm)	Dep. Rate (nm·s^{-1})	Temperature (K)	Composition (at. % ± 1)	T_C (K)
75	0.06	573	Ni$_{50.0}$Mn$_{18.6}$Ga$_{31.4}$	~344

The film was cut into several pieces of approximately "3 mm × 3 mm" for the annealing process. One piece was kept as a reference sample. To promote the dewetting process, the annealing treatment was performed in a furnace (Carbolite) by placing the sample in a quartz tube under a vacuum atmosphere (2×10^{-4} Pa) to avoid the oxidation of the film surface. The selected annealing temperatures (1100 to 1150 K) were reached using a heating rate of ~51 K/min. The samples were then kept at a fixed temperature for the entire annealing time (55 to 165 min). Six annealing conditions were performed as shown in Table 2.

Table 2. Sample annealing conditions and the measured composition of the samples after the heat treatment.

Sample	Annealing Temp. (K)	Annealing Time (min)	Composition (at. % ± 1)
1100_55	1100	55	Ni$_{55.6}$Mn$_{21.0}$Ga$_{23.4}$
1100_110	1100	110	Ni$_{62.8}$Mn$_{13.3}$Ga$_{23.9}$
1100_165	1100	165	Ni$_{71.8}$Mn$_{15.1}$Ga$_{13.1}$
1150_55	1150	55	Ni$_{60.0}$Mn$_{20.4}$Ga$_{19.6}$
1150_110	1150	110	Ni$_{62.0}$Mn$_{22.3}$Ga$_{15.7}$
1150_165	1150	165	Ni$_{62.4}$Mn$_{14.3}$Ga$_{23.3}$

Atomic and magnetic force microscopy (AFM/MFM) images were obtained using a Dimension 3100 scanning probe microscope equipped with a Nanoscope Veeco controller using MESP-V2 magnetic tips in the interleave mode at room temperature. Backscattered electron (BSE) images, as well and material composition, were obtained by a scanning electron microscope (SEM) equipped with energy dispersive X-ray spectroscopy (EDS) and BSE detectors (SEM, FEI Inspect—F). Room-temperature magnetic hysteresis loops were measured using an alternating gradient force magnetometer (AGFM) by applying the magnetic field parallel to the film plane along MgO[100]. The magnetic signal of the sample holder was subtracted. The thermomagnetic curves were obtained using a superconducting quantum interference device (SQUID) magnetometer in the presence of a constant in-plane magnetic field along MgO[100] and the temperature ramp of $dT/dt = 0.5$ K/min.

3. Results and Discussion

Figure 1 summarizes the morphology and the magnetic characteristics of the pristine sample. The AFM images in Figure 1a demonstrate the presence of X-type twins that are characterized by different orientations of the twinning planes (i.e., at 45° degrees to the MgO(001) substrate) [42–47]. The corresponding out-of-plane contribution of the magnetic domains can be visualized in the MFM images as the dark and the bright contrasts (Figure 1b) [42–49]. When cooling from 400 K, the thermomagnetic curves of the pristine sample (Figure 1c) show an abrupt rise in the susceptibility at around 344 K, which is ascribed to crossing the austenitic Curie temperature of the sample [50], where it transforms from the paramagnetic austenitic phase to ferromagnetic austenitic phase. The susceptibility slightly drops afterwards by further cooling, due to martensitic transformation, where the sample transforms from the ferromagnetic austenitic phase to ferromagnetic martensitic phase. The drop of the signal is due to the lower magnetic susceptibility of the martensitic phase compared to the austenitic phase in the applied low magnetic field ($\mu_0 H = 2$ mT) [51]. Upon heating, the reverse transformations takes place, leaving the temperature hysteresis of the structural phase transformation. The pristine sample shows a broad martensitic transition that spans more than 100 K. It can be recognized by looking at the temperature hysteresis (the gap between the cooling and the heating curves). The cooling curve constitutes the upper branch of the gap because of the presence of residual high-susceptibility ferromagnetic austenitic phase in the transition state. Though at room temperature, the transformation is not complete according to the thermomagnetic curves, the magnetic hysteresis loops measured for the pristine sample at room temperature (Figure 1d) show the ferromagnetic characteristics of the material. The saturation magnetization, coercivity and remanent magnetization are $M_S = 48.4$ Am^2/kg, $H_C = 10$ mT and $M_r = 6.2$ Am^2/kg, respectively.

The morphology of the samples was evaluated by AFM after the annealing process (Figure 2). The images show the rough inhomogeneous surface of the samples. The roughness of the annealed samples shows an increase by a factor of up to around 30 with respect to the pristine sample.

The morphology of the annealed samples was evaluated in the larger scale by BSE imaging (Figure 3). This type of imaging directly provides information about the morphology of the samples; moreover, the contrast change can be attributed to the composition variation or/and the variation of the height with respect to the detector. By increasing the annealing time at 1100 K (Figure 3a–c), the holes propagate so that the sample annealed for 165 min shows the largest holes, largest total hole area, and the lowest number of holes (Table 3). However, the samples annealed at 1150 K show a considerable propagation of holes when increasing the annealing time to 110 min, where the holes are integrated constituting ~75% of the scan area (Figure 3g, Table 3). Therefore, the islands are completely isolated, showing the average size of ~0.49 µm^2. In addition, the propagation of the islands and holes seems to follow a symmetric path. Specifically, this is more pronounced for samples 1100_110, 1100_165, and 1150_165. Large–scale BSE images of the samples 1100_110, 1100_165 are provided as examples. We calculated the approximate angles between the symmetric paths that look like lines in large scales (Figure 3d,e). The direction of the lines can be categorized into a pair of orthogonal lines having relative in-plane misalignment of around 22° with respect to each other. Correspondingly, the lines are around 11° misaligned in the plane of the film with respect to the MgO[100] and MgO[010] directions (Figure 3d inset).

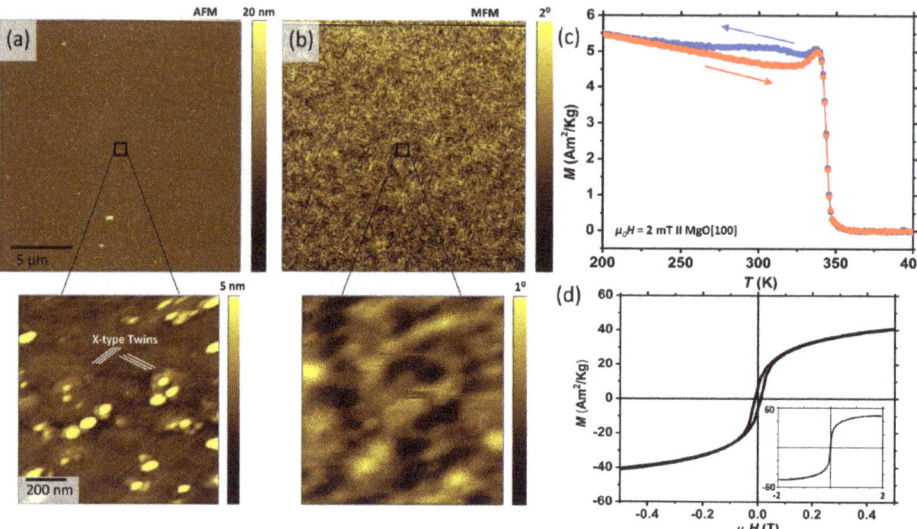

Figure 1. Pristine sample: (**a**) AFM at room temperature, (**b**) MFM at room temperature. (**c**) Thermomagnetic curves for the temperature range of 200–400 K applying low magnetic field ($\mu_0 H = 2$ mT) in the plane of the film along MgO[100]. MgO[100] and MgO[010] directions are along the edges of the AFM/MFM images. (**d**) Magnetic hysteresis loops at room temperature applying external filed along MgO[100] in the plane of the film. The inset shows the full graph.

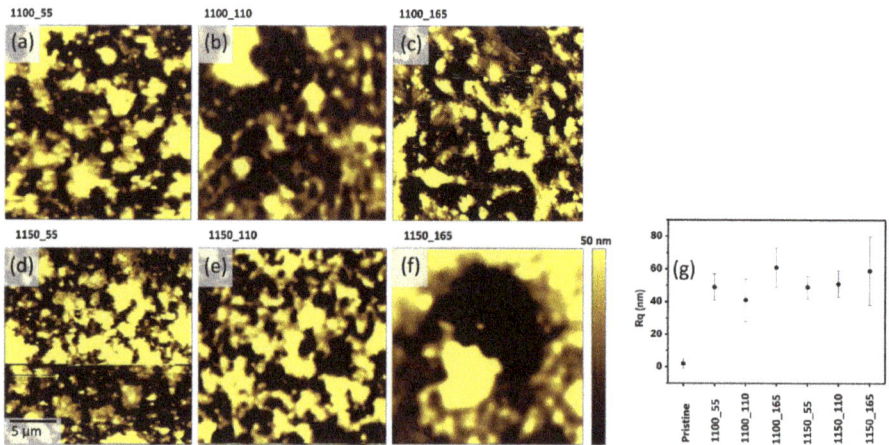

Figure 2. Annealed samples: AFM images of (**a**) 1100_55, (**b**) 1100_110, (**c**) 1100_165, (**d**) 1150_55, (**e**) 1150_110, (**f**) 1150_165. (**g**) Graph showing the roughness of the annealed samples with respect to the pristine sample. The height irregularities compared to the data variance is also provided as the data bars. MgO[100] and MgO[010] directions are along the edges of the AFM images.

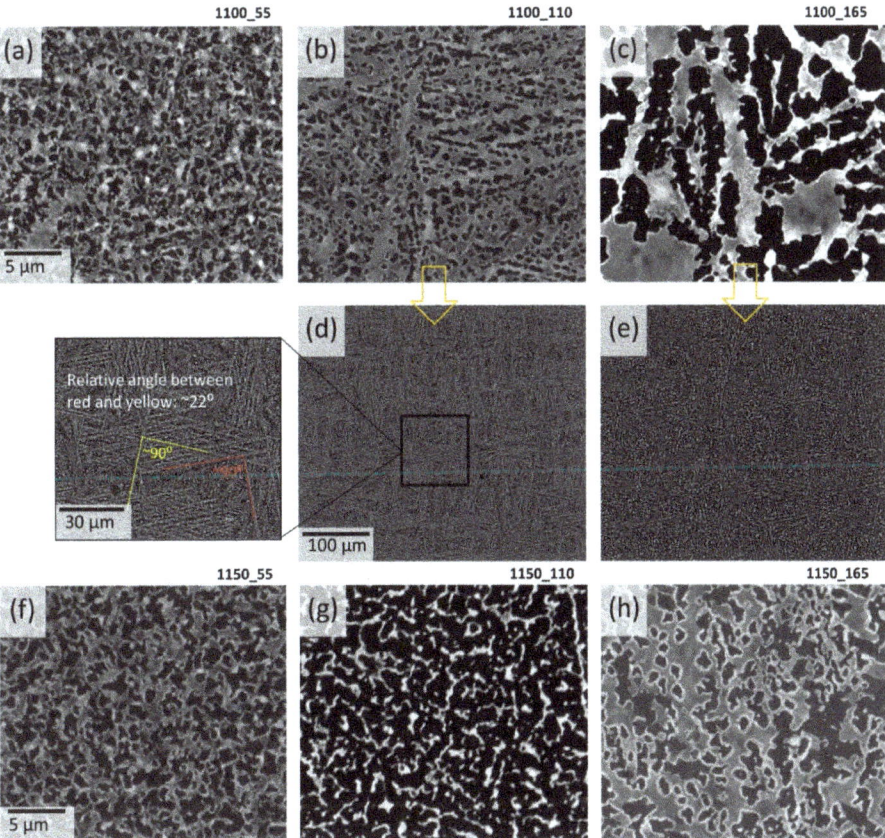

Figure 3. Annealed samples: BSE images of (**a**) 1100_55, (**b**) 1100_110, (**c**) 1100_165, (**d**) large–scale image of 1100_110, (**e**) large–scale image of 1100_165, (**f**) 1150_55, (**g**) 1150_110, (**h**) 1150_165. The inset in (**d**) is the magnified image of the marked area showing the relative orientation of the symmetric paths of islands with respect each other. MgO[100] and MgO[010] directions are along the edges of the BSE images.

Table 3. Morphology characteristics of the annealed samples estimated for Figure 3 (estimation uncertainty ≈ 10%). The average hole number and the average hole size are not applicable (NA) for 1150–110 sample because the holes are integrated (it is not possible anymore to count them as individual components).

Sample	Avg. Hole Number (μm^{-2})	Avg. Hole Size (μm^2)	Total Hole Area (%)
1100_55	1.2	0.33	40
1100_110	1.2	0.24	28
1100_165	0.1	6.26	50
1150_55	0.7	0.55	38
1150_110	NA	NA	75
1150_165	0.4	0.83	37

Room-temperature magnetic hysteresis loops of the samples were measured by applying an external magnetic field up to $\mu_0 H = |0.5|$ T in the plane of the films along MgO[100] (Figure 4). The M_{max} and M_r of the annealed samples show a significant reduction, but the coercivity is in the same range of values compared to the pristine sample. The measured values are provided in Table 4. The hysteresis loops also show that annealing has changed the magnetic behavior of the samples. Annealed samples beside the ferromagnetic component show a paramagnetic component, which raises with the increase in the size of the islands in sample 1100–165.

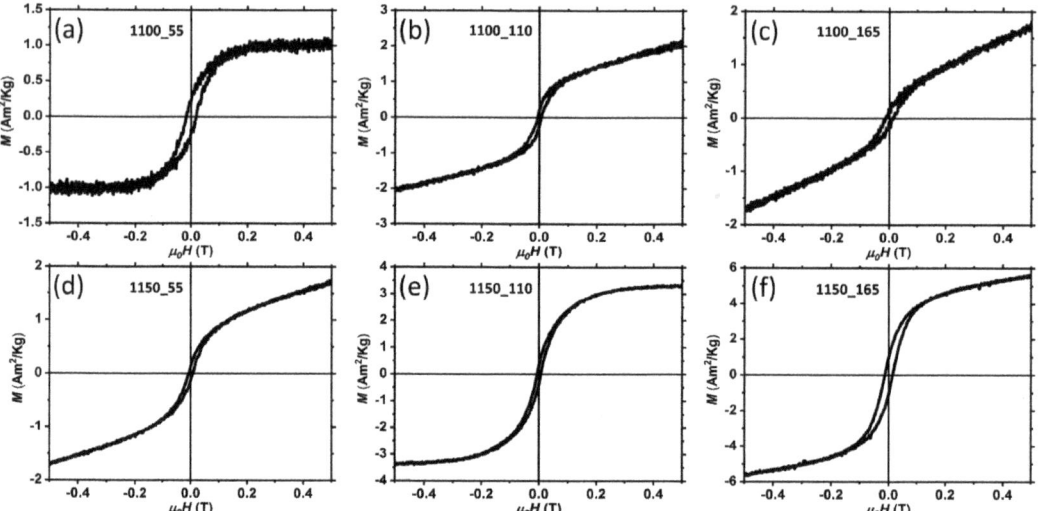

Figure 4. Magnetic hysteresis loops of the annealed samples at room temperature applying external filed along MgO[100] in the plane of the films: (**a**) 1100_55, (**b**) 1100_110, (**c**) 1100_165, (**d**) 1150_55, (**e**) 1150_110, (**f**) 1150_165.

Table 4. Magnetic characteristics of the samples obtained from the magnetic hysteresis loops measured at room temperature.

Sample	M at 0.5 T (Am²/kg)	M_r (Am²/kg)	H_c (mT)
Pristine	40.8	~6.2	~10
1100_55	1.0	0.2	18
1100_110	2.0	0.2	6
1100_165	1.7	0.1	13
1150_55	1.7	0.1	7
1150_110	3.4	0.4	6
1150_165	5.6	1.0	14

Composition measurements gave us a deeper view of the evolution of the samples in the annealing process. Two different approaches have been taken:

1. Obtaining insight into the average percentage of the atomic contents of Ni, Mn and Ga at large–scale (Figure 5a,c, Table 2).
2. EDS mapping on small–scale to obtain information about the local distribution of the atoms in the annealed films (Figure 5b,d).

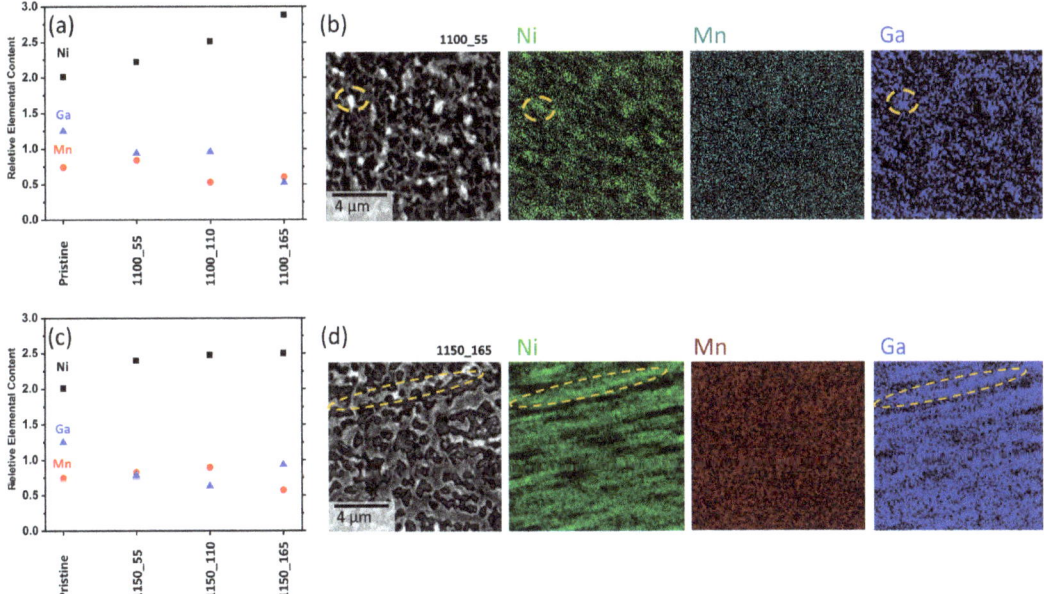

Figure 5. Composition measurements obtained by EDS: (**a**) the relative contents of each of the elements for the pristine and the annealed samples at 1100 K; (**b**) high-resolution elemental mapping of sample 1100_55 (**c**) the relative contents of each of the elements for the pristine and the annealed samples at 1150 K. (**d**) high-resolution elemental mapping of sample 1150_165. Some of the Ni- and Ga-rich areas are highlighted by yellow dash lines.

The first approach for the samples annealed at 1100 K (Figure 5a, Table 2) shows a continuous rise of Ni content at the expense of Ga and then Mn contents as a function of annealing time. This has pushed the composition of the annealed films far away from the stoichiometric Ni_2MnGa. Those results could be due to sublimation of Ga and then Mn at elevated temperatures and high vacuum. In addition to sublimation, our high-magnification composition mapping for 1100–55 sample shows small bright inclusions; one of these has been marked in Figure 5b. The inclusions show higher contents of Ni and Ga with respect to the matrix of the sample, revealing material elemental segregation in the annealing process. One could expect even more significant segregation by increasing the annealing time to 110 and 165 min. The samples annealed at 1150 K (Figure 5c, Table 2) also show a composition variation as a function annealing time. With respect to the samples annealed at 1100 K, the sublimation of Ga is more abrupt, whereas the Mn sublimation occurs only for the longest annealing time.

Thanks to the second approach of composition measurements (EDS mapping in small scale), we can confirm that the segregation process also exists in these samples. The high-magnification composition mapping of 1150_165 sample (Figure 5d) shows that the material islands' propagation lines are richer in Ni and Ga with respect to the matrix of the sample, whereas the Mn content is homogenously distributed. The relatively homogenous distribution of the Mn content in the entire scan area (Figure 5b,d) also shows that the holes appearing in the SEM images do not pass through the entire film.

Finally, we measured the thermomagnetic curves of the samples to evaluate the susceptibility evolution as a function of temperature (Figure S1). Although the annealed samples are magnetic at room temperature, as shown in Figure 4, none of the samples underwent first/second order phase transformations in the investigated range of temperature (200–400 K). This could be due to severe composition variation and the phase segregation of the material in the annealing process. We believe that the magnetic signal of the annealed samples at room temperature could possibly be due the presence of a secondary magnetic phase having T_c above 400 K.

In this work, we used the annealing temperature and time as the driving forces to control the dewetting process. Other parameters such as film thickness, type of gas, gas pressure, and substrate properties, which also contribute to determining the final morphology of the dewetted thin film, were kept constant. The selected temperatures and times successfully promoted the formation of holes and the propagation of holes. However, the magnetic and composition characterizations proved that these annealing parameters used to control the dewetting process are detrimental for these types of ternary compounds that have very sensitive composition-dependent transformations. The annealing parameters resulted in elemental sublimation, phase segregation, and, consequently, termination of first/second order phase transformation of the samples. The next direction of this interesting topic could be to promote the dewetting process at lower temperatures by regulating other parameters (e.g., type of gas, gas pressure, and type of substrate) in order to preserve the composition of the films and, consequently, preserve their peculiar magnetic properties through the dewetting process.

4. Conclusions

Dewetting process was studied in epitaxial Ni-Mn-Ga films deposited on MgO(001). The kinetics of the process have been followed by performing high-temperature annealing in the range of 1100–1150 K for increasing time (55 to 165 min). Scanning electron microscopy images revealed morphology evolution starting from the creation of holes in the continuous layer towards the formation of micro/nanoislands, occurring at increasing temperature and time. The annealing process progressively induced Ga and Mn sublimation accompanied by Ni-Ga segregation. Such a change has reflected into a progressive decrease of magnetic signal. The process completion leads to films containing islands that display a faint magnetic signal without any evidence of first/second order phase transformation.

Supplementary Materials: The following supporting information can be downloaded at: https://www.mdpi.com/article/10.3390/cryst12121826/s1, Figure S1: Thermomagnetic curves of the annealed samples for the temperature range of 200–400 K: (a) 1100_55, (b) 1150_55 applying low magnetic field ($\mu_0 H$ = 2 mT) in the plane of the film along MgO[100], (c) 1150_110, (d) 1150_165 applying magnetic field of ($\mu_0 H$ = 150 mT) in the plane of the film along MgO[100].

Author Contributions: Conceptualization, F.A.; investigation, M.T.G., F.C. (Federica Celegato) and G.B.; resources, M.T.G.; writing—original draft preparation, M.T.G.; writing—review and editing, G.B., F.C. (Francesca Casoli) and P.T.; visualization, M.T.G.; supervision, F.A. and P.T.; funding acquisition, F.A. and P.T. All authors have read and agreed to the published version of the manuscript.

Funding: This research received no external funding.

Data Availability Statement: Not applicable.

Acknowledgments: This work was partially supported by Fondazione Cariparma through the Biomontans project.

Conflicts of Interest: The authors declare no conflict of interest.

References

1. Jani, J.M.; Leary, M.; Subic, A.; Gibson, M.A. A review of shape memory alloy research, applications and opportunities. *Mater. Des.* **2014**, *56*, 1078–1113. [CrossRef]
2. Kohl, M.; Gueltig, M.; Pinneker, V.; Yin, R.; Wendler, F.; Krevet, B. Magnetic shape memory microactuators. *Micromachines* **2014**, *5*, 1135–1160. [CrossRef]
3. Rashidi, S.; Ehsani, M.H.; Shakouri, M.; Karimi, N. Potentials of magnetic shape memory alloys for energy harvesting. *J. Magn. Magn. Mater.* **2021**, *537*, 168112. [CrossRef]
4. Kitanovski, A. Energy applications of magnetocaloric materials. *Adv. Energy Mater.* **2020**, *10*, 1903741. [CrossRef]
5. Bruederlin, F.; Bumke, L.; Chluba, C.; Ossmer, H.; Quandt, E.; Kohl, M. Elastocaloric cooling on the miniature scale: A review on materials and device engineering. *Energy Technol.* **2018**, *6*, 1588–1604. [CrossRef]

6. Kohl, M.; Fechner, R.; Gueltig, M.; Megnin, C.; Ossmer, H. Miniaturization of Shape Memory Actuators. In Proceedings of the 16th International Conference on New Actuators, Bremen, Germany, 25–27 June 2018; pp. 1–9.
7. Rodríguez-Crespo, B.; Salazar, D.; Lanceros-Méndez, S.; Chernenko, V. Development and magnetocaloric properties of Ni (Co)-Mn-Sn printing ink. *J. Alloys Compd.* **2022**, *917*, 165521. [CrossRef]
8. Gao, P.; Tian, B.; Xu, J.; Tong, Y.; Chen, F.; Li, L. Investigation on porous NiMnGa alloy and its composite with epoxy resin. *J. Alloys Compd.* **2022**, *892*, 162248. [CrossRef]
9. Laitinen, V.; Saren, A.; Sozinov, A.; Ullakko, K. Giant 5.8% magnetic-field-induced strain in additive manufactured Ni-Mn-Ga magnetic shape memory alloy. *Scr. Mater.* **2022**, *208*, 114324. [CrossRef]
10. Ituarte, I.F.; Nilsén, F.; Nadimpalli, V.K.; Salmi, M.; Lehtonen, J.; Hannula, S.P. Towards the additive manufacturing of Ni-Mn-Ga complex devices with magnetic field induced strain. *Addit. Manuf.* **2022**, *49*, 102485. [CrossRef]
11. Caputo, M.P.; Berkowitz, A.E.; Armstrong, A.; Müllner, P.; Solomon, C.V. 4D printing of net shape parts made from Ni-Mn-Ga magnetic shape-memory alloys. *Addit. Manuf.* **2018**, *21*, 579–588. [CrossRef]
12. Bhattacharya, K.; DeSimone, A.; Hane, K.F.; James, R.D.; Palmstrøm, C.J. Tents and tunnels on martensitic films. *Mater. Sci. Eng. A* **1999**, *273–275*, 685–689. [CrossRef]
13. Dong, J.W.; Xie, J.Q.; Lu, J.; Adelmann, C.; Palmstrøm, C.J.; Cui, J.; Pan, Q.; Shield, T.W.; James, R.D.; McKernan, S. Shape memory and ferromagnetic shape memory effects in single-crystal Ni_2MnGa thin films. *J. Appl. Phys.* **2004**, *95*, 2593–2600. [CrossRef]
14. Thomas, M.; Heczko, O.; Buschbeck, J.; Lai, Y.W.; McCord, J.; Kaufmann, S.; Schultz, L.; Fähler, S. Stray-Field-Induced Actuation of Free-Standing Magnetic Shape-Memory Films. *Adv. Mater.* **2009**, *21*, 3708–3711. [CrossRef]
15. Campanini, M.; Nasi, L.; Fabbrici, S.; Casoli, F.; Celegato, F.; Barrera, G.; Chiesi, V.; Bedogni, E.; Magén, C.; Grillo, V.; et al. Magnetic Shape Memory Turns to Nano: Microstructure Controlled Actuation of Free-Standing Nanodisks. *Small* **2018**, *14*, 1803027. [CrossRef]
16. Dunand, D.C.; Müllner, P. Size effects on magnetic actuation in Ni-Mn-Ga shape-memory alloys. *Adv. Mater.* **2011**, *23*, 216–232. [CrossRef]
17. Wang, C.; Meyer, J.; Teichert, N.; Auge, A.; Rausch, E.; Balke, B.; Hütten, A.; Fecher, G.H.; Felser, C. Heusler nanoparticles for spintronics and ferromagnetic shape memory alloys. *J. Vac. Sci. Technol. B* **2014**, *32*, 020802. [CrossRef]
18. Hennel, M.; Varga, M.; Frolova, L.; Nalevanko, S.; Ibarra-Gaytán, P.; Vidyasagar, R.; Sarkar, P.; Dzubinska, A.; Galdun, L.; Ryba, T.; et al. Heusler-Based Cylindrical Micro- and Nanowires. *Phys. Status Solidi A* **2022**, *219*, 2100657. [CrossRef]
19. Takhsha Ghahfarokhi, M.; Arregi, J.A.; Casoli, F.; Horký, M.; Cabassi, R.; Uhlíř, V.; Albertini, F. Microfabricated ferromagnetic-shape-memory Heuslers: The geometry and size effects. *Appl. Mater. Today* **2021**, *23*, 101058. [CrossRef]
20. Eichhorn, T.; Hausmanns, R.; Jakob, G. Microstructure of freestanding single-crystalline Ni_2MnGa thin films. *Acta Mater.* **2011**, *59*, 5067–5073. [CrossRef]
21. Lambrecht, F.; Lay, C.; Aseguinolaza, I.R.; Chernenko, V.; Kohl, M. NiMnGa/Si shape memory bimorph nanoactuation. *Shap. Mem. Superelasticity* **2016**, *2*, 347–359. [CrossRef]
22. Schmitt, M.; Backen, A.; Fähler, S.; Kohl, M. Development of ferromagnetic shape memory nanoactuators. In Proceedings of the 2012 12th IEEE International Conference on Nanotechnology (IEEE-NANO), Birmingham, UK, 20–23 August 2012; pp. 1–4.
23. Arivanandhan, G.; Li, Z.; Curtis, S.; Velvaluri, P.; Quandt, E.; Kohl, M. Temperature Homogenization of Co-Integrated Shape Memory—Silicon Bimorph Actuators. *Proceedings* **2020**, *64*, 8.
24. Kohl, M.; Schmitt, M.; Backen, A.; Schultz, L.; Krevet, B.; Fähler, S. Ni-Mn-Ga shape memory nanoactuation. *Appl. Phys. Lett.* **2014**, *104*, 043111. [CrossRef]
25. Lambrecht, F.; Sagardiluz, N.; Gueltig, M.; Aseguinolaza, I.R.; Chernenko, V.A.; Kohl, M. Martensitic transformation in NiMnGa/Si bimorph nanoactuators with ultra-low hysteresis. *Appl. Phys. Lett.* **2017**, *110*, 213104. [CrossRef]
26. Schmitt, M.; Backen, A.; Fähler, S.; Kohl, M. Freely movable ferromagnetic shape memory nanostructures for actuation. *Microelectron. Eng.* **2012**, *98*, 536–539. [CrossRef]
27. Jenkins, C.A.; Ramesh, R.; Huth, M.; Eichhorn, T.; Pörsch, P.; Elmers, H.J.; Jakob, G. Growth and magnetic control of twinning structure in thin films of Heusler shape memory compound Ni_2MnGa. *Appl. Phys. Lett.* **2008**, *93*, 234101. [CrossRef]
28. Mashirov, A.V.; Irzhak, A.V.; Tabachkova, N.Y.; Milovich, F.O.; Kamantsev, A.P.; Zhao, D.; Liu, J.; Kolesnikova, V.G.; Rodionova, V.V.; Koledov, V.V. Magnetostructural Phase Transition in Micro-and Nanosize Ni–Mn–Ga–Cu Alloys. *IEEE Magn. Lett.* **2019**, *10*, 1–4. [CrossRef]
29. Thompson, C.V. Solid-state Dewetting of Thin Films. *Annu. Rev. Mater. Res.* **2012**, *42*, 399–434. [CrossRef]
30. Leroy, F.; Borowik, Ł.; Cheynis, F.; Almadori, Y.; Curiotto, S.; Trautmann, M.; Barbé, J.C.; Müller, P. How to control solid state dewetting: A short review. *Surf. Sci. Rep.* **2016**, *71*, 391–409. [CrossRef]
31. Pierre-Louis, O. Solid-state wetting at the nanoscale. *Prog. Cryst. Growth Charact. Mater.* **2016**, *62*, 177–202. [CrossRef]
32. Motyčková, L. Magnetic Properties of Self-Assembled FeRh Nanomagnets. Master's Thesis, Brno University of Technology, Brno, Czech Republic, 2020.
33. Barrera, G.; Celegato, F.; Cialone, M.; Coïsson, M.; Rizzi, P.; Tiberto, P. Effect of the substrate crystallinity on morphological and magnetic properties of $Fe_{70}Pd_{30}$ nanoparticles obtained by the solid state dewetting. *Sensors* **2021**, *21*, 7420. [CrossRef]
34. Andalouci, A.; Roussigné, Y.; Farhat, S.; Chérif, S.M. Magnetic and magneto-optical properties of assembly of nanodots obtained from solid-state dewetting of ultrathin cobalt layer. *J. Phys. Condens. Matter* **2019**, *31*, 495805. [CrossRef] [PubMed]

35. Barrera, G.; Celegato, F.; Coïsson, M.; Cialone, M.; Rizzi, P.; Tiberto, P. Formation of free-standing magnetic particles by solid-state dewetting of $Fe_{80}Pd_{20}$ thin films. *J. Alloys Compd.* **2018**, *742*, 751–758. [CrossRef]
36. Oh, H.; Pyatenko, A.; Lee, M. A hybrid dewetting approach to generate highly sensitive plasmonic silver nanoparticles with a narrow size distribution. *Appl. Surf. Sci.* **2021**, *542*, 148613. [CrossRef]
37. Esterina, R.; Liu, X.M.; Adeyeye, A.O.; Ross, C.A.; Choi, W.K. Solid-state dewetting of magnetic binary multilayer thin films. *J. Appl. Phys.* **2015**, *118*, 144902. [CrossRef]
38. Bhalla, N.; Jain, A.; Lee, Y.; Shen, A.Q.; Lee, D. Dewetting metal nanofilms—Effect of substrate on refractive index sensitivity of nanoplasmonic gold. *Nanomaterials* **2019**, *9*, 1530. [CrossRef]
39. Song, X.; Liu, F.; Qiu, C.; Coy, E.; Liu, H.; Aperador, W.; Załęski, K.; Li, J.J.; Song, W.; Lu, Z.; et al. Nanosurfacing Ti alloy by weak alkalinity-activated solid-state dewetting (AAD) and its biointerfacial enhancement effect. *Mater. Horiz.* **2021**, *8*, 912–924. [CrossRef]
40. Motyčková, L.; Arregi, J.A.; Staňo, M.; Průša, S.; Částková, K.; Uhlíř, V. Preserving Metamagnetism in Self-Assembled FeRh Nanomagnets. *arXiv* **2022**, arXiv:2209.02469.
41. Lünser, K.; Diestel, A.; Nielsch, K.; Fähler, S. Self-Patterning of Multifunctional Heusler Membranes by Dewetting. *Adv. Mater. Interfaces* **2021**, *8*, 2100966. [CrossRef]
42. Diestel, A.; Neu, V.; Backen, A.; Schultz, L.; Fähler, S. Magnetic domain pattern in hierarchically twinned epitaxial Ni–Mn–Ga films. *J. Phys. Condens. Matter* **2013**, *25*, 266002. [CrossRef]
43. Ranzieri, P.; Fabbrici, S.; Nasi, L.; Righi, L.; Casoli, F.; Chernenko, V.A.; Villa, E.; Albertini, F. Epitaxial Ni–Mn–Ga/MgO(100) thin films ranging in thickness from 10 to 100 nm. *Acta Mater.* **2013**, *61*, 263–272. [CrossRef]
44. Ranzieri, P.; Campanini, M.; Fabbrici, S.; Nasi, L.; Casoli, F.; Cabassi, R.; Buffagni, E.; Grillo, V.; Magén, C.; Celegato, F.; et al. Achieving Giant Magnetically Induced Reorientation of Martensitic Variants in Magnetic Shape-Memory Ni–Mn–Ga Films by Microstructure Engineering. *Adv. Mater.* **2015**, *27*, 4760–4766. [CrossRef] [PubMed]
45. Niemann, R.; Backen, A.; Kauffmann-Weiss, S.; Behler, C.; Rößler, U.K.; Seiner, H.; Heczko, O.; Nielsch, K.; Schultz, L.; Fähler, S. Nucleation and growth of hierarchical martensite in epitaxial shape memory films. *Acta Mater.* **2017**, *132*, 327–334. [CrossRef]
46. Takhsha Ghahfarokhi, M.; Casoli, F.; Fabbrici, S.; Nasi, L.; Celegato, F.; Cabassi, R.; Trevisi, G.; Bertoni, G.; Calestani, D.; Tiberto, P.; et al. Martensite-enabled magnetic flexibility: The effects of post-growth treatments in magnetic-shape-memory Heusler thin films. *Acta Mater.* **2020**, *187*, 135–145. [CrossRef]
47. Takhsha Ghahfarokhi, M.; Nasi, L.; Casoli, F.; Fabbrici, S.; Trevisi, G.; Cabassi, R.; Albertini, F. Following the martensitic configuration footprints in the transition route of Ni-Mn-Ga magnetic shape memory films: Insight into the role of twin boundaries and interfaces. *Materials* **2020**, *13*, 2103. [CrossRef] [PubMed]
48. Takhsha Ghahfarokhi, M.; Chirkova, A.; Maccari, F.; Casoli, F.; Ener, S.; Skokov, K.P.; Cabassi, R.; Gutfleisch, O.; Albertini, F. Influence of martensitic configuration on hysteretic properties of Heusler films studied by advanced imaging in magnetic field and temperature. *Acta Mater.* **2021**, *221*, 117356. [CrossRef]
49. Casoli, F.; Varvaro, G.; Takhsha Ghahfarokhi, M.; Fabbrici, S.; Albertini, F. Insight into the magnetisation process of martensitic Ni–Mn–Ga films: A micromagnetic and vector magnetometry study. *J. Phys. Mater.* **2020**, *3*, 045003. [CrossRef]
50. Albertini, F.; Solzi, M.; Paoluzi, A.; Righi, L. Magnetocaloric properties and magnetic anisotropy by tailoring phase transitions in NiMnGa alloys. *Mater. Sci. Forum* **2008**, *583*, 169–196. [CrossRef]
51. Kamarád, J.; Albertini, F.; Arnold, Z.; Casoli, F.; Pareti, L.; Paoluzi, A. Effect of hydrostatic pressure on magnetization of $Ni_{2+x}Mn_{1-x}Ga$ alloys. *J. Magn. Magn. Mater.* **2005**, *290*, 669–672. [CrossRef]

Article

Hyperbolic Behavior and Antiferromagnetic Order in Rare-Earth Tellurides

Jonathan Gjerde and Radi A. Jishi *

Department of Physics and Astronomy, California State University, Los Angeles, CA 90032, USA
* Correspondence: radi.jishi@calstatela.edu

Abstract: Quasi-2D materials have received much attention in recent years for their unusual physical properties. Among the most investigated of these materials are the rare-earth tellurides, which are primarily studied because they exhibit charge density waves and other quantum phenomena and have a high degree of tunability. In this paper, we examine the optical and magnetic properties of several rare-earth tellurides and find that they are antiferromagnetic materials with hyperbolic dispersion. Hyperbolic materials have very promising applications in sub-diffraction-limit optics, nanolithography, and spontaneous emission engineering, but these applications are hampered by low-quality hyperbolic materials. Rare-earth tellurides may provide insight into solving these issues if their properties can be properly tuned using the large variety of techniques already explored in the literature.

Keywords: rare-earth tellurides; antiferromagnetism; optical properties; hyperbolic materials; density functional theory

1. Introduction

Rare-earth tritellurides (RTe_3) are layered materials that have been widely studied due to the numerous quantum states they exhibit, including charge density waves, magnetic ordering, and superconductivity. In particular, scientists have been interested in the charge density wave (CDW) states of RTe_3 and how CDW states interact and coexist with other quantum states [1]. While other materials exhibit all these quantum states, heightened interest in RTe_3 stems from the fact that their properties can be tuned through the application of pressure [2], varying the rare-earth ion, and intercalation of guest atoms [3] or molecules [4] in the gaps between Te layers. These gaps are referred to as Van der Waals (VdW) gaps because the layers are weakly bound together with Van der Waals forces [5], similar to graphite. In this work, we report first-principles calculations of the magnetic and optical properties of some representative rare-earth tritellurides ($ErTe_3$ and $SmTe_3$) and ditellurides ($ErTe_2$ and $SmTe_2$).

RTe_3 (space group 63: Cmcm) has a weakly orthorhombic structure [6], meaning two lattice parameters are nearly equal (a $\approx$ c $\neq$ b, where b is perpendicular to the layers), with double layers of square Te sheets (sometimes referred to as tellurene). The VdW gap occurs in the space between the two Te layers, which make up the double layer. These Te double layers alternate with double layers of rare-earth telluride (RTe). These two RTe layers can also be viewed as four square layers of a single atom type. From lowest to highest (see Figure 1a), these layers are RE, Te, Te, and RE. However, the distance between the first and second (or third and fourth) layer is very small, typically less than 1 Å, so they are conventionally viewed as a single puckered layer [7]. In these puckered layers, one RE lies at the center of four nearest neighbor Te ions. As can be seen in Figure 1a, the unit cell of RTe_3 has two blocks. The first block consists of an RTe double-layer sandwiched between two Te layers, while the second block is obtained from the first by a translation with the vector $\frac{1}{2}a + \frac{1}{2}b$.

Citation: Gjerde, J.; Jishi, R.A. Hyperbolic Behavior and Antiferromagnetic Order in Rare-Earth Tellurides. *Crystals* **2022**, *12*, 1839. https://doi.org/10.3390/cryst12121839

Academic Editor: Jacek Ćwik

Received: 15 November 2022
Accepted: 14 December 2022
Published: 16 December 2022

Copyright: © 2022 by the authors. Licensee MDPI, Basel, Switzerland. This article is an open access article distributed under the terms and conditions of the Creative Commons Attribution (CC BY) license (https://creativecommons.org/licenses/by/4.0/).

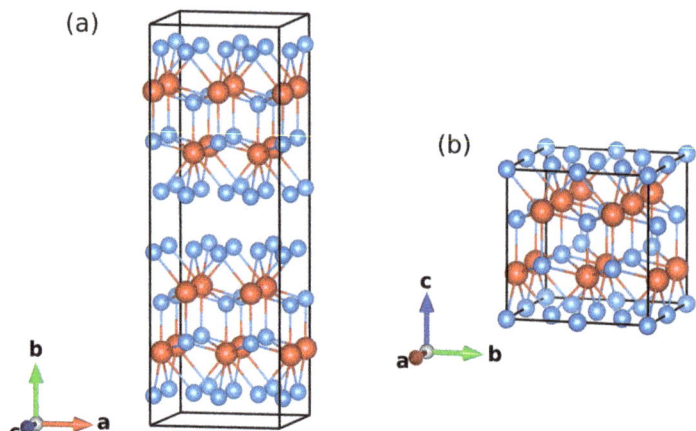

Figure 1. The crystal structures of (**a**) RTe$_3$ and (**b**) RTe$_2$. The red balls represent the rare-earth atoms, while the smaller blue balls represent the tellurium atoms. As the figure shows, RTe$_3$ structure consists of a stack of alternating double RTe layers and double Te layers. In RTe$_2$, double RTe layers alternate with single Te layers.

A closely related group of materials are the tetragonal rare-earth ditellurides (RTe$_2$). RTe$_2$ (see Figure 1b) also contains RTe double layers with a structure similar to that in RTe$_3$, but these alternate with single layers of square Te, so that there is no VdW gap [8]. In RTe$_2$, the long axis, perpendicular to the layers, is conventionally taken as c, while in RTe$_3$ it is b.

Conduction of electrons in RTe$_3$ occurs mainly in the square Te layers of rare-earth tellurides, while the puckered RTe layers do not significantly contribute to the conductivity [9]. This motivated a 2d tight-binding model of the electronic band structure, incorporating the p_x and p_y orbitals of Te [10,11]. The Fermi surface produced by this model is very similar to that measured using angle-resolved photoemission spectroscopy (measured for R = La, Ce, Sm, Gd, Tb, and Dy) [12] and those calculated for LaTe$_3$ using density functional theory (DFT) [13]. This same tight-binding model has also been applied to RTe$_2$ [14], giving somewhat reasonable agreement with ARPES measurements in LaTe$_{1.95}$ and CeTe$_2$ [15]. Some evidence that this model may be appropriate was provided when resistivity measurements showed that CeTe$_2$ was metallic in the plane of the layers and semiconducting perpendicular to the layers [16].

Most rare-earth atoms have unpaired, highly localized 4f electrons, indicating that crystals containing them are likely to have a ferromagnetic or antiferromagnetic ground state. Antiferromagnetism (AFM) occurs when the spins of the magnetic atoms in a material exhibit a regular pattern, but the total magnetic moment adds up to zero, even within a very small volume of the crystal. Unlike ferromagnetism, AFM can lead to several possible arrangements of spin. The long range magnetic orders of RTe$_3$ were investigated by Ru et al. [17] using magnetic susceptibility and specific heat measurements above 1.8 K for various RTe$_3$ compounds. All RTe$_3$ compounds, except those with R = La, Pr, Er, and Tm, were found to exhibit AFM below a maximum Neel temperature (the temperature at which an AFM transition takes place) of about 11 K. La is nonmagnetic, while previous work suggests PrTe$_3$ has a singlet ground state [18]. The remaining structures, ErTe$_3$ and TmTe$_3$, may have an AFM ground state with a Neel temperature below 1.8 K. AFM states have also been observed for GdTe$_2$ below 9.8 K [19] and CeTe$_2$ below 4 K [20], while PrTe$_2$ and LaTe$_2$ again were found to be nonmagnetic. The magnetic and electronic properties of GdTe$_3$ have also been investigated using the VASP DFT code, which confirmed an AFM ground state [21]. The electronic band structures of LaTe$_3$ and NdTe$_3$ have been calculated [22] using the Quantum Espresso DFT code, while those of YTe$_3$ were calculated [3] with VASP. Both VASP [23] and Quantum Espresso [24] use plane waves to expand the wave function of valence electrons, and a pseudopotential to represent the interaction of the valence electrons with core electrons.

Rare-earth tellurides are uniaxial anisotropic materials, meaning symmetry is broken along one of the crystal axes. The interactions of these materials with light depend on the direction of the incident light. This causes the dielectric constant to become a dielectric tensor. Since rare-earth telluride crystals are either tetragonal (RTe$_2$) or almost tetragonal (RTe$_3$), the dielectric tensor can be fully described by two components, namely $\epsilon_{xx} = \epsilon_{yy}$ and ϵ_{zz}, where the principal axis is assumed to be along the z-direction. The dielectric function depends on the frequency of incoming light; the dispersion equation relating the dielectric function components, frequency of light, and wave vectors for uniaxial materials is:

$$\left(\frac{\omega}{c}\right)^2 = \frac{k_x^2 + k_y^2}{\epsilon_{zz}} + \frac{k_z^2}{\epsilon_{xx}} \qquad (1)$$

where k_x, k_y, and k_z are the wave vector components along the x-, y-, and z-axes, respectively. For many materials, this equation implies ellipsoidal surfaces of constant frequency in k-space. However, if one of the dielectric function components, ϵ_{xx} or ϵ_{zz}, is negative while the other component is positive, the isofrequency surfaces become hyperboloids, which are unbounded surfaces in k-space. This means that, in theory, the wave vector can become arbitrarily large in these materials [25]. This has led to research in a number of exciting applications, such as sub-diffraction-limit imaging [26], spontaneous emission engineering [27], and thermal emission engineering [28].

To understand this phenomenon and its application to RTe, we must ask under what circumstances does a material become hyperbolic? A critical feature of a metal is that its dielectric function is negative for some range of frequencies. So the requirement that either ϵ_{xx} or ϵ_{zz} become negative implies that the material must act as a metal in one or two directions, but as a dielectric in the other one or two. One way to accomplish this is by stacking an alternating sequence of metal and dielectric layers. The layers should have a thickness much smaller than the wavelengths of light under consideration, so that the composite material (known as metamaterial) is approximately homogeneous. In this configuration, the material is metallic in the plane of the layers, but not perpendicular to it. This is referred to as a multilayer hyperbolic metamaterial (HMM). Since the material is metallic in two directions, it produces a hyperboloid of one sheet in k-space, referred to as a type 1 HMM. Another option is to take conducting nanowires and insulate them from each other by embedding them in dielectric [29]. Clearly, this makes the material metallic only in the direction along the nanowires. This produces a hyperboloid of two sheets in k-space, which makes it a type 2 HMM (see Figure 2).

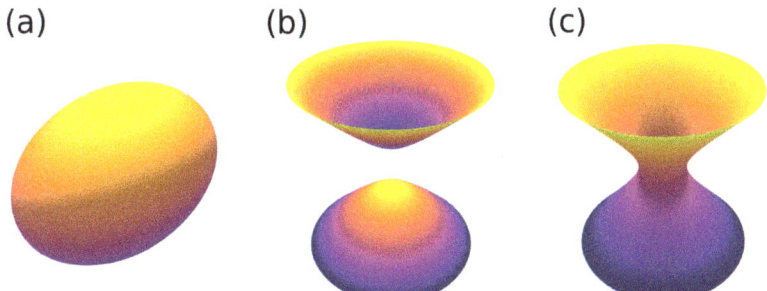

Figure 2. Possible isofrequency surfaces: (a) ellipsoidal, when $\epsilon_{xx} > 0$ and $\epsilon_{zz} > 0$, (b) type 2 hyperbolic, with $\epsilon_{xx} > 0$ and $\epsilon_{zz} < 0$, and (c) type 1 hyperbolic, with $\epsilon_{xx} < 0$ and $\epsilon_{zz} > 0$.

Not all hyperbolic materials need to be engineered in this way. It turns out that there are a number of naturally-occurring materials that meet the requirements for hyperbolic dispersion, including graphite, cuprates, ruthenates, bismuth, sapphire, and hexagonal boron nitride [30]. Graphite, for instance, is a layered material that is known to be able to conduct electrons within the plane of its graphene sheets, but this conduction cannot occur between the layers, because the electrons cannot easily jump across the VdW gap. These are exactly the conditions required for hyperbolic dispersion, and in fact, graphite is hyperbolic for a small range of frequencies in the ultraviolet regime [31]. Cuprates are known to conduct electrons within their copper oxide planes, but not between the planes and, therefore, not in the direction perpendicular to these planes, making them give hyperbolic dispersion [32]. Most of the naturally-occurring hyperbolic materials exhibit hyperbolic behavior in the infrared region [33]. The layered structure of rare-earth tellurides, with conducting tellurium layers separated by nonconducting layers, makes them possible candidates for displaying optical hyperbolic behavior.

The optical properties of RTe$_3$ (with R = La, Ce, Nd, Sm, Gd, Tb, Dy) and RTe$_2$ (with R = La, Ce) have been investigated through reflectivity measurements by Sacchetti and Lavagnini [9,34–36]. These measurements were primarily concerned with investigating the CDW phase and its dependence on pressure. For this reason, only the interactions with incident light perpendicular to the plane of the layers were studied, and the full, two-component, frequency-dependent dielectric tensor remains unknown.

By calculating the total energy of ferromagnetic (FM) and various possible AFM spin configurations, we find that the ground states, in all four compounds, exhibit an antiferromagnetic order. We also performed optical properties calculations, which predict that all four compounds show hyperbolic behavior in a frequency range extending from the infrared into the visible region.

2. Methods

Both the electronic and optical properties were calculated using Wien2k, an all-electron, full-potential, linearized, augmented plane wave DFT code [37]. Wien2k approximates the electronic wave function using a "muffin-tin" method, in which space is divided into two regions. One region consists of the inside of non-overlapping spheres surrounding the nuclei, while the other is the interstitial region between the spheres. Within the spheres, the wave function is expanded in spherical harmonics up to $l_{max} = 10$. In the interstitial region, the wave function is expanded in terms of plane waves up to a maximum wave vector K_{max}, which is chosen so that $K_{max} R_{MT}^{min} = 9$, where R_{MT}^{min} is the smallest muffin-tin radius in the unit cell. The electron density is Fourier expanded in the interstitial region with a wave vector cutoff of $14/a_0$, where a_0 is the Bohr radius. Inside the spheres, it is expanded in a product of radial functions and lattice harmonics with $l_{max} = 6$. The electronic exchange and correlation were treated using the generalized gradient approximation (GGA) and Perdew–Burke–Ernzerhof (PBE) exchange–correlation functionals [38]. In the self-consistent field calculations, integration over the Brillouin zone is replaced by a summation over a mesh of 1800 k-points within the Brillouin zone, and convergence is achieved with a tolerance of 10^{-4} Ry in energy and $10^{-3} e$ in charge. In carrying out the DFT calculations on RTe$_2$ and RTe$_3$ (R = Sm, Er), we used the experimental values of the lattice constants and atomic position coordinates. The lattice parameters are given in Table 1.

Table 1. The lattice constants (Å) of RTe$_3$ and RTe$_2$ (R = Sm, Er).

	a	b	c
SmTe$_3$ [7]	4.334	25.674	4.347
SmTe$_2$ [8]	4.370	4.370	9.000
ErTe$_3$ [7]	4.248	25.275	4.279
ErTe$_2$ [39]	4.248	4.248	8.865

Because of the strong localization of electrons in the 4f orbitals of rare-earth atoms, we add an onsite Coulomb repulsion term, known as a Hubbard U term. Thus, we carry out what is known as a DFT+U calculation of the electronic and optical properties of the rare-earth tellurides. The U values are usually several electron-volts, depending on the type of ion, the localized orbitals, and the chemical environment. In our calculations, we have chosen U = 7 eV for both the samarium and erbium atoms. This is consistent with prior DFT calculations and the generally accepted U values for rare-earth atoms [21,40–42].

The dielectric function, absorption, optical conduction, and related quantities, collectively called the optical properties, describe the interaction of a material with light. These interactions are caused by scattering from one state to another by the perturbing potential of light. Scattering can occur within a band (if the band is partially filled) or between filled bands and partially filled or empty bands. Scattering within a band is approximated using the Drude model of free electrons with damping. The damping is characterized by the Drude term, γ, which varies by material, and may vary depending on direction in anisotropic materials. The intraband contribution to the dielectric function is

$$\epsilon^{intra}(\omega) = 1 - \frac{\omega_p^2}{\omega^2 + i\omega\gamma} \qquad (2)$$

Here, ω_p is the plasma frequency, which, in cgs units, is given by $\sqrt{4\pi n e^2/m}$, where n is the free electron density and m is the effective mass of the electron [43]. It should be noted that in anisotropic materials, such as the rare-earth tellurides studied here, the effective mass is a tensor rather than a scalar; this implies that ω_p is also a tensor. In the calculations reported in this work, we used a value for γ corresponding to an energy of 0.1 eV.

Interband scattering contributes to the permittivity for both metals and dielectrics. In this case, the energy of a photon is absorbed, causing an electron to jump from a valence band to a conduction band. One important consideration is that the photon wave vector is typically much smaller than the width of the Brillouin zone, so that in a transition, the electron wave vector is approximately unchanged. The interband contribution to the imaginary part of the permittivity is given by [44]:

$$Im[\epsilon_{ij}^{inter}(\omega)] = \frac{\hbar^2 e^2}{\pi m^2 \omega^2} \sum_{n,n'} \int_{\vec{k}} \langle n\vec{k}|p_i|n'\vec{k}\rangle \langle n'\vec{k}|p_j|n\vec{k}\rangle \left(f(E_{n\vec{k}}) - f(E_{n'\vec{k}})\right) \delta(E_{n'\vec{k}} - E_{n\vec{k}} - \hbar\omega) \qquad (3)$$

Here, $f(E)$ is the Fermi distribution function, p_i is the ith component of the momentum operator, n and n' are band indices, and $\delta(E)$ is a Dirac delta function. The optical matrix elements $\langle n\vec{k}|p_i|n'\vec{k}\rangle$ are related to the probability amplitude for a transition from band n' to band n. This matrix element leaves the wave vector unchanged since the photon momentum is relatively small. Scattering can only occur from an occupied state to an empty one, which is taken into account by the Fermi functions. Finally, conservation of energy demands that the final energy is equal to the initial energy plus the photon energy, which is guaranteed by the delta function. Knowing the imaginary part of permittivity, the real part may be calculated using the Kramers–Kronig relations. The real and imaginary parts of permittivity are then used to calculate all other optical properties.

Because accurate calculation of the dielectric function requires many more points in the Brillouin zone, a denser mesh of 24 × 24 × 12 k-points is used. Since rare-earth atoms are heavy, we ran these calculations with the spin-orbit coupling contribution added for each material.

3. Results and Discussion

For each material, we calculated the energy of the nonmagnetic, ferromagnetic, and several AFM spin arrangements in order to find the ground state. In the ABAB interlayer AFM configuration, each layer of rare-earth atoms is all spin up or down, with spin direction alternating each layer. ABBA interlayer AFM also has each layer all spin up or down, but alternates spin direction every other layer; i.e., in this arrangement, two adjacent rare-earth layers with one spin direction are followed by two adjacent rare-earth layers with the opposite spin direction. In intralayer AFM, each layer is separately antiferromagnetic. In the stripe configuration, the spins in a row of atoms along a lattice vector are aligned, but the spin direction alternates each row (See Figure 3). In the case of RTe$_3$, we calculated the energies of ABAB interlayer, intralayer, and stripes along both of the two shorter lattice vectors (a and c). For RTe$_2$, we calculated the total energy for the ABAB interlayer, ABBA interlayer, and intralayer AFM configurations. Stripe AFM only applies to RTe$_3$ because the two shorter lattice vectors are slightly different in length, making it possible to break the symmetry along these directions. ABBA interlayer does not apply to RTe$_3$ because the distance between RTe double layers is large, making the interactions between atoms in one double layer with those in the adjacent double layer very weak.

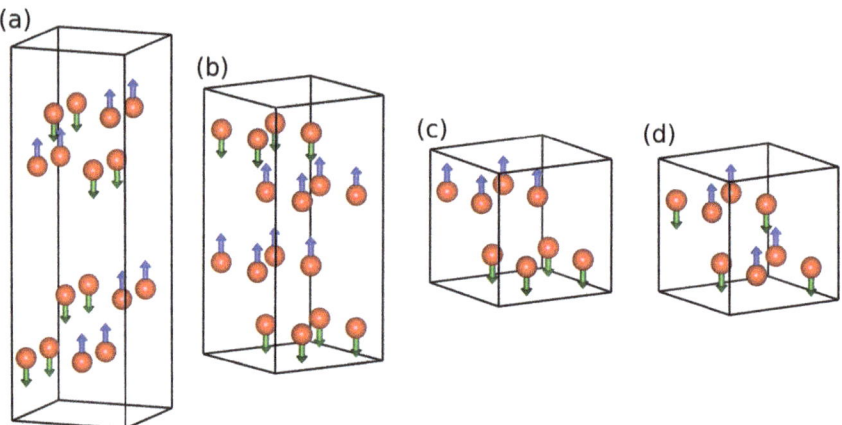

Figure 3. The possible AFM configurations of RTe (with the non-magnetic Te ions hidden): (**a**) stripe (only relevant for RTe$_3$), (**b**) ABBA interlayer (only relevant for RTe$_2$), (**c**) ABAB interlayer, and (**d**) intralayer (c and d apply to both RTe$_3$ and RTe$_2$ structures but are demonstrated here using RTe$_2$).

The calculated total energies of the various configurations are presented in Table 2. As the table shows, the ground state is predicted to be antiferromagnetic for each of the four compounds: SmTe$_2$, SmTe$_3$, ErTe$_2$, and ErTe$_3$. As we noted in the introduction, SmTe$_3$ was indeed found to be antiferromagnetic at very low temperatures, while ErTe$_3$ did not undergo an AFM transition above 1.8 K. Our results, however, suggest that ErTe$_3$ should have an AFM ground state at sufficiently low temperatures. Similarly, we predict that SmTe$_2$ and ErTe$_2$ should also be antiferromagnetic at low temperatures. SmTe$_3$ and ErTe$_3$ were found to have the lowest energy in the AFM stripe configuration, with both compounds favoring stripes along the a-axis, which has the slightly shorter in-plane lattice parameter. Previous DFT calculations of the possible AFM states of GdTe$_3$ also found the ground state to be stripe AFM, though the stripe axis was not specified [21]. The ground states of SmTe$_2$ and ErTe$_2$ exhibited intralayer AFM. In some cases, the energy difference between the ground state and first excited state magnetic configuration was very small, on the order of meV, suggesting that these configurations may be quite unstable. The magnetic moments of Sm in both SmTe$_3$ and SmTe$_2$ were calculated to be approximately 3 μ_B, where μ_B is the Bohr magneton, while the moments of Er in ErTe$_3$ and ErTe$_2$ were about 5.3 μ_B.

We have calculated the band structure for the ground state configuration (stripe AFM) of ErTe$_3$, shown in Figure 4. The bands along the ΓY direction, corresponding to the axis perpendicular to the layers, are notably flat. Since the electron effective mass is related to the curvature of the band, this implies a very large effective mass for the electrons along this direction, indicating that conduction along this direction is very poor, which is in good agreement with experimental observations [11]. Since ω_p^2 is inversely proportional to the electron effective mass, and assuming the z-axis is perpendicular to the rare-earth and tellurium layers, this implies that $\omega_{p,xx}$ will be considerably larger than $\omega_{p,zz}$, as we show later by explicit calculations.

Table 2. The calculated energies (eV), per rare-earth atom, of the various ferromagnetic and antiferromagnetic configurations. The zero point is set to the calculated energy of the nonmagnetic configuration.

	FM	Intra	ABAB	Stripe(a)	Stripe(c)	ABBA
$SmTe_3$	−3.677	−3.670	−3.575	−3.680	−3.671	—
$ErTe_3$	−1.496	−1.496	−1.553	−1.560	−1.554	—
$SmTe_2$	−2.778	−2.826	−2.778	—	—	−2.784
$ErTe_2$	−1.998	−2.594	−2.593	—	—	−2.002

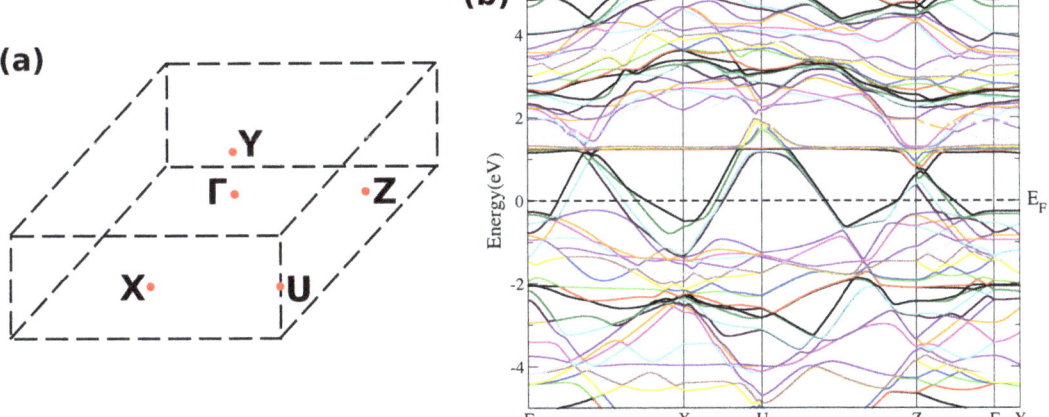

Figure 4. (a) The 3d Brillouin zone of AFM stripe $ErTe_3$, with high symmetry points labeled. Γ is at the center of the Brillouin zone, and Y is at the center of the top face. (b) The energy bands of spin-up electrons in stripe antiferromagnetic $ErTe_3$, plotted along high symmetry directions in the Brillouin zone. Because $ErTe_3$ is AFM, the band structure plot for spin-down electrons is indistinguishable from the plot for spin-up electrons. The Fermi energy is set equal to zero and is shown as a dashed horizontal line. The colors of the bands have no significance; they are used only to make the figure clearer.

The density of states for $ErTe_3$ in the stripe AFM configuration, which we predicted to be the spin state adopted at very low temperatures, is shown in Figure 5a. In this configuration, the unit cell contains four Er and twelve Te atoms; however, the crystal symmetry dictates that there are eight nonequivalent positions. Thus, each atom occupies a position that is equivalent by symmetry to the position occupied by one other atom. Hence, two Er atoms occupy two positions equivalent by symmetry, and one of these atoms is denoted Er1. The other two Er atoms also occupy another two symmetry equivalent positions, and one of these atoms is denoted Er2. A similar situation holds for the 12 Te atoms, where we identify six nonequivalent positions. In Figure 5a we plot the total density of states for stripe AFM $ErTe_3$, along with the partial density derived from the two nonequivalent Er atoms and one of the Te atoms. The plot reveals that states derived from Te orbitals dominate at the Fermi level, indicating that they are the main drivers of electronic conduction. This agrees with theoretical work by DiMasi et al. [11], but is in contrast with a previous calculation of the density of states for $NdTe_3$ [22]. That calculation, using Quantum Espresso, found the density of states for the 4f electrons of Nd to be much larger than that of Te at the Fermi level. They theorized that this was due to hybridization between Nd 4f and Te 5p orbitals. It could be that this disparity is due to differences between the two materials. However, the different result may also be due to the fundamental differences between Wien2k and Quantum Espresso. Wien2k is an all-electron, full potential DFT code, while Quantum Espresso treats core electrons approximately using a pseudopotential.

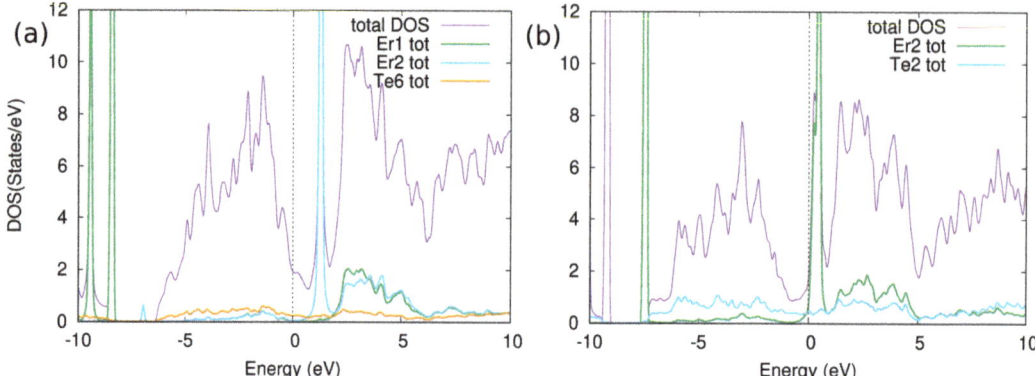

Figure 5. Total and some partial densities of spin-up states for (**a**) stripe antiferromagnetic ErTe$_3$ and (**b**) intralayer antiferromagnetic ErTe$_2$. In the case of stripe AFM ErTe$_3$, the partial densities contributed by two nonequivalent Er atoms and one Te atom are shown, along with the total density of states. In intralayer AFM ErTe$_2$, the contributions of one Er atom and one Te atom are shown, along with the total density of states. The Fermi energy is set at zero energy and is shown as a vertical dotted line.

On the other hand, the density of states for ErTe$_2$ (Figure 5b) shows a somewhat larger contribution from Er than Te at the Fermi level. It seems likely that this is because the quasi-2d character of RTe$_2$ is reduced compared with RTe$_3$ due to having one fewer Te layer separating the RTe layers and to the absence of a VdW gap, across which conduction is extremely difficult. As a result, conduction for RTe$_3$ parallel to the layers is up to 3000 times greater than that perpendicular to the layers [9], while for RTe$_2$, this ratio is between roughly 50 and 100 [15]. This indicates that the tight binding model in RTE$_2$, using only Te 5p orbitals, is perhaps less accurate than in the RTe$_3$ case, though clearly still accurate enough to be useful, given experimental results.

The results in Figure 5 were obtained without including the effect of spin-orbit coupling. Upon including this effect, the only significant change was a downshift in the energy, by about 0.5 eV, of the two peaks below about −8 eV, and a reduction in their height. We therefore concluded that spin-orbit coupling does not lead to a significant change in the electronic properties of these materials. This is consistent with prior RTe DFT calculations, which did not include this interaction and produced results that matched experimental measurements of their electronic properties [3,21,22].

We now consider the optical properties of rare-earth tellurides. As indicated earlier, the dielectric function is the sum of two parts, ϵ^{intra} and ϵ^{inter}. The first part depends on the plasma frequency which, due to the anisotropy of the crystals under study, is a tensor. Taking the axis perpendicular to the layers as the z-axis, we have $\epsilon_{xx} \approx \epsilon_{yy} \neq \epsilon_{zz}$ for the weakly orthorhombic (almost tetragonal) RTe$_3$ crystals, and $\epsilon_{xx} = \epsilon_{yy} \neq \epsilon_{zz}$ for tetragonal RTe$_2$ crystals. Since we are interested in the optical properties of rare-earth tellurides at room temperature, and since the AFM order sets in only at very low temperatures, we calculate the optical properties of these materials in the non-magnetic phase. Expressed in energy units, the calculated plasma frequencies $\omega_{p,xx}$ and $\omega_{p,zz}$ are respectively given by 3.11 eV and 0.49 eV for SmTe$_3$, 4.95 eV and 1.03 eV for ErTe$_3$, 4.98 eV and 1.56 eV for SmTe$_2$, and 5.07 eV and 1.60 eV for ErTe$_2$. The large difference between the values of $\omega_{p,xx}$ and $\omega_{p,zz}$ is a consequence of the large anisotropy in these materials.

The calculated real parts of the components ϵ_{xx} and ϵ_{zz} of the dielectric functions are shown in Figure 6, while the corresponding imaginary parts are shown in Figure 7. The plots show that there are energy ranges where the real parts of ϵ_{xx} and ϵ_{zz} have opposite signs; this indicates that in those ranges of the spectrum, the tellurides display hyperbolic dispersion. The imaginary parts were found to be rather large in the hyperbolic range, suggesting that losses would be large if these materials were used in devices. Large losses have been a stumbling block in utilizing hyperbolic materials to their full potential [45]. However, one of the main reasons rare-earth tellurides have been so widely studied is their high degree of tunability. Perhaps these losses could be greatly reduced by tuning the number of charge carriers, their mobility, or other properties using intercalation, substitution, vacancies, or pressure. In particular, it has been shown that pressure has a large impact on the optical properties of CeTe$_3$, as does varying the rare-earth ion, which can be viewed as applying chemical pressure [34]. Another study demonstrated the possibility of altering the Fermi surface and band structure of YTe$_3$ by intercalating Pd atoms in the VdW gap [3]. This would certainly affect the optical properties by altering the available interband transitions, while introducing charge carriers would impact the Drude permittivity. Significant Te vacancies would similarly affect the available band transitions and carrier densities. Substitution of Te atoms with Sb in RTe$_2$ is another possible tuning method that has been experimentally demonstrated [14]. Further study is required to determine the exact effects of these modifications on the full dielectric tensor.

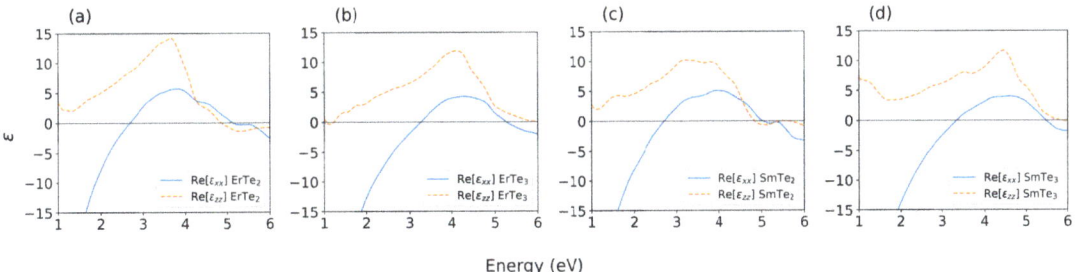

Figure 6. Real parts of the components of the dielectric function parallel (ϵ_{xx}) and perpendicular (ϵ_{zz}) to the layers for (**a**) ErTe$_2$, (**b**) ErTe$_3$, (**c**) SmTe$_2$, and (**d**) SmTe$_3$.

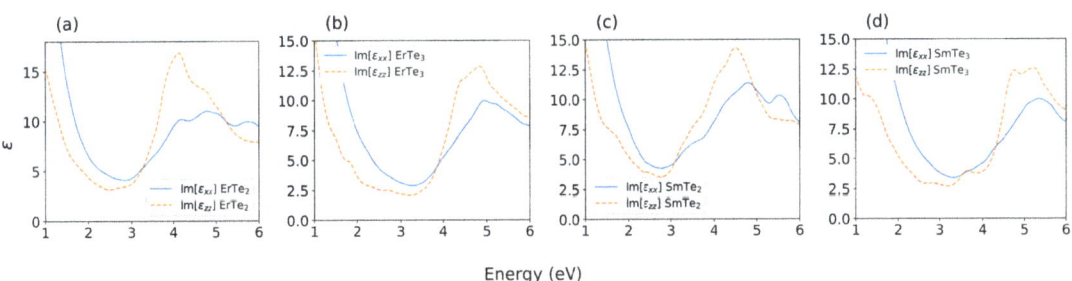

Figure 7. Imaginary parts of the components of the dielectric function parallel (ϵ_{xx}) and perpendicular (ϵ_{zz}) to the layers for (**a**) ErTe$_2$, (**b**) ErTe$_3$, (**c**) SmTe$_2$, and (**d**) SmTe$_3$.

Our calculations predict that all crystals studied in this work (SmTe$_2$, ErTe$_2$, SmTe$_3$, and ErTe$_3$) display hyperbolic dispersion in the infrared to visible frequency ranges, as can be seen in Figure 8. They all exhibited type 1 hyperbolic behavior for some range of energies. This behavior is a consequence of the large anisotropy attending the electronic properties of these materials, with conductivity in the plane of the layers being much greater than that normal to the layers. For a very limited range of frequencies, however, SmTe$_2$ and ErTe$_2$ were also found to exhibit type 2 hyperbolic behavior in the ultraviolet.

This further demonstrates that metallic behavior perpendicular to the layers is much more possible in RTe$_2$, since electrons are not insulated by a VdW gap.

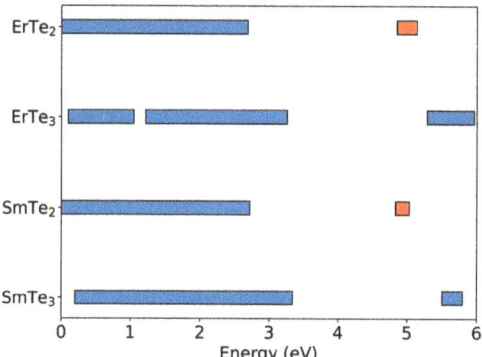

Figure 8. The energy ranges in which RTe crystals exhibit hyperbolic dispersion. The hyperbolic behavior of type 1 is shown in blue and type 2 in red.

In conclusion, we have calculated the particular AFM ground states of several rare-earth tellurides, along with their electronic and optical properties, providing deeper insight into how these properties differ between RTe$_2$ and RTe$_3$. We predict that both classes of materials are hyperbolic in the technologically important visible and infrared range, and suggest several feasible methods for altering the optical properties to suit the needs of the experimenter.

Author Contributions: Conceptualization, R.A.J.; validation, R.A.J.; investigation, R.A.J. and J.G.; resources, R.A.J.; writing—original draft preparation, J.G.; writing—review and editing, R.A.J.; visualization, J.G. All authors have read and agreed to the published version of the manuscript.

Funding: This work has been partially supported by the National Science Foundation with Awards HRD-1547723 and HRD-2112554.

Institutional Review Board Statement: Not applicable.

Data Availability Statement: Data will be made available upon reasonable request.

Conflicts of Interest: The authors declare no conflict of interest.

References

1. Yumigeta, K.; Qin, Y.; Li, H.; Blei, M.; Attarde, Y.; Kopas, C.; Tongay, S. Advances in Rare-Earth Tritelluride Quantum Materials: Structure, Properties, and Synthesis. *Adv. Sci.* **2021**, *8*, 2004762. [CrossRef] [PubMed]
2. Sacchetti, A.; Condron, C.L.; Gvasaliya, S.N.; Pfuner, F.; Lavagnini, M.; Baldini, M.; Toney, M.F.; Merlini, M.; Hanfland, M.; Mesot, J.; et al. Pressure-Induced Quenching of the Charge-Density-Wave State in Rare-Earth Tritellurides Observed by X-Ray Diffraction. *Phys. Rev. B* **2009**, *79*, 201101. [CrossRef]
3. He, J.B.; Wang, P.P.; Yang, H.X.; Long, Y.J.; Zhao, L.X.; Ma, C.; Yang, M.; Wang, D.M.; Shangguan, X.C.; Xue, M.Q.; et al. Superconductivity in Pd-Intercalated Charge-Density-Wave Rare Rarth Poly-Tellurides RETe$_n$. *Supercond. Sci. Technol.* **2016**, *29*, 065018. [CrossRef]
4. Malliakas, C.D.; Kanatzidis, M.G. Charge Density Waves in the Square Nets of Tellurium of AMRETe$_4$ (A = K, Na; M = Cu, Ag; RE = La, Ce). *J. Am. Chem. Soc.* **2007**, *129*, 10675. [CrossRef] [PubMed]
5. Lei, S.; Lin, J.; Jia, Y.; Gray, M.; Topp, A.; Farahi, G.; Klemenz, S.; Gao, T.; Rodolakis, F.; McChesney, J.L.; et al. High Mobility in a Van der Waals Layered Antiferromagnetic Metal. *Sci. Adv.* **2020**, *6*, eaay6407. [CrossRef]
6. Norling, B.K.; Steinfink, H. The Crystal Structure of Neodymium Tritelluride. *Inorg. Chem.* **1966**, *5*, 1488. [CrossRef]
7. Malliakas, C.D.; Kanatzidis, M.G. Divergence in the Behavior of the Charge Density Wave in RETe$_3$ (RE = Rare-Earth Element) with Temperature and RE Element. *J. Am. Chem. Soc.* **2006**, *128*, 12612. [CrossRef]
8. Pardo, M.P.; Flahaut, J.; Domange, L. Chimie Minerale-Les Tellurures des Elements des Terres Rares de Formule Generale MTe$_2$. *C. R. Hebd. Seances L'Academie des Sci.* **1963**, *256*, 953.

9. Sacchetti, A.; Degiorgi, L.; Giamarchi, T.; Ru, N.; Fisher, I.R. Chemical Pressure and Hidden One-Dimensional Behavior in Rare-Earth Tri-Telluride Charge-Density Wave Compounds. *Phys. Rev. B* **2006**, *74*, 125115. [CrossRef]
10. Brouet, V.; Yang, W.L.; Zhou, X.J.; Hussain, Z.; Ru, N.; Shin, K.Y.; Fisher, I.R.; Shen, Z.X. Fermi Surface Reconstruction in the CDW State of CeTe$_3$ Observed by Photoemission. *Phys. Rev. Lett.* **2004**, *93*, 126405. [CrossRef]
11. DiMasi, E.; Foran, B.; Aronson, M.C.; Lee, S. Quasi-Two-Dimensional Metallic Character of Sm$_2$Te$_5$ and SmTe$_3$. *Chem. Mater.* **1994**, *6*, 1867. [CrossRef]
12. Brouet, V.; Yang, W.L.; Zhou, X.J.; Hussain, Z.; Moore, R.G.; He, R.; Lu, D.H.; Shen, Z.X.; Laverock, J.; Dugdale, S.B.; et al. Angle-Resolved Photoemission Study of the Evolution of Band Structure and Charge Density Wave Properties in RTe$_3$ (R = Y, La, Ce, Sm, Gd, Tb, and Dy). *Phys. Rev. B* **2008**, *77*, 235104. [CrossRef]
13. Pariari, A.; Koley, S.; Roy, S.; Singha, R.; Laad, M.S.; Taraphder, A.; Mandal, P. Interplay Between Charge Density Wave Order and Magnetic Field in the Nonmagnetic Rare-Earth Tritelluride LaTe$_3$. *Phys. Rev. B* **2021**, *104*, 155147. [CrossRef]
14. DiMasi, E.; Foran, B.; Aronson, M.C.; Lee, S. Stability of Charge-Density Waves Under Continuous Variation of Band Filling in LaTe$_{2-x}$Sb$_x$ ($0 \leq x \leq 1$). *Phys. Rev. B* **1996**, *54*, 13587. [CrossRef] [PubMed]
15. Shin, K.Y.; Brouet, V.; Ru, N.; Shen, Z.X.; Fisher, I.R. Electronic Structure and Charge-Density Wave Formation in LaTe$_{1.95}$ and CeTe$_{2.00}$. *Phys. Rev. B* **2005**, *72*, 085132. [CrossRef]
16. Kwon, Y.S.; Min, B.H. Anisotropic Transport Properties in RTe$_2$ (R: La, Ce, Pr, Sm and Gd). *Physica B Condens. Matter* **2000**, *281–282*, 120. [CrossRef]
17. Ru, N.; Chu, J.-H.; Fisher, I.R. Magnetic Properties of the Charge Density Wave Compounds RTe$_3$ (R = Y, La, Ce, Pr, Nd, Sm, Gd, Tb, Dy, Ho, Er, and Tm). *Phys. Rev. B* **2008**, *78*, 012410. [CrossRef]
18. Iyeiri, Y.; Okumura, T.; Michioka, C.; Suzuki, K. Magnetic Properties of Rare-Earth Metal Tritellurides RTe$_3$ (R = Ce, Pr, Nd, Gd, Dy). *Phys. Rev. B* **2003**, *67*, 144417. [CrossRef]
19. Shin, Y.S.; Han, C.W.; Min, B.H.; Lee, H.J.; Choi, C.H.; Kim, Y.S.; Kim, D.L.; Kwon, Y.S. Anisotropic Magnetization in RTe$_2$ (R: Ce, Pr, Gd and Sm). *Phys. B Condens. Matter* **2000**, *291*, 225. [CrossRef]
20. Kwon, Y.S.; Park, T.S.; Lee, K.R.; Kim, J.M.; Haga, Y.; Suzuki, T. Transport and Optical Properties of CeTe$_2$. *J. Magn. Magn. Mater.* **1995**, *140–144*, 1173. [CrossRef]
21. Xu, Z.; Ji, S.-H.; Tang, L.; Wu, J.; Li, N.; Cai, X.; Chen, X. Molecular Beam Epitaxy Growth and Electronic Structures of Monolayer GdTe$_3$. *Chin. Phys. Lett.* **2021**, *38*, 077102. [CrossRef]
22. Hong, Y.; Wei, Q.; Liang, X.; Lu, W. Origin and Strain Tuning of Charge Density Wave in LaTe$_3$. *Phys. B Condens. Matter* **2022**, *639*, 413988. [CrossRef]
23. Kresse, G.; Furthmüller, J. Efficient Iterative Schemes for Ab Initio Total-Energy Calculations Using a Plane-Wave Basis Set. *Phys. Rev. B* **1996**, *54*, 11169. [CrossRef] [PubMed]
24. Giannozzi, P.; Baroni, S.; Bonini, N.; Calandra, M.; Car, R.; Cavazzoni, C.; Ceresoli, D.; Chiarotti, G.L.; Cococcioni, M.; Dabo, I.; et al. QUANTUM ESPRESSO: A Modular and Open-Source Software Project for Quantum Simulations of Materials. *J. Phys. Condens. Matter* **2009**, *21*, 395502. [CrossRef] [PubMed]
25. Ferrari, L.; Wu, C.; Lepage, D.; Zhang, X.; Liu, Z. Hyperbolic Metamaterials and Their Applications. *Prog. Quantum Electron.* **2015**, *40*, 1. [CrossRef]
26. Jacob, Z.; Alekseyev, L.V.; Narimanov, E. Optical Hyperlens: Far-Field Imaging Beyond the Diffraction Limit. *Opt. Express* **2006**, *14*, 8247. [CrossRef]
27. Kidwai, O.; Zhukovsky, S.V.; Sipe, J.E. Effective-Medium Approach to Planar Multilayer Hyperbolic Metamaterials: Strengths and Limitations. *Phys. Rev. A* **2012**, *85*, 053842. [CrossRef]
28. Guo, Y.; Cortes, C.L.; Molesky, S.; Jacob, Z. Broadband Super-Planckian Thermal Emission from Hyperbolic Metamaterials. *Appl. Phys. Lett.* **2012**, *101*, 131106. [CrossRef]
29. Guo, Z.; Jiang, H.; Chen, H. Hyperbolic Metamaterials: From Dispersion Manipulation to Applications. *J. Appl. Phys.* **2020**, *127*, 071101. [CrossRef]
30. Korzeb, K.; Gajc, M.; Pawlak, D.A. Compendium of Natural Hyperbolic Materials. *Opt. Express* **2015**, *23*, 25406. [CrossRef]
31. Sun, J.; Zhou, J.; Li, B.; Kang, F. Indefinite Permittivity and Negative Refraction in Natural Material: Graphite. *Appl. Phys. Lett.* **2011**, *98*, 101901. [CrossRef]
32. Nee, T.J. Anisotropic Optical Properties of YBa$_2$Cu$_3$O$_7$. *Appl. Phys.* **1992**, *71*, 6002. [CrossRef]
33. Sun, J.; Litchinitser, N.M.; Zhou, J. Indefinite by Nature: From Ultraviolet to Terahertz. *ACS Photonics* **2014**, *1*, 293. [CrossRef]
34. Sacchetti, A.; Arcangeletti, E.; Perucchi, A.; Baldassarre, L.; Postorino, P.; Lupi, S.; Ru, N.; Fisher, I.R.; Degiorgi, L. Pressure Dependence of the Charge-Density-Wave Gap in Rare-Earth Tritellurides. *Phys. Rev. Lett.* **2007**, *98*, 026401. [CrossRef] [PubMed]
35. Lavagnini, M.; Sacchetti, A.; Degiorgi, L.; Shin, K.Y.; Fisher, I.R. Optical Properties of the Ce and La Ditelluride Charge Density Wave Compounds. *Phys. Rev. B* **2007**, *75*, 205133. [CrossRef]
36. Lavagnini, M.; Sacchetti, A.; Degiorgi, L.; Arcangeletti, E.; Baldassarre, L.; Postorino, P.; Lupi, S.; Perucchi, A.; Shin, K.Y.; Fisher, I.R. Pressure Dependence of the Optical Properties of the Charge-Density-Wave Compound LaTe$_2$. *Phys. Rev. B* **2008**, *77*, 165132. [CrossRef]
37. Blaha, P.; Schwarz, K.; Tran, F.; Laskowski, R.; Madsen, G.K.H.; Marks, L.D. WIEN2k: An APW+lo Program for Calculating the Properties of Solids. *J. Chem. Phys.* **2020**, *152*, 074101. [CrossRef]

38. Perdew, J.P.; Burke, K.; Ernzerhof, M. Generalized Gradient Approximation Made Simple. *Phys. Rev. Lett.* **1996**, *77*, 3865. [CrossRef]
39. Cannon, J.F.; Hall, H.T. High-Pressure, High-Temperature Syntheses of Selected Lanthanide-Tellurium Compounds. *Inorg. Chem.* **1970**, *9*, 1639. [CrossRef]
40. Xu, G.; Wang, J.; Felser, C.; Qi, X.L.; Zhang, S.C. Quantum Anomalous Hall Effect in Magnetic Insulator Heterostructure. *Nano Lett.* **2015**, *15*, 2019. [CrossRef]
41. Kozub, A.L.; Shick, A.B.; Máca, F.; Kolorenč, J.; Lichtenstein, A.I. Electronic Structure and Magnetism of Samarium and Neodymium Adatoms on Free-Standing Graphene. *Phys. Rev. B* **2016**, *94*, 125113. [CrossRef]
42. Shick, A.B.; Shapiro, D.S.; Kolorenc, J.; Lichtenstein, A.I. Magnetic Character of Holmium Atom Adsorbed on Platinum Surface. *Sci. Rep.* **2017**, *7*, 1. [CrossRef] [PubMed]
43. Ashcroft, N.; Mermin, N. *Solid State Phys*, 1st ed.; Harcourt: Fort Worth, TX, USA, 1976; pp. 16–18.
44. Ambrosch-Draxl, C.; Sofo, J.O. Linear Optical Properties of Solids Within the Full-Potential Linearized Augmented Planewave Method. *Comput. Phys. Commun.* **2006**, *175*, 1. [CrossRef]
45. Naik, G.V.; Saha, B.; Liu, J.; Saber, S.M.; Stach, E.A.; Irudayaraj, J.M.K.; Sands, T.D.; Shalaev, V.M.; Boltasseva, A. Epitaxial Superlattices with Titanium Nitride as a Plasmonic Component for Optical Hyperbolic Metamaterials. *Proc. Natl. Acad. Sci. USA* **2014**, *111*, 7546. [CrossRef] [PubMed]

Article

Boundary Effect and Critical Temperature of Two-Band Superconducting FeSe Films

Chenxiao Ye [1], Jiantao Che [2] and Hai Huang [2,*]

[1] School of Nuclear Science and Engineering, North China Electric Power University, Beijing 102206, China
[2] Department of Mathematics and Physics, North China Electric Power University, Beijing 102206, China
* Correspondence: huanghai@ncepu.edu.cn

Abstract: Based on two-band Bogoliubov–de Gennes theory, we study the boundary effect of an interface between a two-gap superconductor FeSe and insulator (or vacuum). New boundary terms are introduced into two-band Ginzburg–Landau formalism, which modifies the boundary conditions for the corresponding order parameters of superconductor. The theory allows for a mean-field calculation of the critical temperature suppression with the decrease in FeSe film thickness. Our numerical results are in good agreement with the experimental data observed in this material.

Keywords: two-band superconductor; boundary term; critical temperature; FeSe film

1. Introduction

The discovery of iron-based materials with a high superconducting transition temperature has triggered great interest for both fundamental studies and practical applications in this field [1]. Among these superconductors, the FeSe system has a simple crystal structure, clean superconducting phase and low toxicity, making it an appealing candidate for studying the superconducting properties of Fe-based compounds [2]. The FeSe layers consist of square lattices of Fe atoms with tetrahedrally coordinated covalent bonds to the Se anions, and the lattice constant perpendicular to the layered plane is about 0.55 nm [3]. The Fermi surface of this compound consists of one electron and one-hole thin cylinders through Shubnikov–de Haas oscillations [4]. At around 100 K, FeSe shows a structural phase transition from tetragonal to orthorhombic without an accompanying magnetic phase transition and becomes superconducting below 8 K [2,5–7]. The picture of two-gap superconductivity has been clearly confirmed by the scanning tunneling microscopy measurements, multiple Andreev reflection spectroscopy, specific heat measurements and other experiments [8–12].

High-quality, superconducting thin films have an important role in applications and basic research of superconductivity. In this respect, preparing high-quality thin-film samples not only satisfies the demands for some measurements of basic physical properties but also provides suitable bases for making tunneling junctions, which determines several important superconducting parameters, such as gap value and paring symmetry [13]. Previous studies examining the thickness dependence of FeSe have been limited to measurements on thin films grown using techniques such as molecular beam expitaxy [14,15], pulsed laser deposition [16] and radio-frequency sputtering [17], all of which require well-optimised growth protocols. An alternative to the growth of thin films is to create devices by mechanical exfoliation of high-quality single crystals. With this method, a series of FeSe superconducting films with thicknesses ranging from 470 to 2.2 nm was obtained on the Si/SiO$_2$ substrate [18]. A dramatic depression of T_c has been observed when the thickness is smaller than 27 nm. Farrar et al. have also observed similar behavior: a sharp decrease in superconductivity occurs at $d \lesssim 25$ nm [19]. One possible origin for this T_c suppression could be the increase in disorder scattering with reducing thickness [19,20]. Another alternative possibility is due to the interaction between FeSe thin film and the

substrate [21]. However, up to now, there is still no consensus on the explanation of the experimental data mentioned above.

In this paper, we propose that the T_c dependence on the film thickness is due to the influence of the boundary effect between the two-band superconductor and the insulator (or vacuum). We first introduce the appropriate boundary conditions for the Ginzburg–Landau (GL) order parameters at the superconductor–insulator interface. We then give a microscopic analysis of these new boundary terms based on two-band Bogoliubov–de Gennes theory. Our theoretical result is consistent with the experimental data of FeSe films, which suggests the boundary effect is an important factor for the understanding of superconducting properties in this iron-based compound.

The rest of this article is structured as follows. In Section 2, we first review two-band Bogoliubov–de Gennes theory and GL equations, and then give a microscopic derivation of proper boundary conditions for the superconductor-insulator interface. In Section 3, we perform the calculations on the film-thickness-dependence of critical temperature for the FeSe compound in the context of GL theory. Finally, we conclude this article in Section 4.

2. Theoretical Scheme

Based on the previous literature [22–31], we can write the Hamiltonian of a two-band superconductor as

$$H = \sum_{i\sigma} c_{i\sigma}^\dagger(r)\hat{h}(r)c_{i\sigma}(r) - \sum_{ii'} g_{ii'} c_{i\uparrow}^\dagger(r) c_{i\downarrow}^\dagger(r) c_{i'\downarrow}(r) c_{i'\uparrow}(r), \tag{1}$$

where $i, i' = 1, 2$ are the band indices and $\sigma = \uparrow, \downarrow$ is the spin index. $\hat{h}(r)$ is the single-particle Hamiltonian of the normal metal, and $g_{ii'}$ are the electron–phonon interaction constants with $g_{12} = g_{21}$.

We can introduce the gap functions as

$$\Psi_i(r) = -\sum_{i'} g_{ii'} \langle c_{i'\downarrow}(r) c_{i'\uparrow}(r) \rangle \tag{2}$$

and transform the Hamiltonian into the mean-field form

$$H_{eff} = \sum_{i\sigma} c_{i\sigma}^\dagger(r)\hat{h}(r)c_{i\sigma}(r) + \sum_i [\Psi_i(r) c_{i\uparrow}^\dagger(r) c_{i\downarrow}^\dagger(r) + \text{H.c.}]. \tag{3}$$

This effective Hamiltonian can be diagonalized by means of the Bogoliubov transformation with b and $b^\dagger$ as the annihilation and creation operators of quasi-particle excitations:

$$c_{i\uparrow}(r) = \sum_k [u_{ik}(r) b_{ik\uparrow} - v_{ik}^*(r) b_{ik\downarrow}^\dagger] \tag{4}$$

and

$$c_{i\downarrow}(r) = \sum_k [u_{ik}(r) b_{ik\downarrow} + v_{ik}^*(r) b_{ik\uparrow}^\dagger] \tag{5}$$

where k is the wave vector. As a result, the effective Hamiltonian can be written as

$$H_{eff} = E_g + \sum_{ik\sigma} E_{ik} b_{ik\sigma}^\dagger b_{ik\sigma}. \tag{6}$$

E_g is the ground state energy, and E_{ik} is the energy of the excitation.

Using the commutator $[c_{i\sigma}(r), H_{eff}]$, together with Equations (4)–(6), we can obtain the Bogoliubov–de Gennes equations for a two-band superconductor [32–35]

$$\begin{pmatrix} \hat{h} & \Psi_i(r) \\ \Psi_i^*(r) & -\hat{h}^* \end{pmatrix} \begin{pmatrix} u_{ik}(r) \\ v_{ik}(r) \end{pmatrix} = E_{ik} \begin{pmatrix} u_{ik}(r) \\ v_{ik}(r) \end{pmatrix}. \tag{7}$$

From Equation (6), we can also obtain $\langle b_{ik\uparrow}^\dagger b_{ik\uparrow}\rangle = f(E_{ik})$ with $f(E_{ik}) = [1+\exp(E_{ik}/k_BT)]^{-1}$. Then, with Equation (2), we can transform the self-consistent gap equations into

$$\Psi_i(r) = \sum_{i'k} g_{ii'} v_{i'k}^*(r) u_{i'k}(r) \times [1 - 2f(E_{i'k})]. \tag{8}$$

In analogy with the single-band case [36], for small gap functions Ψ_i, we can obtain the linearized form of self-consistency conditions from Equations (7) and (8) as

$$\Psi_i(r) = \sum_{i'} \int K_{ii'}(r,r')\Psi_{i'}(r')dr' \tag{9}$$

with the kernel

$$K_{ii'}(r,r') = g_{ii'} k_B T \sum_{kk'} \sum_\omega \frac{\Phi_{i'k}^*(r')\Phi_{i'k'}^*(r')\Phi_{i'k}(r)\Phi_{i'k'}(r)}{(\varepsilon_{i'k} - i\hbar\omega)(\varepsilon_{i'k'} + i\hbar\omega)}. \tag{10}$$

$\Phi_{i'k}(r)$ is defined as the normal-state eigenfunction of the electron; $\hat{h}\Phi_{i'k} = \varepsilon_{i'k}\Phi_{i'k}$. The frequency $\omega = (2\nu + 1)\pi k_B T/\hbar$, where ν is an integer.

With the explicit expressions of the kernels in the bulk system and the addition of nonlinear terms to the gap equations, we can obtain the two-band GL equations from Equation (9) as [27]

$$\alpha_1(T)\Psi_1 + \beta_1|\Psi_1|^2\Psi_1 - \gamma_1\nabla^2\Psi_1 - \epsilon_{12}\Psi_2 = 0 \tag{11}$$

and

$$\alpha_2(T)\Psi_2 + \beta_2|\Psi_2|^2\Psi_2 - \gamma_2\nabla^2\Psi_2 - \epsilon_{12}\Psi_1 = 0 \tag{12}$$

with the GL parameters

$$\alpha_{1,2} = N_{1,2}\left[\frac{\lambda_{22,11}}{\lambda} - \frac{1}{\lambda_{max}} - \ln\left(\frac{T_{c0}}{T}\right)\right], \quad \beta_i = \frac{7\zeta(3)N_i}{16\pi^2(k_B T_{c0})^2}, \tag{13}$$

$$\gamma_i = \frac{7\zeta(3)\hbar^2 N_i v_{Fi}^2}{16\pi^2(k_B T_{c0})^2} \text{ and } \epsilon_{12} = \frac{N_1 \lambda_{12}}{\lambda} = \frac{N_2 \lambda_{21}}{\lambda}. \tag{14}$$

$\lambda_{ii'} = g_{ii'}N_{i'}$ with $N_{i'}$ being the density of states at the Fermi level for each band; $\lambda = \lambda_{11}\lambda_{22} - \lambda_{12}\lambda_{21}$ and $\lambda_{max} = \frac{1}{2}\left[(\lambda_{11} + \lambda_{22}) + \sqrt{(\lambda_{11} - \lambda_{22})^2 + 4\lambda_{12}\lambda_{21}}\right]$ are the determinant and the largest eigenvalue of λ-matrix, respectively. T_{c0} is the bulk critical temperature, and v_{Fi} is the average Fermi velocity for each band.

In the spatially homogeneous case, we can neglect the gradient γ-terms. Equations (11) and (12) will yield the gap equations at $T = T_{c0}$:

$$\begin{pmatrix} \lambda_{11} & \lambda_{12} \\ \lambda_{21} & \lambda_{22} \end{pmatrix}\begin{pmatrix} \Psi_1 \\ \Psi_2 \end{pmatrix} = \lambda_{max}\begin{pmatrix} \Psi_1 \\ \Psi_2 \end{pmatrix}, \tag{15}$$

which obviously give a consistent result.

Meanwhile, we can write down the boundary conditions for the two-band GL theory at the interface between the two-band superconductor and insulator (or vacuum) as

$$\nabla\Psi_i \cdot s|_S = -\sum_{i'} A_{ii'}\Psi_{i'} \tag{16}$$

with S, the boundary coordinate, and $A_{ii'}$, as some constants. From Equation (16), we can see that the ordinary Neumann boundary condition corresponds to $A_{ii'} = 0$. However, we will show that the boundary effect induced by these A-terms is important for the understanding of the T_c suppression in FeSe thin films.

At this stage, we would also like to point out the boundary effect and different interband interactions in multi-band superconductors have already been extensively studied in the literature. Babaev et al. presented a microscopic study on the behavior of the order parameters for two-band superconductors based on the free boundary condition [37]. With the Neumann boundary condition, the stable edge states and the dynamic response of such states to an external applied current have been investigated in the time-dependent Ginzburg–Landau formalism for the two-band mesoscopic superconductors [38]. Aguirre et al. also discussed the effects of different interband interactions on the vortex states by solving the two-band Ginzburg–Landau equations with the Neumann boundary condition [39,40]. However, to explain the suppression of critical temperature with the decrease in film thickness, we need to conduct detailed microscopic analysis and derive the correct boundary terms based on the two-band Bogoliubov–de Gennes theory for the superconductor FeSe.

Now, we try to derive these boundary A-terms in Equation (16) from the two-band Bogoliubov–de Gennes theory. In all cases, we assume that there is no current flowing through the boundary. The equation to be solved reads

$$\Psi_i(s) = \sum_{i'} \int K_{ii'}(s,s') \Psi_{i'}(s') ds' \tag{17}$$

where s measures the normal distance from the boundary. For simplicity, we set the cross-section of the boundary as 1. $K_{ii'}(s,s')$ is defined by Equation (10), and due to the existence of interface $\Psi_i(s)$ will decrease exponentially in the insulating regime.

Following the procedure suggested by de Gennes [36], we suppose that the form of gap functions close to the surface behaves as

$$\Psi_i(s) = \Psi_{i0} + \left(\sum_{i'} A_{ii'} \Psi_{i'0} \right) s \tag{18}$$

with Ψ_{i0} being the gap function at the boundary and $s > 0$ inside the superconductor. It is easy to see that the boundary condition in Equation (16) follows naturally from Equation (18). However, beyond the scale of the coherence length from the boundary, the linear dependence definitely becomes invalid. Ψ_i will then have a negative curvature and reach the BCS value deep in the superconductor.

If we introduce $K^0_{ii'}(s,s')$ as the kernel of gap functions in the bulk metal, we can then transform Equation (17) as

$$\Psi_i(s) - \sum_{i'} \int K^0_{ii'}(s,s') \Psi_{i'}(s') ds'$$
$$= -\sum_{i'} \int [K^0_{ii'}(s,s') - K_{ii'}(s,s')] \Psi_{i'}(s') ds' \equiv -\sum_{i'} H_{ii'}(s). \tag{19}$$

From Equations (11) and (12) with the higher order β-terms omitted, while also noting that $K^0_{ii'}(s,s') = K^0_{ii'}(s-s')$ due to the translational symmetry, we can read out the Laplace transformation of $K^0_{ii'}$ close to the critical temperature as

$$K^0_{ii'}(p) = \frac{\lambda_{ii'}}{\lambda_{\max}} + \frac{\lambda_{ii'} \gamma_{i'}}{N_{i'}} p^2. \tag{20}$$

By plugging Equation (20) into Equation (19), we can get

$$\Psi_i(p) - \sum_{i'} (\lambda_{ii'}/\lambda_{\max}) \Psi_{i'}(p) - \sum_{i'} (\lambda_{ii'} \gamma_{i'}/N_{i'}) p^2 \Psi_{i'}(p) = -\sum_{i'} H_{ii'}(p). \tag{21}$$

$\Psi_i(p)$ and $H_{ii'}(p)$ are the Laplace transformations of $\Psi_i(s)$ and $H_{ii'}(s)$, respectively. Since the first two terms of the left-handed side in Equation (21) can be approximately canceled out according to Equation (15), we have

$$\sum_{i'}(\lambda_{ii'}\gamma_{i'}/N_{i'})p^2\Psi_{i'}(p) = \sum_{i'} H_{ii'}(p). \quad (22)$$

We can see that both sides in Equation (22) take the main contribution from the boundary region. Notice that the Laplace transformation of the gap functions in Equation (18) takes the form

$$\Psi_i(p) = \frac{\Psi_{i0}}{p} + \sum_{i'} \frac{A_{ii'}\Psi_{i'0}}{p^2}. \quad (23)$$

Then, at $p \to 0$, we will obtain from Equation (22)

$$\sum_{i'i''}(\lambda_{ii'}\gamma_{i'}/N_{i'})A_{i'i''}\Psi_{i''0} = \sum_{i'} H_{ii'}(p=0). \quad (24)$$

According to de Gennes' analysis [36,41], from the sum rules

$$\int K^0_{ii'}(s,s')ds' = \frac{\lambda_{ii'}}{\lambda_{max}} \quad \text{and} \quad \int K_{ii'}(s,s')ds' = \frac{\lambda_{ii'}N_{i'}(s)}{\lambda_{max}N_{i'}} \quad (25)$$

with $N_{i'}(s)$ as the local density of states at the Fermi surface, we can write the Laplace transformation of the kernel difference at $p \to 0$ as

$$H_{ii'}(p=0) = \int H_{ii'}(s)ds = \frac{\lambda_{ii'}\Psi_{i'0}}{\lambda_{max}} \int \frac{\Psi_{i'}(s)}{\Psi_{i'0}}\left[1 - \frac{N_{i'}(s)}{N_{i'}}\right]ds. \quad (26)$$

$\Psi_{i'}(s)/\Psi_{i'0}$ approaches zero in the insulating region, and is of the order of one in the metallic region. $N_{i'}(s)/N_{i'}$ also passes from $0 \to 1$ a few atoms from the boundary. Therefore, the integrand in Equation (26) is nonvanishing only in a width of the order of the lattice constant a. We can then estimate $H_{ii'}(p=0)$ as

$$H_{ii'}(p=0) = \frac{\lambda_{ii'}a}{\lambda_{max}}\Psi_{i'0}. \quad (27)$$

By combining Equation (24) with Equation (27), we can finally obtain

$$A_{ii} = \frac{N_i a}{\gamma_i \lambda_{max}} \quad \text{and} \quad A_{12} = A_{21} = 0. \quad (28)$$

With these formulae, we successfully demonstrate the microscopic origin of boundary conditions in Equation (16).

Based on two-band GL theory, we can write the supercurrent at the boundary S as

$$J_S = -ie\hbar \sum_{i=1}^{2} \frac{1}{m_i}[\Psi_i^*(\nabla\Psi_i \cdot s)|_S - \Psi_i(\nabla\Psi_i^* \cdot s)|_S]. \quad (29)$$

According to our boundary conditions, we have the supercurrent

$$J_S = -\frac{ie\hbar}{m_1}[\Psi_1^*(-A_{11}\Psi_1) - \Psi_1(-A_{11}\Psi_1^*)] - \frac{ie\hbar}{m_2}[\Psi_2^*(-A_{22}\Psi_2) - \Psi_2(-A_{22}\Psi_2^*)] = 0, \quad (30)$$

which definitely gives a consistent result.

3. Critical Temperature of FeSe Films in Ginzburg–Landau Theory

In this section, we try to understand the film-thickness-dependence of T_c for FeSe based on the boundary effect mentioned above. We suppose that the film extends from $z = -d/2$ to $z = d/2$ and the film thickness is d.

From Equations (11) and (12), two-band GL equations can be written as

$$\begin{pmatrix} \hat{H}_{11} & \hat{H}_{12} \\ \hat{H}_{21} & \hat{H}_{22} \end{pmatrix} \begin{pmatrix} \Psi_1(r) \\ \Psi_2(r) \end{pmatrix} = 0 \qquad (31)$$

with

$$\hat{H}_{ii} = -\gamma_i \nabla^2 + \alpha_i(T) \qquad (32)$$

and

$$\hat{H}_{12} = \hat{H}_{21} = -\epsilon_{12}. \qquad (33)$$

Note that close to the critical temperature, the magnitudes of order parameters are small, and the higher order β-terms can be neglected.

Similarly to the single-band case [41], we set the form of gap functions for the superconducting film as

$$\begin{pmatrix} \Psi_1(z) \\ \Psi_2(z) \end{pmatrix} = \begin{pmatrix} \eta_1 \cos(k_1 z) \\ \eta_2 \cos(k_2 z) \end{pmatrix} \qquad (34)$$

with η_i as the constant. Then, from the boundary conditions

$$\left. \frac{d\Psi_i}{dz} \right|_{z=\pm d/2} = \mp A_{ii} \Psi_i, \qquad (35)$$

we have k_i, satisfying

$$k_i \tan\left(\frac{k_i d}{2}\right) = A_{ii}. \qquad (36)$$

Let us introduce

$$H_{ii'} = \langle \Psi_i | \hat{H}_{ii'} | \Psi_{i'} \rangle = \int_{-d/2}^{d/2} \Psi_i(z) \hat{H}_{ii'} \Psi_{i'}(z) dz. \qquad (37)$$

We can transform Equation (31) into

$$\begin{pmatrix} H_{11} & H_{12} \\ H_{21} & H_{22} \end{pmatrix} \begin{pmatrix} \eta_1 \\ \eta_2 \end{pmatrix} = 0 \qquad (38)$$

with

$$H_{ii} = \left[\gamma_i k_i^2 + \alpha_i(T) \right] \left[\frac{d}{2} + \frac{\sin(k_i d)}{2 k_i} \right] \qquad (39)$$

and

$$H_{12} = H_{21} = -\epsilon_{12} \left[\frac{\sin(\frac{k_1 d}{2} + \frac{k_2 d}{2})}{k_1 + k_2} + \frac{\sin(\frac{k_1 d}{2} - \frac{k_2 d}{2})}{k_1 - k_2} \right]. \qquad (40)$$

The critical temperature of the two-band superconducting film will be determined by the condition

$$H_{11} H_{22} - H_{12} H_{21} = 0 \qquad (41)$$

at $T = T_c$, which can be explicitly written as

$$\prod_{i=1,2} \left[\gamma_i k_i^2 + \alpha_i(T_c) \right] \left[\frac{d}{2} + \frac{\sin(k_i d)}{2 k_i} \right] = \epsilon_{12}^2 \left[\frac{\sin(\frac{k_1 d}{2} + \frac{k_2 d}{2})}{k_1 + k_2} + \frac{\sin(\frac{k_1 d}{2} - \frac{k_2 d}{2})}{k_1 - k_2} \right]^2. \qquad (42)$$

For the two-band superconductor FeSe, we have $T_{c0} \approx 8$ K [5] and the average lattice constant $a \approx 0.55$ nm [3]. The density of states at the Fermi level for each band are $N_1 = 0.30$ and $N_2 = 0.65$ eV^{-1}, respectively, [8,42]. From the numerical work in reference [43], we can get $\lambda_{11} = g_{11} N_1 = 0.30$ and the ratio $g_{11}:g_{12}:g_{22} = 1:0.30:0.37$. Then, we have $\lambda = 0.054$, $\lambda_{max} = 0.41$ and $\epsilon_{12} = 1.1$ eV^{-1}. With the average Fermi velocities $v_{F1} = 3.7$ and $v_{F2} = 4.4$ in units of 10^{13} nm·s^{-1} [44], we get $\gamma_1 = 20$ nm^2·eV^{-1} and $\gamma_2 = 61$ nm^2·eV^{-1} from Equation (14). Then, from Equation (28), we can obtain the characteristic length scales $A_{11} = (50 \text{ nm})^{-1}$ and $A_{22} = (70 \text{ nm})^{-1}$. For a given film thickness d, we first get k_i from Equation (36). By plugging it into Equation (42), the critical temperature T_c as a function of d can be calculated numerically and then plotted in Figure 1. It is shown that the critical temperature keeps gradually decreasing with the decreases in d and $T_c \approx 3$ K when the film thickness is reduced to $d = 3$ nm. From Figure 1, we can see that our theoretical results fit the experimental data on Si/SiO$_2$ substrate well.

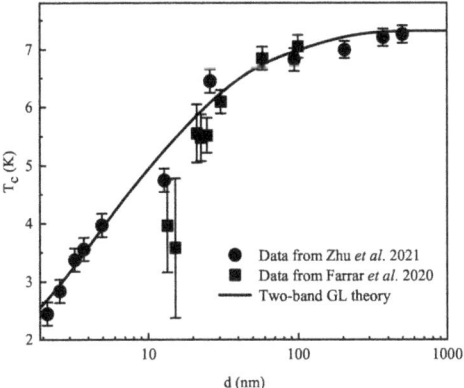

Figure 1. The critical temperature as a function of FeSe film thickness. The experimental data are taken from Ref. [18] (circles) and Ref. [19] (squares) respectively.

At this point, we can also discuss the single-band superconducting system of niobium [45,46], which shows similar behavior to FeSe. The critical temperature of Nb films gradually decreases with a reduction in the thickness from 300 to about 50 nm. A much stronger dependence of the critical temperature on thickness is observed for films thinner than 50 nm. We would also like to apply our theoretical scheme with the similar boundary condition $\nabla \Psi \cdot s|_S = -A \Psi$ to understand this physical property for single-band superconductor Nb. With the lattice constant $a = 0.33$ nm, the Fermi velocity $v_F = 3.9$ in units of 10^{13} nm·s^{-1}, the density of states at the Fermi level $N = 1.5$ eV^{-1} and the dimensionless interaction parameter $\lambda = 0.32$ for niobium [47,48], we can obtain $\gamma = \frac{7\zeta(3)\hbar^2 N v_F^2}{16\pi^2(k_B T_{c0})^2} = 80$ nm^2·eV^{-1} with the bulk critical temperature $T_{c0} = 9.4$ K and $A = \frac{Na}{\gamma\lambda} = (52 \text{ nm})^{-1}$. Thus, with this characteristic length scale, our theoretical scenario can qualitatively explain the strong T_c suppression around 50 nm in the single-band superconducting Nb films.

4. Conclusions

In conclusion, we introduced the appropriate boundary conditions in two-band GL theory at the interface between two-gap superconductor and insulator (or vacuum). We also gave a microscopic derivation of these boundary terms based on two-band Bogoliubov–de Gennes formalism. For the two-band superconductor FeSe, we obtained the characteristic length scales of the boundary effect as 50 and 70 nm. The theory can perfectly explain the dramatic suppression of T_c when the film thickness is reduced to the same order of these length scales. Our investigation thus suggests that the boundary effect induced by these new terms may play an important role in the research of some iron-based superconducting films.

Author Contributions: Conceptualization, C.Y., J.C. and H.H.; methodology, C.Y., J.C. and H.H.; software, C.Y.; validation, C.Y., J.C. and H.H.; formal analysis, C.Y., J.C. and H.H.; investigation, C.Y., J.C. and H.H.; resources, H.H.; data curation, C.Y., J.C. and H.H.; writing—original draft preparation, C.Y.; writing—review and editing, C.Y., J.C. and H.H.; visualization, C.Y.; supervision, J.C. and H.H.; project administration, H.H. All authors have read and agreed to the published version of the manuscript.

Funding: This research received no external funding.

Data Availability Statement: Not applicable.

Conflicts of Interest: The authors declare no conflict of interest.

References

1. Kamihara, Y.; Watanabe, T.; Hirano, M.; Hosono, H. Iron-Based layered superconductor La[$O_{1-x}F_x$]FeAs (x=0.05-0.12) with T_c = 26 K. *J. Am. Chem. Soc.* **2008**, *130*, 3296–3297. [CrossRef] [PubMed]
2. Hsu, F.C.; Luo, J.Y.; Yeh, K.W.; Chen, T.K.; Huang, T.W.; Wu, P.M.; Lee, Y.C.; Huang, Y.L.; Chu, Y.Y.; Yan, D.C.; et al. Superconductivity in the PbO-type structure α-FeSe. *Proc. Natl. Acad. Sci. USA* **2008**, *105*, 14262. [CrossRef] [PubMed]
3. Schneider, R.; Zaitsev, A.G.; Fuchs, D.; Löhneysen, H.V. Excess conductivity and Berezinskii-Kosterlitz-Thouless transition in superconducting FeSe thin films. *J. Phys. Condens. Matter* **2014**, *26*, 455701. [CrossRef] [PubMed]
4. Terashima, T.; Kikugawa, N.; Kiswandhi, A.; Choi, E.S.; Brooks, J.S.; Kasahara, S.; Watashige, T.; Ikeda, H.; Shibauchi, T.; Matsuda, Y.; et al. Anomalous Fermi surface in FeSe seen by Shubnikov-de Haas oscillation measurements. *Phys. Rev. B* **2014**, *90*, 144517. [CrossRef]
5. McQueen, T.M.; Williams, A.J.; Stephens, P.W.; Tao, J.; Zhu, Y.; Ksenofontov, V.; Casper, F.; Felser, C.; Cava, R.J. Tetragonal-to-orthorhombic structural phase transition at 90 K in the superconductor $Fe_{1.01}$Se. *Phys. Rev. Lett.* **2009**, *103*, 057002. [CrossRef]
6. McQueen, T.M.; Huang, Q.; Ksenofontov, V.; Felser, C.; Xu, Q.; Zandbergen, H.; Hor, Y.S.; Allred, J.; Williams, A.J.; Qu, D.; et al. Extreme sensitivity of superconductivity to stoichiometry in $Fe_{1+\delta}$Se. *Phys. Rev. B* **2009**, *79*, 014522. [CrossRef]
7. Watson, M.D.; Kim, T.K.; Haghighirad, A.A.; Davies, N.R.; McCollam, A.; Narayanan, A.; Blake, S.F.; Chen, Y.L.; Ghannadzadeh, S.; Schofield, A.J.; et al. Emergence of the nematic electronic state in FeSe. *Phys. Rev. B* **2015**, *91*, 155106. [CrossRef]
8. Wang, Q.Y.; Li, Z.; Zhang, W.H.; Zhang, Z.C.; Zhang, J.S.; Li, W.; Ding, H.; Ou, Y.B.; Deng, P.; Chang, K.; et al. Interface-induced high-temperature superconductivity in single unit-cell FeSe films on $SrTiO_3$. *Chin. Phys. Lett.* **2012**, *29*, 037402. [CrossRef]
9. Chareev, D.; Osadchii, E.; Kuzmicheva, T.; Lin, J.Y.; Kuzmichev, S.; Volkova, O.; Vasiliev, A. Single crystal growth and characterization of tetragonal $FeSe_{1-x}$ superconductors. *Cryst. Eng. Commun.* **2013**, *15*, 1989. [CrossRef]
10. Lin, J.Y.; Hsieh, Y.S.; Chareev, D.A.; Vasiliev, A.N.; Parsons, Y.; Yang, H.D. Coexistence of isotropic and extended s-wave order parameters in FeSe as revealed by low-temperature specific heat. *Phys. Rev. B* **2011**, *84*, 220507(R). [CrossRef]
11. Schneider, R.; Zaitsev, A.G.; Fuchs, D.; Hott, R. Anisotropic field dependence of the electronic transport in superconducting FeSe thin films. *Supercond. Sci. Technol.* **2020**, *33*, 075011. [CrossRef]
12. Sun, Y.; Zhang, W.H.; Xing, Y.; Li, F.S.; Zhao, Y.F.; Xia, Z.C.; Wang, L.L.; Ma, X.C.; Xue, Q.K.; Wang, J. High temperature superconducting FeSe films on $SrTiO_3$ substrates. *Sci. Rep.* **2014**, *4*, 6040. [CrossRef] [PubMed]
13. Chen, T.K.; Luo, J.Y.; Ke, C.T.; Chang, H.H.; Huang, T.W.; Yeh, K.W.; Chang, C.C.; Hsu, P.C.; Wu, C.T.; Wang, M.J.; et al. Low-temperature fabrication of superconducting FeSe thin films by pulsed laser deposition. *Thin Solid Films* **2010**, *519*, 1540. [CrossRef]
14. Tan, S.Y.; Zhang, Y.; Xia, M.; Ye, Z.Y.; Chen, F.; Xie, X.; Peng, R.; Xu, D.F.; Fan, Q.; Xu, H.C.; et al. Interface-induced superconductivity and strain-dependent spin density waves in $FeSe/SrTiO_3$ thin films. *Nat. Mater.* **2013**, *12*, 634. [CrossRef] [PubMed]
15. Wang, Q.Y.; Zhang, W.H.; Zhang, Z.C.; Sun, Y.; Xing, Y.; Wang, Y.Y.; Wang, L.L.; Ma, X.C.; Xue, Q.K.; Wang, J. Thickness dependence of superconductivity and superconductor-insulator transition in ultrathin FeSe films on $SrTiO_3$ (001) substrate. *2D Mater.* **2015**, *2*, 044012. [CrossRef]
16. Nabeshima, F.; Imai, Y.; Hanawa, M.; Tsukada, I.; Maeda, A. Enhancement of the superconducting transition temperature in FeSe epitaxial thin films by anisotropic compression. *Appl. Phys. Lett.* **2013**, *103*, 172602. [CrossRef]
17. Schneider, R.; Zaitsev, A.G.; Fuchs, D.; Löhneysen, H.V. Superconductor-insulator quantum phase transition in disordered FeSe thin films. *Phys. Rev. Lett.* **2012**, *108*, 257003. [CrossRef]
18. Zhu, C.S.; Lei, B.; Sun, Z.L.; Cui, J.H.; Shi, M.Z.; Zhuo, W.Z.; Luo, X.G.; Chen, X.H. Evolution of transport properties in FeSe thin flakes with thickness approaching the two-dimensional limit. *Phys. Rev. B* **2021**, *104*, 024509. [CrossRef]
19. Farrar, L.S.; Bristow, M.; Haghighirad, A.A.; McCollam, A.; Bending, S.J.; Coldea, A.I. Suppression of superconductivity and enhanced critical field anisotropy in thin flakes of FeSe. *NPJ Quantum Mater.* **2020**, *5*, 29. [CrossRef]

20. Böhmer, A.E.; Taufour, V.; Straszheim, W.E.; Wolf, T.; Canfield, P.C. Variation of transition temperatures and residual resistivity ratio in vapor-grown FeSe. *Phys. Rev. B* **2016**, *94*, 024526.
21. Phan, G.N.; Nakayama, K.; Sugawara, K.; Sato, T.; Urata, T.; Tanabe, Y.; Tanigaki, K.; Nabeshima, F.; Imai, Y.; Maeda, A.; et al. Effects of strain on the electronic structure, superconductivity, and nematicity in FeSe studied by angle-resolved photoemission spectroscopy. *Phys. Rev. B* **2017**, *95*, 224507. [CrossRef]
22. Suhl, H.; Matthias, B.T.; Walker, L.R. Bardeen-Cooper-Schrieffer theory of superconductivity in the case of overlapping bands. *Phys. Rev. Lett.* **1959**, *3*, 552–554. [CrossRef]
23. Moskalenko, V.A. Superconductivity in metals with overlapping energy bands. *Fiz. Metal. Metalloved* **1959**, *8*, 2518–2520.
24. Tilley, D.R. The Ginzburg-Landau equations for pure two band superconductors. *Proc. Phys. Soc.* **1964**, *84*, 573. [CrossRef]
25. Tilley, D.R. The Ginsburg-Landau equations for anisotropic alloys. *Proc. Phys. Soc.* **1965**, *86*, 289. [CrossRef]
26. Gurevich, A. Enhancement of the upper critical field by nonmagnetic impurities in dirty two-gap superconductors. *Phys. Rev. B* **2003**, *67*, 184515. [CrossRef]
27. Zhitomirsky, M.E.; Dao, V.H. Ginzburg-Landau theory of vortices in a multigap superconductor. *Phys. Rev. B* **2004**, *69*, 054508. [CrossRef]
28. Dao, V.H.; Zhitomirsky, M.E. Anisotropy of the upper critical field in MgB$_2$: The two-gap Ginzburg-Landau theory. *Eur. Phys. J. B* **2005**, *44*, 183. [CrossRef]
29. Gurevich, A. Limits of the upper critical field in dirty two-gap superconductors. *Physica C* **2007**, *456*, 160. [CrossRef]
30. Silaev, M.; Babaev, E. Microscopic theory of type-1.5 superconductivity in multiband systems. *Phys. Rev. B* **2011**, *84*, 094515. [CrossRef]
31. Silaev, M.; Babaev, E. Microscopic derivation of two component Ginzburg-Landau model and conditions of its applicability in two-band systems. *Phys. Rev. B* **2012**, *85*, 134514. [CrossRef]
32. Zhang, L.F.; Covaci, L.; Milošević, M.V.; Berdiyorov, G.R.; Peeters, F.M. Unconventional vortex states in nanoscale superconductors due to shape-induced resonances in the inhomogeneous cooper-pair condensate. *Phys. Rev. Lett.* **2012**, *109*, 107001. [CrossRef] [PubMed]
33. Zhang, L.F.; Covaci, L.; Milošević, M.V.; Berdiyorov, G.R.; Peeters, F.M. Vortex states in nanoscale superconducting squares: The influence of quantum confinement. *Phys. Rev. B* **2013**, *88*, 144501. [CrossRef]
34. Zhang, L.F.; Becerra, V.F.; Covaci, L.; Milošević, M.V. Electronic properties of emergent topological defects in chiral *p*-wave superconductivity. *Phys. Rev. B* **2016**, *94*, 024520. [CrossRef]
35. Zhang, L.F.; Covaci, L.; Milošević, M.V. Topological phase transitions in small mesoscopic chiral *p*-wave superconductors. *Phys. Rev. B* **2017**, *96*, 224512. [CrossRef]
36. de Gennes, P.G. *Superconductivity of Metals and Alloys*; Westview Press: New York, NY, USA, 1966.
37. Benfenati, A.; Samoilenka, A.; Babaev, E. Boundary effects in two-band superconductors. *Phys. Rev. B* **2021**, *103*, 144512. [CrossRef]
38. Gonçalves, W.C.; Sardella, E.; Becerra, V.F.; Milošević, M.V.; Peeters, F.M. Numerical solution of the time dependent Ginzburg-Landau equations for mixed (*d*+*s*)-wave superconductors. *J. Math. Phys.* **2014**, *55*, 041501. [CrossRef]
39. Aguirre, C.; Martins, Q.D.; Barba-Ortega, J. Vortices in a superconducting two-band disk: Role of the Josephson and bi-quadratic coupling. *Physica C* **2021**, *581*, 1353818. [CrossRef]
40. Aguirre, C.; de Arruda, A.; Faúndez, J.; Barba-Ortega, J. ZFC process in 2+1 and 3+1 multi-band superconductor. *Physica B* **2021**, *615*, 413032. [CrossRef]
41. Ketterson, J.B.; Song, S.N. *Superconductivity*; Cambridge University Press: Cambridge, UK, 1999.
42. Subedi, A.; Zhang, L.J.; Singh, D.J.; Du, M.H. Density functional study of FeS, FeSe, and FeTe: Electronic structure, magnetism, phonons, and superconductivity. *Phys. Rev. B* **2008**, *78*, 134514. [CrossRef]
43. Chen, Y.J.; Zhu, H.P.; Shanenko, A.A. Interplay of Fermi velocities and healing lengths in two-band superconductors. *Phys. Rev. B* **2020**, *101*, 214510. [CrossRef]
44. Tamai, A.; Ganin, A.Y.; Rozbicki, E.; Bacsa, J.; Meevasana, W.; King, P.D.C.; Caffio, M.; Schaub, R.; Margadonna, S.; Prassides, K.; et al. Strong electron correlations in the normal state of the iron-based FeSe$_{0.42}$Te$_{0.58}$ superconductor observed by angle-resolved photoemission spectroscopy. *Phys. Rev. Lett.* **2010**, *104*, 097002. [CrossRef] [PubMed]
45. Ilin, K.S.; Vitusevich, S.A.; Jin, B.B.; Gubin, A.I.; Klein, N.; Siegel, M. Peculiarities of the thickness dependence of the superconducting properties of thin Nb films. *Physica C* **2004**, *408–410*, 700. [CrossRef]
46. Gubin, A.I.; Ilin, K.S.; Vitusevich, S.A.; Siegel, M.; Klein, N. Dependence of magnetic penetration depth on the thickness of superconducting Nb thin films. *Phys. Rev. B* **2005**, *72*, 064503. [CrossRef]
47. Chen, L.Z.; Wang, X.C.; Wen, Y.H.; Zhu, Z.Z. Jahn-Teller effect in Nb planar atomic sheet. *Acta Phys. Sin.* **2007**, *56*, 2920. [CrossRef]
48. Li, C.Z.; Li, C.; Wang, L.X.; Wang, S.; Liao, Z.M.; Brinkman, A.; Yu, D.P. Bulk and surface states carried supercurrent in ballistic Nb-Dirac semimetal Cd$_3$As$_2$ nanowire-Nb junctions. *Phys. Rev. B* **2018**, *97*, 115446. [CrossRef]

Disclaimer/Publisher's Note: The statements, opinions and data contained in all publications are solely those of the individual author(s) and contributor(s) and not of MDPI and/or the editor(s). MDPI and/or the editor(s) disclaim responsibility for any injury to people or property resulting from any ideas, methods, instructions or products referred to in the content.

Article

First-Principles Calculations of Structural and Mechanical Properties of Cu–Ni Alloys

Yun Wei [1], Ben Niu [1], Qijun Liu [1], Zhengtang Liu [2] and Chenglu Jiang [3,*]

[1] School of Physical Science and Technology, Southwest Jiaotong University, Chengdu 610031, China
[2] State Key Laboratory of Solidification Processing, Northwestern Polytechnical University, Xi'an 710072, China
[3] College of Water Conservancy and Hydropower Engineering, Sichuan Agricultural University, Ya'an 625014, China
* Correspondence: juul@sicau.edu.cn

Abstract: Nanostructured Cu–Ni alloys have become the focus of public attention due to their better corrosion resistance and high hardness in experimental measurements. First-principles calculation based on the density functional theory (DFT) has been confirmed as an effective tool and used to illustrate the mechanical properties of these alloys. In this paper, the DFT has been employed to calculate the mechanical properties of Cu–Ni alloys, including bulk modulus, shear modulus, Young's modulus, anisotropic index, Poisson's ratio, average velocity, and B/G. We find that the Ni-rich Cu–Ni alloys have relatively higher mechanical parameters, and the Cu-rich alloys have smaller mechanical parameters, which is consistent with previous experiments. This provides an idea for us to design alloys to improve alloy strength.

Keywords: first-principles calculations; Cu–Ni alloys; mechanical properties

1. Introduction

Copper and copper-based alloys are commonly used in manufacturing, meaning they are indispensable metals in society. Cu is known to have excellent electrical conductivity, good ductility, and high thermal conductivity, which is an advantage when it is deformed in terms of the geometry structure of devices and microwires [1]. In order to improve the hardness of Cu to meet certain applications, researchers suggest reliable mechanisms that cause mechanical hardening: (1) solid solution strengthening, (2) precipitation hardening [2].

One of the most attractive Cu-based alloys are Cu–Ni binary compounds. Cu–Ni alloys are considered to possess good hardness and corrosion resistance [2–4]. Some Cu–Ni alloys have good electrocatalytic [5] and electrical [6] properties in conducting electricity. Other interesting characteristics of Cu–Ni alloys, such as thermal conductivity reduction in nanostructuration [4] and ferromagnetic features in Ni-rich Cu–Ni deposits [7], are revealed. The deposited films of nanostructured Cu-rich Cu–Ni have better corrosion protection than pure Cu, and films have a high hardness [2]. The mechanical properties are widely revealed in experiments for nanostructured Cu–Ni films; they find that Young's modulus increases with the increasing Ni concentration, and Young's modulus decreases with the decreasing Cu concentration [8–10].

Recently, the Cu–X alloys for Cu–Al, Cu–Sc, Cu–Zn, and Cu–Ce were discussed in formation energy by VASP, which shows that Cu–Ce displays high mechanical energy and strong bonds [11]. The charge density difference and Mulliken population of Mg–Gd–Cu alloys were calculated by first-principles calculations, where the Cu–Gd terminated interface is probably the most stable interface [12]. Cu–Zn$_4$ alloy has a high Young's modulus with lower elastic anisotropy, which is rationally recommended for improving the stiffness of Al–Cu–Zn alloys based on first-principles calculations [13]. It can be seen that the first-principles calculation is one of the most important tools to compute mechanical properties, aiming to improve stiffness. Cu–Ni-based alloys have high elasticity and high

electrical conductivity [14–16], with the addition of Co improving the conductivity [14] and the addition of Sn enhancing the mechanical stability of alloys [15,16]. The high stiffness and corrosion protection found in experiments with Cu–Ni alloys has brought them wide attention. Those features of the mechanical properties of Cu–Ni alloys can also be obtained from calculations, but have been less reported previously.

Therefore, we consider first-principles calculations to discuss the mechanical properties of Cu–Ni alloys in nanostructuration, trying to support and promote the understanding of these alloys.

2. Computational Details

First-principles calculations based on density functional theory (DFT) are executed by the CASTEP code [17], where the GGA-PBE is chosen as the exchange-correlation energy functional and potential [18]. Total energy for cutoff is set to 440 eV to ensure the energy tolerances. K-point mesh for Monkhorst–Pack grid is set to $12 \times 12 \times 12$; the actual spacing for mesh is 0.004 nm^{-1}. In geometry optimization processing, the detailed convergence tolerance for atom is evinced by setting energy to 5.0×10^{-6} eV/atom, max. force to 0.01 eV/Å, max. stress to 0.02 GPa, and max. displacement to 5.0×10^{-5} nm, where the max. iteration is 100 to ensure the energy convergence. The Broyden–Fletcher–Goldfarb–Shanno (BFGS) method is applied to mathematical algorithm for convergence calculation [19].

As shown in Figure 1, three Cu–Ni alloys are implanted in this paper, containing $Cu_{0.5}Ni_{0.5}$, $Ni_{0.92}Cu_{0.08}$, and $Cu_{0.95}Ni_{0.05}$ with same space group $Fm\bar{3}m$. Red points for k-point and green dash for Brillouin edge are marked, respectively. The vectors demonstrate the direction in reciprocal space by $\vec{g}1$, $\vec{g}2$, $\vec{g}3$. The results of geometry optimization are listed in Table 1. It can be seen that the large Cu concentration has the larger lattice parameters, and the small atom percentage of Ni concentration has the smaller lattice parameters, which are consistent with experiments. Lattice parameters error is less than 2% compared with XRD data [2,9,10]. Figure 2 shows the simulated XRD patterns of Cu–Ni alloys based on pseudo-Voigt peak shape profile [20], where the peak broadening accounting for instrument broadening is described by Caglioti equation [21] by the U = 0.05, V = 0.002, and W = 0.002; the [UVW] is the direction of incident of the electron beam. It can be seen that the XRD patterns of Cu–Ni alloys are similar; the peak for (111) and (200) reflections shift towards larger angles as the Ni contents increase. The peak position and peak tendency for Cu–Ni alloys are in great agreement with experiments [2,9,10], conforming to the observed particle shapes and crystallite sizes. Hence, these structures are considered to further compute the mechanical properties with GGA-PBE functional.

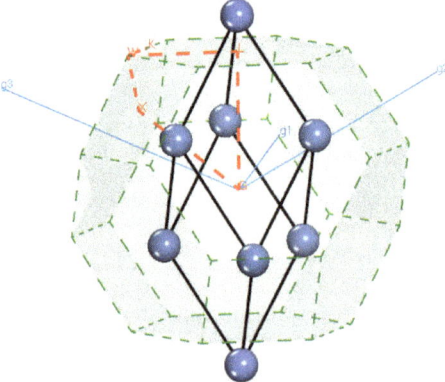

Figure 1. Primitive cell of Cu–Ni; the blue atom corresponds to Cu and Ni; the contents of atom are $Cu_{0.5}Ni_{0.5}$, $Ni_{0.92}Cu_{0.08}$, and $Cu_{0.95}Ni_{0.05}$. Red line and point refer to integration path and k-point in Brillouin zone. $\vec{g}1$, $\vec{g}2$, $\vec{g}3$ are reciprocal vector.

Table 1. Optimized lattice parameter for Cu–Ni alloys with experiments [2,9,10].

Composition	Lattice Parameter a (nm)	Density (g·cm^{-3})	Volume (nm^3)
$Cu_{0.08}Ni_{0.92}$	0.35302	8.9223	0.010999
$Cu_{0.5}Ni_{0.5}$	0.35557	9.0315	0.011239
$Cu_{0.95}Ni_{0.05}$	3.6174	8.8827	0.011834
$Cu_{0.97}Ni_{0.03}$	0.35836 [2]		
$Cu_{0.46}Ni_{0.54}$	0.3560 2 [2]		
$Cu_{0.55}N_{0.45}$	0.3561 [9]		
$Cu_{0.13}Ni_{0.87}$	0.3534 [9]		
$Cu_{100}Ni_0$	0.36148 [10]		
Cu_0Ni_{100}	0.35232 [10]		

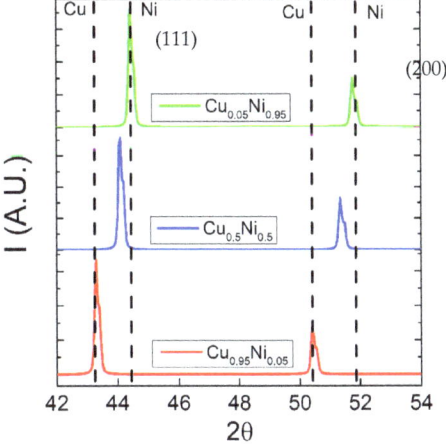

Figure 2. Simulated XRD patterns of Cu–Ni alloy crystal in the 2θ = 42–54° range. The dashed lines indicate the position of the (111) and (200) fcc reflection for pure copper and pure nickel, the copper (Cu) is used for anode type, λ = 0.1540562 nm.

3. Mechanical Properties

The elastic constants are important to understand the physical properties and engineering applications [22–24]. Calculated elastic constant $C_{\rho\sigma}$ can be solved by energy density u with tensors $s_\rho s_\sigma$, which was first noted by Born and Huang [25]:

$$u = \frac{1}{2}\sum_{\rho\sigma} C_{\rho\sigma} s_\rho s_\sigma \quad (1)$$

where the ρ, σ are tensor notations in Voigt's way. Then, the famous Hooke's law can be listed:

$$S_\rho = \frac{\partial u}{\partial s_\rho} = \sum_\sigma C_{\rho\sigma} s_\sigma \quad (2)$$

The S_ρ is elastic stress. When the elastic constants are obtained, the mechanical properties of the crystal can be solved by the elastic constant matrix. The Voigt, Reuss, and Hill [26–28] methods are popular approaches for mechanical properties including bulk modulus B, shear modulus G, average velocity v_m, Young's modulus E, ratio B/G, and Poisson's ratio v. In the cubic system, the equation can be simple:

$$B_V = \frac{C_{11} + 2C_{12}}{3} \quad (3)$$

$$G_V = \frac{C_{11} - C_{12} + 3C_{44}}{5} \tag{4}$$

$$B_R = \frac{1}{3S_{11} + 6S_{12}} \tag{5}$$

Average value for Hill's way is widely used to describe the mechanical properties, i.e.:

$$B = B_{VRH} = \frac{B_V + B_R}{2} \tag{6}$$

$$G = G_{VRH} = \frac{G_V + G_R}{2} \tag{7}$$

The average velocity for the crystal can be solved by [13]:

$$v_m = \left[\frac{1}{3}\left(\frac{2}{v_t^3} + \frac{1}{v_l^3}\right)\right]^{-1/3} \tag{8}$$

$$v_l = \left(\frac{3B + 4G}{3\rho}\right)^{1/2} \tag{9}$$

$$v_t = \left(\frac{G}{\rho}\right)^{1/2} \tag{10}$$

Other mechanical properties for Young's modulus E, anisotropic index A_U, and Poisson's ratio v are expressed [13]:

$$E = \frac{9BG}{3B + G} \tag{11}$$

$$A_U = 5\frac{G_v}{G_R} - \frac{B_v}{B_R} - 6 \tag{12}$$

$$v = \frac{3B - 2G}{2(3B + G)} \tag{13}$$

The ratio of B/G has been an adopted criterion to judge the ductility and brittleness of solids; if $B/G > 1.75$, the solid behaves in a ductile manner, otherwise it behaves in a brittle manner [29]. Based on DFT, the energy density u can be easily obtained. The calculated mechanical properties are shown in Table 2, solving from the Equations (3) to (13).

Table 2. Calculated mechanical properties of Cu–Ni alloys, $C_{\rho\sigma}$ is elastic constant, B is bulk modulus, G is shear modulus, v_m is average velocity, E is Young's modulus, and v is Poisson's ratio.

	C_{11}	C_{12}	C_{44}	B	G	E	A_U	v	v_m	B/G
$Cu_{0.05}Ni_{0.95}$	241.1	163.0	110.0	189.0	72.7	193.3	1.405	0.330	3145.7	2.60
$Cu_{0.5}Ni_{0.5}$	165.2	89.4	102.2	114.6	68.7	171.7	1.278	0.250	3002.2	1.67
$Cu_{0.95}Ni_{0.05}$	154.8	137.9	71.4	143.5	32.1	89.6	7.884	0.396	2011.8	4.47
$Cu_{0.97}Ni_{0.03}$						111 [2]				
$Cu_{0.46}Ni_{0.54}$						163 [2]				
$Cu_{0.55}Ni_{0.45}$						172 [9]				
$Cu_{0.13}Ni_{0.87}$						195 [9]				
$Cu_{100}Ni_{0}$						158 [10]				

Firstly, for different works, the Cu-rich Cu content in Cu–Ni alloys has a smaller Young's modulus in their respective results, whatever the simulation or experimental measurements. Our calculated Young's modulus is less small than that of these other experiments, which is caused by the slight difference in the approaching process. It is reasonable and meaningful to further discuss the mechanical properties in calculation for deep consideration.

In Figure 3a, elastic constants refer to the deformation resistance of the crystal in anisotropy. In Voigt's way, the subscript of C_{11} represents the deformation resistance in the x-axis for Cu–Ni alloys. The calculated elastic constant for C_{11} reduces with the increasing Cu content, indicating that the deformation resistance in the x-axis for Cu–Ni is becoming weak. Similarly, C_{44} represents the deformation resistance in the x,y-axis becoming weak, while C_{12} represents the deformation resistance in the y-axis to stress in the x-axis decreasing first, then increasing. The Cu content in the crystal has a great effect on deformation resistance in the x-axis. In Figure 3b, the popular moduli for materials are shown in different Cu content. Bulk modulus has a similar tendency as C_{12}, showing the total deformation resistance of alloys. Shear modulus G and Young's modulus E decrease in Cu-rich Cu content, predicting the fluctuation of hardness.

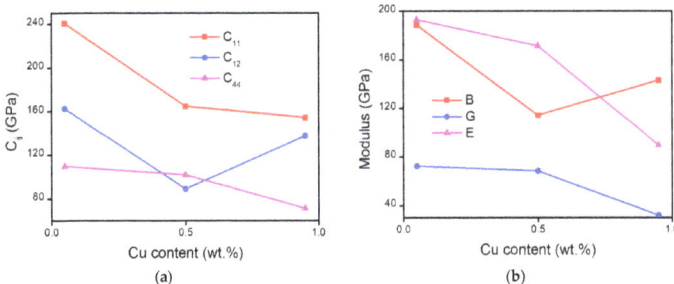

Figure 3. Cu content-dependent mechanical properties: (a) elastic constants; (b) bulk modulus B, Shear modulus G, and Young's modulus E.

In Figure 4a, for the red curve, there is a typical transition from ductile to brittle. In Cu-rich $Cu_{0.95}Ni_{0.05}$ and Ni-rich $Cu_{0.05}Ni_{0.85}$ alloys, they are greatly ductile with the $B/G > 1.75$; meanwhile, the Cu-rich alloys have the higher ductility. On the other hand, the $Cu_{0.5}Ni_{0.5}$ alloys are brittle due to the $B/G = 1.67$. It shows that the right ratio for the binary compound of Cu–Ni systems is beneficial to improve the brittleness of pure metal. For the blue curve, the anisotropic index has the same tendency, where the $Cu_{0.5}Ni_{0.5}$ alloy with the smallest numerical index presented the lowest anisotropy. The low anisotropy with higher Young's modulus shows that the $Cu_{0.5}Ni_{0.5}$ alloy has a great stiffness.

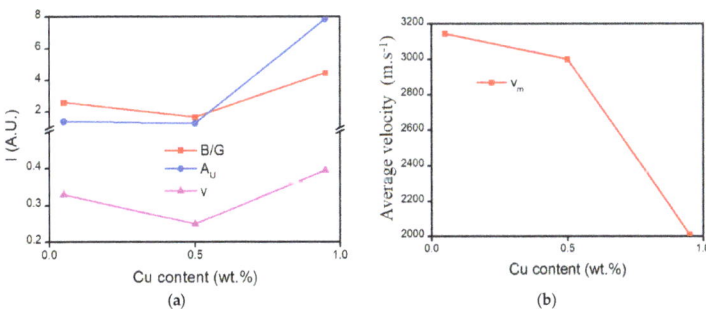

Figure 4. Cu content-dependent brittleness and stiffness: (a) criterion for ductile and brittle B/G, anisotropic index A_U, and Poisson's ratio v; (b) average velocity v_m.

For Poisson's ratio v in Figure 4a, we can obtain the same results as B/G. When the Poisson's ratio $v > 0.26$, the alloys are ductile; otherwise, it behaves in a brittle way [13]. That is, the $Cu_{0.5}Ni_{0.5}$ alloy is brittle. In Figure 4b, the average velocity decreases with the increasing Cu content; the significant drop can be found in the $Cu_{0.5}Ni_{0.5}$ position. It shows that the $Cu_{0.5}Ni_{0.5}$ alloys have a good elastic strength by loading elastic wave velocity, almost comparable with pure copper.

Elastic constant can be applied in a mechanical stability criterion, to reveal the mechanical stability of a crystal. For the cubic system, the criterion can be written simply [30]:

$$C_{11} - C_{12} > 0, \; C_{11} + 2C_{12} > 0, \; C_{44} > 0 \tag{14}$$

We put the elastic constants from Table 2 in the mechanical stability criterion, and the detailed results are shown in Figure 5; it can be seen that the alloys are mechanically stable.

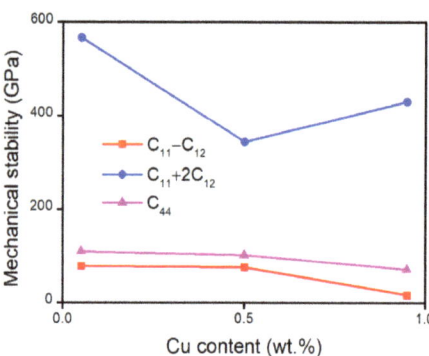

Figure 5. Mechanical stability of Cu–Ni alloys.

4. Conclusions

First-principles calculations-based density functional theory is used in this paper; the lattice crystal and mechanical properties for Cu–Ni alloys are good, illustrated in the stiffness compared with the experiment for nanostructured Cu–Ni films. Firstly, we employ the pseudo-Voigt way to simulate the XRD results, showing the crystal is a reliable structure. Then, for the mechanical properties we computed, the main results are included as a list.

1. Based on DFT, we calculated elastic constants, bulk modulus, shear modulus, Young's modulus, anisotropic index A_U, Poisson's ratio v, average velocity, and B/G in this paper. It improves and supports the results for the experiment.
2. Cu-rich and Ni-rich Cu–Ni alloys are ductile; the Ni-rich alloy has the highest uniaxial deformation resistance due to having the largest Young's modulus.
3. $Cu_{0.5}Ni_{0.5}$ as the most suitable binary compound is predicted to have great stiffness in the Cu–Ni system, due to the brittleness and low anisotropy.

Author Contributions: Conceptualization, C.J.; software, Z.L.; formal analysis, Y.W. and B.N.; resources, Q.L.; writing—original draft preparation, Y.W.; writing—review and editing, C.J. All authors have read and agreed to the published version of the manuscript.

Funding: This research received no external funding.

Data Availability Statement: Data available on request from the authors.

Conflicts of Interest: The authors declare no conflict of interest.

References

1. Peterson, J.; Honnell, K.; Greeff, C.; Johnson, J.; Boettger, J.; Crockett, S. Global equation of state for copper. *AIP Conf. Proc.* **2012**, *1426*, 763.
2. Varea, A.; Pellicer, E.; Pané, S.; Nelson, B.J.; Suriñach, S.; Baró, M.D.; Sort, J. Mechanical Properties and Corrosion Behaviour of Nanostructured Cu-rich CuNi Electrodeposited Films. *Int. J. Electrochem. Sci.* **2012**, *7*, 1288.
3. Metikoš-Huković, M.; Babić, R.; Rončević, I.Š.; Grubač, Z. Corrosion resistance of copper–nickel alloy under fluid jet impingement. *Desalination* **2011**, *276*, 228. [CrossRef]
4. Manzano, C.; Caballero-Calero, O.; Tranchant; Bertero, E.; Cervino-Solana, P.; Martin-Gonzales, M.; Philippe, L. Thermal conductivity reduction by nanostructuration in electrodeposited CuNi alloys. *J. Mater. Chem. C* **2021**, *9*, 3447. [CrossRef]

5. Durivault, L.; Brylev, O.; Reyter, D.; Sarrazin, M.; Bélanger, D.; Roué, L. Cu–Ni materials prepared by mechanical milling: Their properties and electrocatalytic activity towards nitrate reduction in alkaline medium. *J. Alloys Compd.* **2007**, *432*, 323. [CrossRef]
6. Hur, S.; Kim, D.; Kang, B.; Yoon, S. The Structural and Electrical Properties of CuNi Thin-Film Resistors Grown on AlN Substrates for Π -Type Attenuator Application. *J. Electrochem. Soc.* **2005**, *152*, G472. [CrossRef]
7. Chen, M.; Ma, E.; Hemker, K.J.; Sheng, H.; Wang, Y.; Cheng, X. Deformation twinning in nanocrystalline aluminum. *Science* **2003**, *23*, 1275. [CrossRef]
8. Shen, T.; Koch, C.; Tsui, T.; Pharr, G. On the elastic moduli of nanocrystalline Fe, Cu, Ni, and Cu–Ni alloys prepared by mechanical milling/alloying. *J. Mater. Res.* **1995**, *10*, 2892. [CrossRef]
9. Pellicer, E.; Varea, A.; Pané, S.; Nelson, B.J.; Menéndez, E.; Estrader, M.; Suriñach, S.; Baró, M.D.; Nogués, J.; Sort, J. Nanocrystalline Electroplated Cu–Ni: Metallic Thin Films with Enhanced Mechanical Properties and Tunable Magnetic Behavior. *Adv. Funct. Mater.* **2010**, *20*, 983. [CrossRef]
10. Wang, C.; Bhuiyan, M.E.H.; Moreno, S.; Minary-Jolandan, M. Alloy with Controlled Composition from a Single Electrolyte Using Co-Electrodeposition. *Appl. Mater. Interfaces* **2020**, *12*, 18683. [CrossRef]
11. Hui, J.; Zhang, X.; Yang, G.; Liu, T.; Liu, W. First-principles study of de-twinning in a FCC alloy. *J. Solid State Chem.* **2021**, *293*, 121765. [CrossRef]
12. Hao, Y.; Chen, X.; Chen, B. The microstructure and property of lamellar interface in ternary Mg–Gd–Cu alloys: A combined experimental and first-principles study. *J. Mater. Sci.* **2021**, *56*, 9470. [CrossRef]
13. Iwaoka, H.; Hircosawa, S. First-principles calculation of elastic properties of Cu-Zn intermetallic compounds for improving the stiffness of aluminum alloys. *Comput. Mater. Sci.* **2020**, *174*, 109479. [CrossRef]
14. Wang, Z.; Li, J.; Fan, Z.; Zhang, Y.; Hui, S.; Peng, L.; Huang, G.; Xie, H.; Mi, X. Effects of Co Addition on the Microstructure and Properties of Elastic Cu-Ni-Si-Based Alloys for Electrical Connectors. *Materials* **2021**, *14*, 1996. [CrossRef] [PubMed]
15. Yang, M.; Hu, X.; Li, X.; Li, Z.; Zheng, Y.; Li, N.; Dong, C. Microstructure and electrical contact behavior of Al_2O_3-Cu/30W3SiC (0.5 Y_2O_3) composites. *J. Mater. Res. Technol.* **2022**, *17*, 1246. [CrossRef]
16. Li, Z.; Cheng, Z.; Li, X.; Hu, Y.; Li, N.; Zheng, Y.; Shao, Y.; Liu, R.; Dong, C. Enthalpic interaction promotes the stability of high elastic Cu-Ni-Sn alloys. *J. Alloys Compd.* **2022**, *896*, 163068. [CrossRef]
17. Segall, M.; Lindan, P.; Probert, M.; Pickard, C.; Hasnip, P.; Clark, S.; Payne, M. First-principles simulation: Ideas, illustrations and the CASTEP code. *J. Phys. Condens. Matter* **2002**, *14*, 2717. [CrossRef]
18. Perdew, J.; Burke, K.; Ernzerhof, M. Generalized Gradient Approximation Made Simple. *Phys. Rev. Lett.* **1996**, *77*, 3865. [CrossRef]
19. Pfrommer, B.G.; Cote, M.; Louie, S.G.; Cohen, M.L. Relaxation of crystals with the quasi-Newton method. *J. Comput. Phys.* **1997**, *131*, 233. [CrossRef]
20. Dasgupta, P. On Use of Pseudo-Voigt Profiles in Diffraction Line Broadening Analysis. *Fizika A* **2000**, *9*, 61.
21. Caglioti, G.; Paoletti, A.; Ricci, F. Choice of Collimators for Crystal Spectrometers for Neutron Diffraction, Nuclear Instruments and Methods. *Nucl. Instrum.* **1958**, *3*, 223. [CrossRef]
22. Katsura, T.; Tange, Y. A simple derivation of the Birch–Murnaghan equations of state (EOSs) and comparison with EOSs derived from other definitions of finite strain. *Minerals* **2019**, *9*, 745. [CrossRef]
23. Kapahi, A.; Udaykumar, H. Dynamics of void collapse in shocked energetic materials: Physics of void–void interactions. *Shock. Waves* **2013**, *23*, 537. [CrossRef]
24. Liu, W.; Liu, Q.; Zhong, M.; Gan, Y.; Liu, F.; Li, X.; Tang, B. Predicting impact sensitivity of energetic materials: Insights from energy transfer of carriers. *Acta Mater.* **2022**, *236*, 118137. [CrossRef]
25. Born, M.; Huang, K. *Dynamical Theory of Crystal Lattices*; Oxford University Press: London, UK, 1954.
26. Phacheerak, K.; Thanomngam, P. Pressure Dependence of Structural and Elastic Properties of Na_2O: First-Principles Calculations. *Intergrated Ferroelectr.* **2022**, *224*, 256. [CrossRef]
27. Yang, C.; Duan, Y.; Yu, J.; Peng, M.; Zheng, S.; Li, M. Elastic anisotropy and thermal properties of MBN (M = Al, Ga) systems using first-principles calculations. *Vacuum* **2023**, *207*, 111626. [CrossRef]
28. Zhou, Y.; Lin, Y.; Wang, H.; Dong, Q.; Tan, J. First-principles study on the elastic anisotropy and thermal properties of Mg–Y compounds. *J. Phys. Chem. Solids* **2022**, *171*, 111034. [CrossRef]
29. Wu, Y.; Ma, L.; Zhou, X.; Duan, Y.; Shen, L.; Peng, M. Insights to electronic structures, elastic properties, fracture toughness, and thermal properties of $M_{23}C_6$ carbides. *Int. J. Refract. Met. Hard Mater.* **2022**, *109*, 105985. [CrossRef]
30. Gao, J.; Liu, Q.; Jiang, C.; Fan, D.; Zhang, M.; Liu, F.; Tang, B. Criteria of Mechanical Stability of Seven Crystal Systems and Its Application: Taking Silica as an Example. *Chin. J. High Press. Phys.* **2022**, *36*, 051101.

Disclaimer/Publisher's Note: The statements, opinions and data contained in all publications are solely those of the individual author(s) and contributor(s) and not of MDPI and/or the editor(s). MDPI and/or the editor(s) disclaim responsibility for any injury to people or property resulting from any ideas, methods, instructions or products referred to in the content.

Article

GPU-Based Cellular Automata Model for Multi-Orient Dendrite Growth and the Application on Binary Alloy

Jingjing Wang, Hongji Meng *, Jian Yang and Zhi Xie

School of Information Science and Engineering, Northeastern University, Shenyang 110819, China
* Correspondence: menghonhji@ise.neu.edu.cn

Abstract: To simulate dendrite growth with different orientations more efficiently, a high-performance cellular automata (CA) model based on heterogenous central processing unit (CPU)+ graphics processing unit (GPU) architecture has been proposed in this paper. Firstly, the decentered square algorithm (DCSA) is used to simulate the morphology of dendrite with different orientations. Secondly, parallel algorithms are proposed to take full advantage of many cores by maximizing computational parallelism. Thirdly, in order to further improve the calculation efficiency, the task scheduling scheme using multi-stream is designed to solve the waiting problem among independent tasks, improving task parallelism. Then, the present model was validated by comparing its steady dendrite tip velocity with the Lipton–Glicksman–Kurz (LGK) analytical model, which shows great agreement. Finally, it is applied to simulate the dendrite growth of the binary alloy, which proves that the present model can not only simulate the clear dendrite morphology with different orientations and secondary arms, but also show a good agreement with the in situ experiment. In addition, compared with the traditional CPU model, the speedup of this model is up to $158\times$, which provides a great acceleration.

Keywords: GPU-CA model; multi-orient dendrite; parallel algorithm; speedup; task schedule scheme

Citation: Wang, J.; Meng, H.; Yang, J.; Xie, Z. GPU-Based Cellular Automata Model for Multi-Orient Dendrite Growth and the Application on Binary Alloy. *Crystals* **2023**, *13*, 105. https://doi.org/10.3390/cryst13010105

Academic Editor: Yuying Wu

Received: 27 November 2022
Revised: 25 December 2022
Accepted: 31 December 2022
Published: 6 January 2023

Copyright: © 2023 by the authors. Licensee MDPI, Basel, Switzerland. This article is an open access article distributed under the terms and conditions of the Creative Commons Attribution (CC BY) license (https://creativecommons.org/licenses/by/4.0/).

1. Introduction

Dendrite is an important component of solidification structure, whose morphological characteristics directly affect the performance of an alloy. Therefore, studying the growth of dendrite is helpful for researchers to optimize the process parameters and improve the performance of alloys. Experimental methods for obtaining the microstructure during solidification have been proposed, such as the electronic probe [1], synchrotron X-ray [2,3], etc. that have been developed, but cannot be applied to product at a large scale due to the expensive cost and harsh environment. Moreover, these methods are unable to show the dynamic evolution process of dendrite growth.

Over the past few decades, a considerable number of numerical models have grown up around the theme of simulating the evolution process of dendrite morphology during solidification, such as level set [4], monte carlo (MC) [5], cellular automata (CA) [6–8], phase field (PF) [9], etc. Among them, the phase field model (PF) and CA model are the most widely used. Due to complex physical equations and massive calculations, the PF model is only applied to small-scale simulations. Whereas the CA model can perform large-scale simulations and reveal the evolution of micro-mesoscopic dendrites, such as the transformation of columnar crystals into equiaxed crystals (CET) [10], the behavior of dendrite deflection in fluids [6–8], and the formation of silicon facet dendrite [11]. Influenced by grid layout and capture rules, the traditional CA model can only simulate dendrites that grow along the x axis or 45° with the x axis. However, during the actual solidification process of the alloy, dendrites grow with different preferred orientations, which is caused by the solute profile and can lead to macro-segregation in the cast product [12]. Therefore, it is very important for the model to be able to simulate the dendrite with different orientations. Researchers have proposed various methods to achieve multi-orient dendrite simulation, such as defining block cells [13], establishing random ZigZag neighbor cell capture rules [14], etc. These methods can only simulate a dendrite with the same orientation at the same time due to complex algorithms. Hence, Rappaz and Gandin developed the decentered square algorithm (DCSA) [15], which can simulate dendrites with different orientations simultaneously, and it has been widely used to simulate the growth of columnar and equiaxed dendrites [16–19]. However, these models suffered from velocity anisotropy, which means velocities along different orientations are asymmetrical and have multiple interfaces. Then, Luo et al. [20] modified the velocity of the interface cell through the local solute equilibrium and improved the phenomenon of multiple interfaces. Wang et al. [21]. determined the length of the square half diagonal by the preferred orientation to ensure sharpness of the interface and introduced GF to reduce velocity anisotropy.

However, the DCSA rule dynamically determines the capture position and condition by tracking the dendrite tip, which is very complex. Massive calculations of the CA model incorporating the DCSA rule make large-scale simulations very time-consuming. It is, therefore, necessary to speed up the calculation. Some methods have attempted to improve calculation efficiency, such as adaptive meshes [22,23], parallelizing the computer program using MPI technology [24], and parallel computing technology based on serial arithmetic [25]. These accelerators are CPU-based, so the speedup is not good enough for the number of cores, being limited by the hardware architecture. Unlike the CPU, GPU is characterized by integrating thousands of computer cores inside, which makes it the most effective way to achieve acceleration. With the development of GPU in general purpose computing, it has been used in various fields [26,27]. Over the past two decades, the GPU-based acceleration method has become a hotspot in computational and material science [28–30] and has made great progress in dendrite growth simulation [31–34]. However, most of these contributions focused on the PF model, which is still unable to simulate dendrite growth at large scale, even when accelerated by using GPU.

This paper proposes a high-performance CA model incorporating the DCSA capture rule to simulate dendrite growth with different orientations more efficiently. To make full use of the hardware resources and improve the calculation efficiency, parallel algorithms and a task scheduling scheme using multi-stream are proposed. Compared to the traditional CPU-CA model, this model can achieve great acceleration with a speedup of 158×. In addition, the steady-state tip velocity predicted by this work shows great agreement with the analytical value of the LGK model. The simulation results of the binary alloy by this model not only show dendrites with different orientations, but also agree well with the experimental result in situ. This model will be a promising tool for studying the dendrite growth during the large-scale solidification.

The following chapters will cover the establishment, solution, and verification of the model in detail.

2. Materials and Methods

2.1. Description of Numerical Model

2.1.1. Heat Transfer Model

Equation (1) is used to describe heat transfer during alloy solidification.

$$\rho c_p \frac{\partial T}{\partial t} = \frac{\partial}{\partial x}\left(\lambda \frac{\partial T}{\partial x}\right) + \frac{\partial}{\partial y}\left(\lambda \frac{\partial T}{\partial y}\right) \tag{1}$$

where T is the temperature, f_s is the solid fraction, ρ is the density, c_p is the equivalent specific heat, and L is the latent heat.

2.1.2. Solute Distribution Model

The solute distribution model consists of solute redistribution at the interface and solute diffusion across the entire domain. For one thing, the solute redistribution is determined by Equations (2) and (3) according to local equilibrium.

$$C_s = kC_l \tag{2}$$

$$dC = C_l(1-k)\Delta f_s \tag{3}$$

where dC is the residual solute expelled by the interface cell, and k is the coefficient of equilibrium solute partition ($k < 1$ in the present work).

First, the expelled solute will be discharged into the remaining liquid of the current cell. Then, the current C_l will be compared to the equilibrium solute concentration C_l^*; the part beyond C_l^* will be discharged to adjacent liquid cells. C_l^* is calculated by Equation (4).

$$C_l^* = C_0 + \frac{T - T_0}{m_l} + \frac{\Gamma Ka(\hat{n})}{m_l} \tag{4}$$

where T is the actual temperature, T_0 is the initial temperature, C_0 is the initial solute concentration, m_l is the liquidus slope, and $a(\hat{n})$ is the function of interface anisotropy.

If the current cell is completely solidified, then all the excluded solute will be discharged to the adjacent liquid cells. How much solute will be discharged into an adjacent liquid cell is designed by weight, as shown in Equation (5).

$$w_i = \frac{C_i^0 - C_l}{\sum_{j=0}^{N}(C_j^0 - C_l)} \tag{5}$$

where $C_i^0 - C_l$ is the solute difference between the interface cell and the adjacent liquid cell i, N is the number of liquid neighbours.

For another, the solute diffusion coefficient in liquid is three to four orders of magnitude larger than that in solid, so diffusion in solid phase is neglected in this paper. Equation (6) shows the diffusion of a solute in liquid and interface.

$$\frac{\partial C_l}{\partial t} = \frac{\partial}{\partial x}\left(D_i \frac{\partial C_l}{\partial x}\right) + \frac{\partial}{\partial y}\left(D_i \frac{\partial C_l}{\partial y}\right) \tag{6}$$

where D_i is the solute diffusion coefficient, which is governed by Equation (7).

$$D_i = f_s D_s + (1 - f_s)D_l \tag{7}$$

where D_l and D_s are the diffusion coefficient in liquid and solid, respectively.

2.1.3. CA Model

In the traditional CA model, only fully solidified cells ($f_s = 1$) can capture their liquid neighbors. This capture rule is severely affected by the grid layout, so it can only simulate dendrites growing along the x axis and 45° with the x axis. However, the growth orientation of dendrite is chaotic in the actual solidification process of the alloy. In order to simulate dendrite growth closer to reality, the decentered square rule for capturing has been adopted by the present CA model. As shown in Figure 1, the key to the improved algorithms is dynamically calculating the nucleation position and nucleation conditions, according to the position where the interface cell (parent cell) reaches the liquid cell. All cells in the entire computing domain are initialized to liquid ($f_s = 0$). As the temperature decreases, the one who receives the nucleation conditions will receive the chance to nucleate. The nucleation probability is calculated by Equation (8) combined with random probability.

$$\frac{dn}{d\Delta T} = \frac{n_{max}}{\sqrt{2}\exp\left[-\frac{1}{2}\left(\frac{\Delta T-\Delta T_N}{\Delta T_\sigma}\right)\right]} \quad (8)$$

where ΔT is the undercooling, n_{max} is the maximum density of nuclei given by the integral of undercooling (from 0 to ΔT) of Equation (8), ΔT_N is the average nucleation undercooling, and ΔT_σ is the standard deviation. Once a cell is nucleated, its solid fraction will be changed from 0 to 1, and a decentered square with an angle θ between diagonal and x axis will be located at its center. Additionally, four corners of this square penetrate the nearest liquid neighbors, capturing them in the corners and changing them into interface cells. Captured cells will inherit the growth properties of the parent cell and grow along the diagonal to capture its liquid neighbors until it is completely solidified.

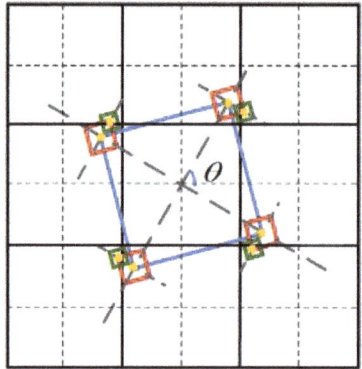

Figure 1. The diagram of the decentered algorithm (θ is the angle between diagonal and x axis).

In order to simulate the dendrite morphology in detail, the local level rule method is used to calculate the increment of the solid fraction [35], wherein Δf_s of each time step is determined by Equation (9). The length of the primary arm of the dendrite is updated by Equation (10).

$$\Delta f_s = \frac{(C_l^* - C_l)}{(C_l^*(1-k))}GF \quad (9)$$

$$l(t + \Delta t) = l(t) + \Delta f_s l_{max} \quad (10)$$

l_{max} is the maximum half-diagonal length determined by Equation (11).

$$l_{max} = l\left(\frac{\Delta x}{\max(\sin\theta, \cos\theta)}\right) \quad (11)$$

Δx is the size of the cell, and GF [36] is the geometry factor related to the state of neighbors, which is introduced to avoid multiple interfaces and is limited to no more than 1.

$$GF = \min\left\{1, G\left(\sum_{m=1}^{4} s_m^I + \sqrt{2}\sum_{m=1}^{4} s_m^{II}\right)\right\} \quad (12)$$

where G is an adjustable factor, s_m^I and s_m^{II} are the state of the nearest neighbor and the second nearest neighbor, respectively, and both of them have two values, as shown in Equation (13).

$$s_m^I, s_m^{II} = \begin{cases} 0, & f_s < 1 \\ 1, & f_s = 1 \end{cases} \quad (13)$$

2.2. Parallel Solver Based on GPUq

To simulate the morphology of dendrite with clequeqear secondary arms, the cell size is generally divided into 1~2 μm. There will be huge calculations when conducting large-scale simulations, which is time-consuming using CPU. In addition, the purpose of the simulation is to optimize the process parameters by performing many groups of experiments. Therefore, it is necessary to accelerate the calculation. This paper designs a GPU-based parallel solver for acceleration.

2.2.1. GPU-CA Framework

The GPU-CA frame used in this work is shown in Figure 2, in which each slice has the cell state of the cross-section corresponding to this moment.

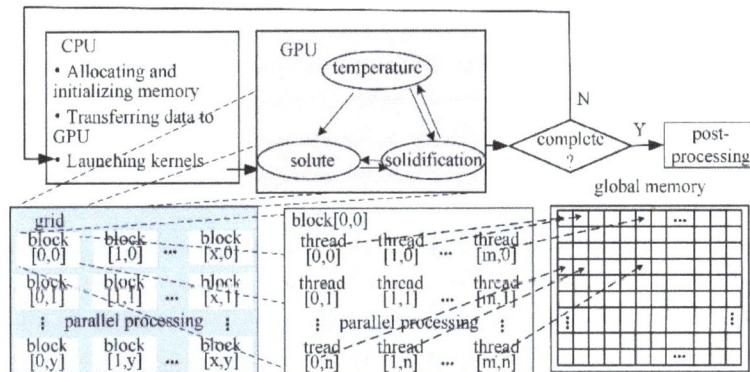

Figure 2. GPU-CA architecture.

In this heterogeneous framework, the CPU is responsible for allocating memory, initializing, and transferring the initial state to the global memory on the GPU via the PCIe bus. Then, all the threads on the GPU execute instructions simultaneously in single-instruction-multiple-data (SIMD) mode until the required calculations are completed. Finally, data are transferred from GPU memory to CPU memory via the PCIe bus for post-processing. This work uses Pascal GP100 GPU (embedded with 3584 CUDA cores) as the accelerator, where millions of threads can execute instructions simultaneously, and each thread corresponds to a cell.

2.2.2. Implementation by CUDA C

This work adopts CUDA as the programming model, because it allows us to execute applications on heterogeneous computing systems by simply annotating code with a small set of extensions to the C programming language. In addition, the CUDA programming model provides the exposed memory hierarchy, in which global memory is the most widely used, because it has a large memory space. This work chose global memory and, based on it, designed the structure of array (SOA) to organize data, because the SOA is more liable to maximize the efficiency of accessing the global memory.

1. Task scheduling scheme

CUDA provides streams to achieve concurrency between kernels, where kernels are executed sequentially in the same stream, simultaneously in different streams. To increase parallelism, this paper divides the problem into eight small tasks, each corresponding to a kernel function, as shown in Table 1.

Table 1. Task assignment in GPU-CA model.

Kernel Functions	Computation Task
capture	capturing neighbors
calD	solute diffusion coefficient
solute_Dif	solute diffusion
schange	solute redistribution in interface
get_T	temperature distribution
calB	equilibrium solute
growth	velocity and the arm length
backup	storing data for the next slice

All kernels are related, but at some point, some kernels are independent. For example, before kernel solute_Dif is scheduled, kernels capture, schange, calD are independent. To solve the waiting problem among independent tasks, this paper puts all kernels into two streams based on the kernel dependency, as shown in Figure 3. It is noted that all operations in the non-default stream are non-blocking with respect to the host thread. Thus, we need to synchronize the host with operations running in a stream. The overall flow of the numerical simulation is shown in Figure 3, where kernel function solute_Dif depends on values of C_L and D calculated by calD and schange, respectively. In this case, the host has to wait until the calD and schange finish their calculation before solute_Dif is scheduled. This work uses the cudaDeviceSynchronize(void) function to achieve synchronization between the host and operations running in a stream.

2 Parallel algorithms

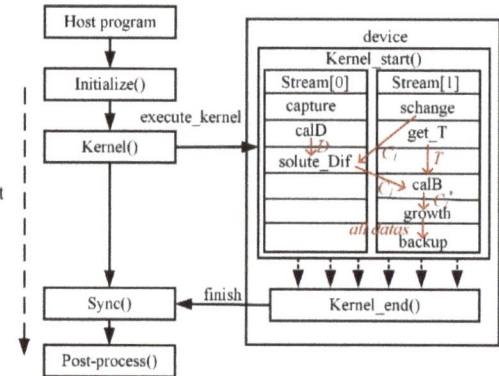

Figure 3. The flow-chart of the program model.

Global memory can be accessed on the device from any SM throughout the application's lifetime. When multiple threads write data to the same address at the same time, there will be a 'data race' that can lead to an undefined error. The CA model in this work adopted the Moore neighbor, in which each cell has eight neighbors. Therefore, there may be more than one neighbor discharging solute into a liquid cell or capturing it at the same time. Mapping to the programming model, there are multiple threads that attempt to modify the data in the same address, introducing a 'data race'. This paper avoided this phenomenon by preventing the cell from 'actively' discharging solute to neighboring cells or capturing it. For example, the solute redistribution algorithm adopts a kernel function and a device function to turn 'active' to 'passive', as shown in Figure 4. There are two steps to complete the solute redistribution process. The pseudo-code is listed in Algorithm 1.

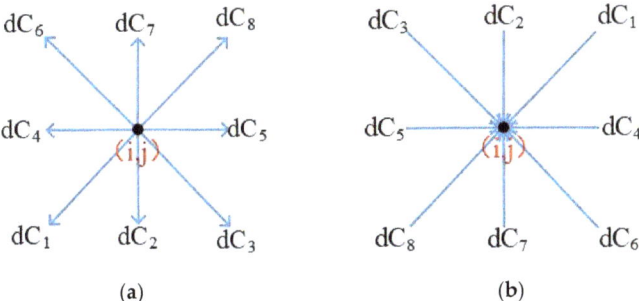

Figure 4. Avoiding data race ((a) store (b) read->change).

First, calculate how much solute will be discharged to each liquid neighbor and store it in the intermediate variables, lines 2–7 in Algorithm 1.

Then, read the data stored by the first step from the neighbors and sum it to the solute of the current cell changed, lines 8–11 in Algorithm 1.

Algorithm 1 parallel solute redistribution algorithm

1   id : assign thread to cell
2   for all interface cells do
3      $dC \leftarrow$ diacharged solute
4   if$(dC > 0)$
5         $dC_i \leftarrow$ store the solute dischared to cells
6   end
7   end
8   for all liquid and interface cells do
9   temp $\leftarrow$ read dC_i from neighbors
10    $C_l \leftarrow C_l +$ temp//add temp to C_l
11 end

This algorithm can avoid the 'data race' and make all cells in the calculation domain independent.

3. Results and Discussion

3.1. Validated by the LGK Model

This model was validated by comparing the steady velocity of a free dendrite crystal in an undercooled melt with the analytical value of the LGK model. This work defines the velocity of the dendrite tip through the cell size and time interval as the cell stays in the interface state, and the steady state of dendrite growth is determined as the solute at the boundary opposite the dendrite tip reaches 1.01 times the initial value [37,38]. The physical parameters used in this paper are shown in Table 2.

Table 2. Physical properties [39,40].

Property and Symbol	Fe–0.6C	Fe–5.3Si
Initial composition, C_0 (wt.%)	0.6	5.3
Liquidus temperature T_l (K)	1763.37	1732.87
Liquidus slope, m_l (K%$^{-1}$)	−80	−7.6
Solute partition coefficient, k_0	0.34	0.77
Solute diffusion coefficient in liquid D_l (m^2·s^{-1})	2×10^{-9}	$8.0 \times 10^{-4} \exp(-29{,}943.23/T)$
Solute diffusion coefficient in solid D_s (m^2·s^{-1})	5×10^{-10}	$8.0 \times 10^{-8} \exp(-29{,}943.23/T)$
Gibbs-Thomson coefficient, (m·K)	1.9×10^{-7}	1.9×10^{-7}

This test was performed in a square domain with a size of 400 μm × 400 μm, which is uniformly divided into 400 × 400 cells with the size of 1 μm × 1 μm. At the beginning of solidification, a nucleus with a preferred orientation of 0° was placed in the center of the domain.

Figure 5 shows the evolution of the transient velocity of the dendrite tip as a function of time calculated by the present CA model under the undercooling of 8K and the steady velocity calculated by the LGK model. It can be seen that tip velocity of the dendrite decreases dramatically in the transient growth period; then, the decreasing tendency becomes mild in the steady growth period. Finally, the velocity is stable at 56.33 μm/s, which is very close to the analytical value of 55.90 μm/s calculated by the LGK model.

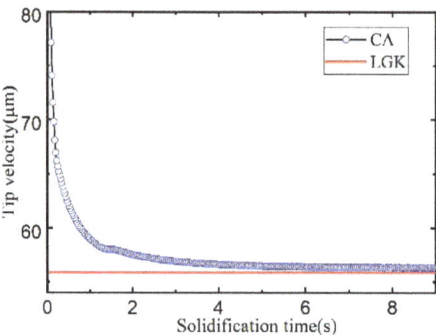

Figure 5. Comparison of the steady tip velocity at a constant undercooling $\Delta T = 8K$.

3.2. Model Capability

This model is applied to simulate the dendrite growth of binary alloy. Firstly, a group of dendrites of Fe–0.6C alloy are simulated to evaluate the ability to simulate the dendrites with different orientations. Secondly, it is applied to simulate multi-orient dendrites' growth of Fe–5.3Si alloy [41], and the simulated result is compared with the in situ observation. Finally, the performance of the parallel solution is analyzed by comparing the calculation time with the traditional CPU calculation.

3.2.1. Single Dendrite of Fe–0.6C Alloy

A group of dendrites of Fe–0.6C with different orientations are simulated in this part, where the orientation ranges from 0° to 90° due to the four-fold symmetry of the dendrite. Simulations are also performed on a square domain with a size of 400 μm × 400 μm that is divided into 400 × 400 cells. Figure 6 shows the simulation results at 0.6 s with a constant undercooling of 8K, which shows that the present model can simulate clear dendrite morphology with not only different orientations but also second dendrite arms.

In addition, as shown in Figure 7, the primary arm of dendrites with different orientations of 0°, 30°, and 60° are almost the same length, indicating that growth velocities with different orientations are in good symmetry.

3.2.2. Multi-Dendrites of Fe–5.3Si Alloy

The development of synchrotron X-ray technology has enabled the observation of solidification in metallic alloys, which provides a powerful tool to verify the model. Yasuda et al. [41] performed in situ observation of Fe–5.3Si wt.% alloy, which shows a clear morphology of dendrite, including secondary dendrite arms. To validate this model, this paper simulated multi-orient dendrites' growth of Fe–5.3Si alloy under the same experimental condition as the in situ observation, which set the nuclei positions and orientations corresponding to the in situ experimental observation before the simulation began. Figure 8a,b depict the simulated dendrite morphology and solute profile, respectively.

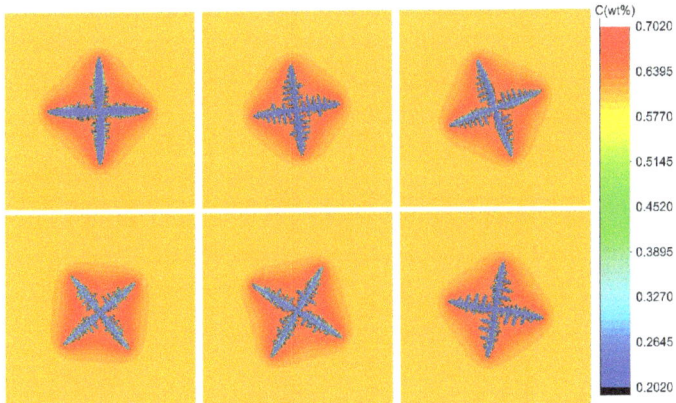

Figure 6. Solute profile and morphology of the equiaxed dendrite at the melt undercooling of 8K as the orientation is (**a**) 0°, (**b**) 5°, (**c**) 20°, (**d**) 40°, (**e**) 60°, and (**f**) 80°.

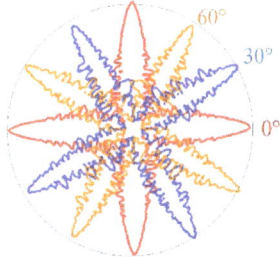

Figure 7. Velocity symmetry in different orientations.

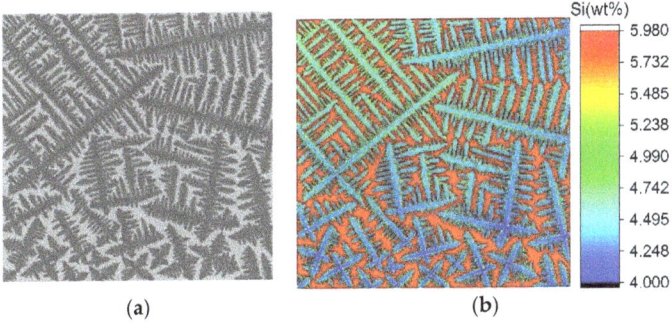

Figure 8. Muti-dendrite morphology of Fe–5.3Si alloy simulated by present model (**a**) morphology, and (**b**) solute profile.

Simulation results show that the bottom dendrite is less developed than the upper because there are more dendrites at the bottom than at the upper, which agrees well with the in situ observation by synchrotron X-ray imaging [41]. For one thing, the growth of many dendrites will make the surrounding solute enriched, which will restrict the development of dendrites. For another, the more dendrites, the less space for each of them. In addition, Figure 8b shows the segregation between dendrites and within a dendrite, which satisfies well the non-equilibrium solidification theory [42]. Therefore, this model can simulate the growth of the dendrite in the actual casting process and reveal the phenomenon of segregation.

3.2.3. Acceleration Performance

The block size configured when launching a kernel function means a lot to performance because of its impact on latency hiding, memory efficiency, and occupancy, etc. A group of numerical experiments were carried out to find the optimal execution configuration according to the grid and block size guidelines [43]. Further, parts of the representative numerical simulation results are shown in Table 3, which shows that the configuration of (32,4) can obtain optimal performance. Accordingly, the following numerical experiments are performed under this configuration.

Table 3. Time elapsed with different kernel configuration.

Kernel Configuration	Time Elapsed (s)
(128,1)	37.94
(128,2)	37.64
(128,4)	38.85
(64,4)	37.59
(64,2)	37.53
(62,1)	38.13
(32,8)	37.70
(32,4)	37.23
(32,2)	37.85

Speedup, the ratio between the CPU and GPU computational time, is used to evaluate the performance of the present GPU-CA model. The same numerical simulations were performed both on the Intel(R) Xeon(R) CPU E5-2680 v4 @ 2.40 GHz and Pascal GP 100 GPU. All simulations are carried out on the assumption that solidification has been started with 0.6 s, which equals 600,000 steps in the case of $\Delta t = 0.0001$ s. Figure 9 shows the computation time on the CPU and GPU as well as the speedup, which shows that this model can achieve great acceleration. In addition, the speedup increases with the number of grids, which is up to 158×, but the tendency to increase becomes low from a certain point. This can be explained by limited hardware resources, and before it is fully utilized, performance will be significantly improved when the grid number is increased. Once occupancy reaches its maximum, performance may be limited by additional scheduling overhead when increasing the grid number again.

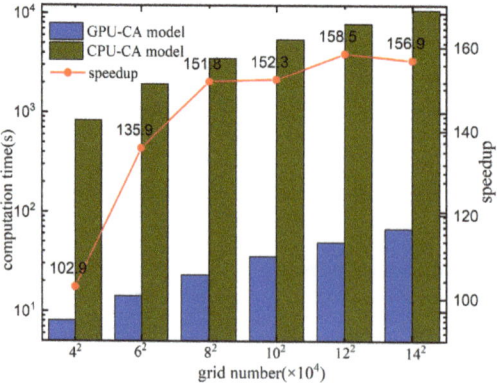

Figure 9. Computational performance.

4. Conclusions

This paper aims to develop a high-performance CA model incorporating the decentered square capture rule to simulate dendrite growth with different orientations more efficiently. The calculation efficiency of the CA model based on GPU has been improved by the proposed parallel algorithms adapted to the many-core architecture of GPU and the task scheduling scheme using multi-stream, and it is validated by comparing the calculated value with the analytical value by the LGK model. By applying this model to simulate the single dendrite of the Fe–0.6C alloy in different orientations and the multi-orient dendrite growth of the Fe–5.3Si alloy, the following conclusions can be drawn:

(1) The steady dendrite tip velocity calculated by this CA model agrees well with the analytical LGK model.
(2) The present model can simulate dendrite morphologies with a random orientation of the Fe–0.6C alloy and can maintain velocity symmetry under different orientations.
(3) The simulation result of Fe–5.3Si not only matches well with the in situ experiment but can also reveal the segregation existing between and within dendrite. This model can be used to simulate multi-dendrite growth in actual casting.
(4) Compared to traditional CPU calculation, this work can achieve noticeable acceleration, and the speedup increases with the number of grids, which is up to 158×.

This GPU-based model can greatly accelerate the calculation, which will be a promising tool in the field of studying dendrite growth during large-scale solidification.

Author Contributions: Methodology, writing—original draft preparation, J.W.; formal analysis, H.M. and J.Y.; funding acquisition and supervision, Z.X. and H.M. All authors have read and agreed to the published version of the manuscript.

Funding: This work was funded by National Natural Science Foundation of China, grant number No.51634002 and No.61703084 and Fundamental Research Funds for the Central Universities, grant number N224001-8.

Data Availability Statement: The data applied in this research are available from the authors upon request.

Conflicts of Interest: The authors declare no conflict of interest.

References

1. Domitner, J.; Kharicha, A.; Grasser, M.; Ludwig, A. Reconstruction of Three-Dimensional Dendritic Structures based on the Investigation of Microsegregation Patterns. *Steel Res. Int.* **2010**, *81*, 644–651. [CrossRef]
2. Guo, E.Y.; Shuai, S.; Kazantsev, D.; Karagadde, S.; Phillion, A.B.; Jing, T.; Li, W.Z.; Lee, P.D. The influence of nanoparticles on dendritic grain growth in Mg alloys. *Acta Mater.* **2018**, *152*, 127–137. [CrossRef]
3. Liss, K.D.; Garbe, U.; Li, H.J.; Schambron, T.; Almer, J.D.; Yan, K. In Situ Observation of Dynamic Recrystallization in the Bulk of Zirconium Alloy. *Adv. Eng. Mater.* **2009**, *11*, 637–640. [CrossRef]
4. Osher, S.; Fedkiw, R.P. Level set methods: An overview and some recent results. *J. Comput. Phys.* **2001**, *169*, 463–502. [CrossRef]
5. Rodgers, T.M.; Madison, J.D.; Tikare, V. Simulation of metal additive manufacturing microstructures using kinetic Monte Carlo. *Comput. Mater. Sci.* **2017**, *135*, 78–89. [CrossRef]
6. Wei, L.; Lin, X.; Wang, M.; Huang, W.D. Cellular automaton simulation of the molten pool of laser solid forming process. *Acta. Phys. Sin.-Chi. Ed.* **2015**, *64*, 018103. [CrossRef]
7. Bai, Y.; Wang, Y.; Zhang, S.; Wang, Q.; Li, R. Numerical Model Study of Multiple Dendrite Motion Behavior in Melt Based on LBM-CA Method. *Crystals* **2020**, *10*, 70. [CrossRef]
8. Wang, Q.; Wang, Y.; Zhang, S.; Guo, B.; Li, C.; Li, R. Numerical Simulation of Three-Dimensional Dendrite Movement Based on the CA–LBM Method. *Crystals* **2021**, *11*, 1056. [CrossRef]
9. Zhang, X.F.; Zhao, J.Z. Effect of forced flow on three dimensional dendritic growth of al-cu alloys. *Acta. Met. Sin.* **2012**, *48*, 615–620. [CrossRef]
10. Wang, W.; Wang, Z.; Yin, S.; Luo, S.; Zhu, M. Numerical simulation of solute undercooling influenced columnar to equiaxed transition of Fe-C alloy with cellular automaton. *Comput. Mater. Sci.* **2019**, *167*, 52–64. [CrossRef]
11. Ma, W.; Li, R.; Chen, H. Three-Dimensional CA-LBM Model of Silicon Facet Formation during Directional Solidification. *Crystals* **2020**, *10*, 669. [CrossRef]
12. SenGupta, A.; Santillana, B.; Sridhar, S.; Auinger, M. Dendrite growth direction measurements: Understanding the solute advancement in continuous casting of steel. *IOP Conf. Ser. Mater. Sci. Eng.* **2019**, *529*, 012065. [CrossRef]

13. Beltran-Sanchez, L.; Stefanescu, D.M. Growth of solutal dendrites: A cellular automaton model and its quantitative capabilities. *Met. Mater. Trans. A* **2003**, *34*, 367–382. [CrossRef]
14. Wei, L.; Lin, X.; Wang, M.; Huang, W. A cellular automaton model for the solidification of a pure substance. *Appl. Phys. A Mater.* **2010**, *103*, 123–133. [CrossRef]
15. Rappaz, M.; Gandin, C.A. Probabilistic modelling of microstructure formation in solidification proc. *Acta Mater.* **1993**, *41*, 345–360. [CrossRef]
16. Wang, W.; Lee, P.D.; McLean, M. A model of solidification microstructures in nickel-based superalloys: Predicting primary dendrite spacing selection. *Acta Mater.* **2003**, *51*, 2971–2987. [CrossRef]
17. Yuan, L.; Lee, P.D. Dendritic solidification under natural and forced convection in binary alloys: 2D versus 3D simulation. *Model. Simul. Mater. Sci.* **2010**, *18*, 055008. [CrossRef]
18. Zhao, Y.; Chen, D.F.; Long, M.J.; Arif, T.T.; Qin, R.S. A Three-Dimensional Cellular Automata Model for Dendrite Growth with Various Crystallographic Orientations During Solidification. *Met. Mater. Trans. B* **2014**, *45*, 719–725. [CrossRef]
19. Chen, R.; Xu, Q.; Liu, B. A Modified Cellular Automaton Model for the Quantitative Prediction of Equiaxed and Columnar Dendritic Growth. *J. Mater. Sci. Technol.* **2014**, *30*, 1311–1320. [CrossRef]
20. Luo, S.; Zhu, M.Y. A two-dimensional model for the quantitative simulation of the dendritic growth with cellular automaton method. *Compute Mater. Sci.* **2013**, *71*, 10–18. [CrossRef]
21. Wang, W.L.; Luo, S.; Zhu, M.Y. Development of a CA-FVM Model with Weakened Mesh Anisotropy and Application to Fe–C Alloy. *Crystals* **2016**, *6*, 147. [CrossRef]
22. Wei, L.; Lin, X.; Wang, M.; Huang, W.D. Orientation selection of equiaxed dendritic growth by three-dimensional cellular automaton model. *Phys. B* **2012**, *407*, 2471–2475. [CrossRef]
23. Provatas, N.; Greenwood, M.; Athreya, B.; Goldenfeld, N.; Dantzig, J. Multiscale modeling of solidification: Phase-field methods to adaptive mesh refinement. *Int. J. Mod. Phys. B* **2005**, *19*, 4525–4565. [CrossRef]
24. Jelinek, B.; Eshraghi, M.; Felicelli, S.; Peters, J.F. Large-scale parallel lattice Boltzmann-cellular automaton model of two-dimensional dendritic growth. *Comput. Phys. Commun.* **2014**, *185*, 939–947. [CrossRef]
25. Feng, W.M.; Xu, Q.Y.; Liu, B.C. Microstructure simulation of aluminum alloy using parallel computing technique. *ISIJ Int.* **2002**, *42*, 702–707. [CrossRef]
26. Campos, R.S.; Lobosco, M.; dos Santos, R.W. A GPU-based heart simulator with mass-spring systems and cellular automaton. *J. Supercomput.* **2014**, *69*, 1–8. [CrossRef]
27. Yam-Uicab, R.; Lopez-Martinez, J.; Trejo-Sanchez, J.; Hidalgo-Silva, H.; Gonzalez-Segura, S. A fast Hough Transform algorithm for straight lines detection in an image using GPU parallel computing with CUDA-C. *J. Supercomput.* **2017**, *73*, 4823–4842. [CrossRef]
28. Aoki, T.; Ogawa, S.; Yamanaka, A. Multiple-GPU Scalability of Phase-Field Simulation for Dendritic Solidification Progress in nuclear science and technology. *Prog. Nucl. Sci. Technol.* **2011**, *2*, 639–642.
29. Takaki, T.; Rojas, R.; Ohno, M.; Shimokawabe, T.; Aoki, T. GPU phase-field lattice Boltzmann simulations of growth and motion of a binary alloy dendrite. *IOP Conf. Ser. Mater. Sci. Eng.* **2015**, *84*, 012066. [CrossRef]
30. Yang, C.; Xu, Q.; Liu, B. GPU-accelerated three-dimensional phase-field simulation of dendrite growth in a nickel-based superalloy. *Comput. Mater. Sci.* **2017**, *136*, 133–143. [CrossRef]
31. Sakane, S.; Takaki, T.; Rojas, R.; Ohno, M.; Shibuta, Y.; Shimokawabe, T.; Aoki, T. Multi-GPUs parallel computation of dendrite growth in forced convection using the phase-field-lattice Boltzmann model. *J. Cryst. Growth* **2017**, *474*, 154–159. [CrossRef]
32. Yang, C.; Xu, Q.Y.; Liu, B.C. Primary dendrite spacing selection during directional solidification of multicomponent nickel-based superalloy: Multiphase-field study. *J. Mater. Sci.* **2018**, *53*, 9755–9770. [CrossRef]
33. Sakane, S.; Takaki, T.; Ohno, M.; Shimokawabe, T.; Aoki, T. GPU-accelerated 3D phase-field simulations of dendrite competitive growth during directional solidification of binary alloy. *IOP Conf. Ser. Mater. Sci. Eng.* **2015**, *84*, 012063. [CrossRef]
34. Kao, A.; Krastins, I.; Alexandrakis, M.; Shevchenko, N.; Eckert, S.; Pericleous, K. A Parallel Cellular Automata Lattice Boltzmann Method for Convection-Driven Solidification. *JOM* **2019**, *71*, 48–58. [CrossRef]
35. Wang, T.M.; Wei, J.J.; Wang, X.D.; Yao, M. Progress and Application of Microstructure Simulation of Alloy Solidification. *Acta Met. Sin.* **2018**, *54*, 193–203.
36. Shin, Y.H.; Hong, C.P. Modeling of dendritic growth with convection using a modified cellular automaton model with a diffuse interface. *ISIJ Int.* **2002**, *42*, 359–367. [CrossRef]
37. Wang, J.J.; Meng, H.J.; Yang, J.; Xie, Z. A fast method based on GPU for solidification structure simulation of continuous casting billets. *J. Comput. Sci.* **2021**, *48*, 101265. [CrossRef]
38. Beltran, S.L.; Stefanescu, D.M. A quantitative dendrite growth model and analysis of stability concepts. *Met. Mater. Trans. A* **2004**, *35a*, 2471–2485. [CrossRef]
39. Wang, W.L.; Ji, C.; Luo, S.; Zhu, M.Y. Modeling of Dendritic Evolution of Continuously Cast Steel Billet with Cellular Automaton. *Met. Mater. Trans. B* **2018**, *49*, 200–212. [CrossRef]
40. Nastac, L. Numerical modeling of solidification morphologies and segregation patterns in cast dendritic alloys. *Acta Mater.* **1999**, *47*, 4253–4262. [CrossRef]
41. Yasuda, H.; Yamamoto, Y.; Nakatsuka, N.; Yoshiya, M.; Nagira, T.; Sugiyama, A.; Ohnaka, I.; Uesugi, K.; Umetani, K. In situ observation of solidification phenomena in Al-Cu and Fe-Si-Al alloys. *Int. J. Cast Met. Res.* **2009**, *22*, 15–21. [CrossRef]

42. Kurz, W.; Fisher, D.J. *Fundamentals of Solidification*, 3rd ed.; Trans Tech Publication: Aedermannsdorf, Switzerland, 1992; pp. 71–92.
43. Cheng, J.; Crossman, M.; Mckercher, T. *Professional CUDA C Programming*; John Wiley & Sons, Inc.: Indianapolis, Indiana, 2014; p. 96.

Disclaimer/Publisher's Note: The statements, opinions and data contained in all publications are solely those of the individual author(s) and contributor(s) and not of MDPI and/or the editor(s). MDPI and/or the editor(s) disclaim responsibility for any injury to people or property resulting from any ideas, methods, instructions or products referred to in the content.

Article

Direct Evidence for Phase Transition Process of VC Precipitation from (Fe,V)₃C in Low-Temperature V-Bearing Molten Iron

Lei Cao [1,2], Desheng Chen [1,*], Xiaomeng Sang [3], Hongxin Zhao [1], Yulan Zhen [1], Lina Wang [1], Yahui Liu [1], Fancheng Meng [1] and Tao Qi [1,4]

[1] National Engineering Research Center of Green Recycling for Strategic Metal Resources, Institute of Process Engineering, Chinese Academy of Sciences, Beijing 100190, China
[2] School of Chemical Engineering, University of Chinese Academy of Sciences, Beijing 100190, China
[3] Hebei Zhongke Tongchuang Vanadium & Titanium Technology Co., Ltd., Hengshui 053099, China
[4] Ganjiang Innovation Academy, Chinese Academy of Sciences, Ganzhou 341119, China
* Correspondence: dshchen@ipe.ac.cn

Abstract: V-bearing molten iron was obtained by adding Na_2CO_3 in the smelting process of vanadium titanomagnetite at low temperature. Two forms of V-rich carbides ((Fe,V)₃C, VC) were detected in the V-bearing pig iron products. Once the smelting temperature was above 1300 °C, most of the V in the raw ore was reduced into molten iron. Owning to the high content of V, the unsteady (Fe,V)₃C solid solution decomposed along with the precipitation of graphite and VC during the solidification process. The presence of VC cluster and VC precursor in (Fe,V)₃C was detected by transmission electron microscopy, which confirmed the possibility of this transition process at the atomic perspective. The transformation dramatically affected the compositions and properties of V-bearing pig iron and had important guiding significance for the actual production process.

Keywords: phase transition; precipitation; VC; (Fe,V)₃C; vanadium-bearing molten iron

1. Introduction

Vanadium is an important strategic metal. In 2021, 89% of the global vanadium output came from vanadium titanomagnetite [1]. V-bearing molten iron is a key intermediate product, and its composition mainly depends on the raw material and smelting process. Under traditional blast furnace smelting conditions, the reduction temperature is about 1500 °C [2,3]. A large number of high-melting-point carbides such as TiC, TiN, VC, and VN are formed in the molten iron [4], leading to a series of production difficulties and quality problems [5–7]. Wang [8] studied the thermodynamic properties such as the solubility and activity coefficient of V and Ti elements in the molten iron. He [9] and Zhang [10] revealed the influence mechanism of molten iron fluidity by studying viscous flow properties focusing on the carbides of titanium. Hou [11] found through thermodynamic calculation that the precipitation amount of VC and VN in V-bearing molten iron increased sharply at 1200–1250 °C, which was consistent with the change trend of hot metal viscosity. The formation of V-rich carbonitride particles was observed by a confocal laser scanning high-temperature microscope. However, the precipitation process of V-rich carbides has not been thoroughly studied.

The research on carbides of vanadium was mainly concentrated in the field of high-V medium-/low-carbon steel [12–14], focusing on the morphology and distribution of carbides in the alloy structure and its modification control measures. Kesri [15] studied the phase diagram of Fe–C–V ternary system and determined the thermodynamic stable regions of four phases: austenite (A, γ-Fe), ferrite (F, α-Fe), VC$_x$ type carbide, and M₃C type carbide in the solidification process, in which the equilibrium temperature between the VC$_x$ and M$_3$C stable regions was 1120 °C. Kawalec [16] and Fras [17] calculated the eutectic transformation process of VC in white cast iron, proposed the concept of the degree of eutectic saturation (S$_c$) in high-vanadium cast iron, and characterized the microstructure of eutectic VC. Recently, microalloyed steel with Nb, V, and Ti has received more and more attention with the increasing demand for high-quality steel [18,19]. The precipitation mechanism of the second-phase particles and their precipitation strengthening effect on the alloy structure have been studied extensively [20,21]. Pan [22] found five forms of V carbides in V-microalloyed medium-carbon steel. Liu [23] studied the effect of V content on VC precipitation in PD3 steel. Sayed [24] studied the effect of the heat-treatment parameters on the random vanadium precipitation in low-carbon steel. Javier [25] proposed a model to simulate the precipitation of V(CN). Li [26] revealed the hetero-nucleation effect of V(CN) during $\gamma \rightarrow \alpha$ phase change. Wang [27] observed an intermediate coherent crystal structure within the α-Fe matrix through TEM, revealing a two-step nucleation and growth mechanism of V(CN) in microalloyed steels. However, all the above studies focused on the eutectoid process of steel. The precipitation process and mechanism of VC in V-bearing molten iron in the solidification process have not been fully explored.

Recently, a novel low-temperature smelting technology was developed by adding soda [28] or alkali [29] in the reduction smelting process of vanadium titanomagnetite. Chen's research [30,31] showed that the oxide components in the slag were transformed into highly active sodium salt. This slag system has a low melting point and high fluidity, and it displays excellent impurity removal performance. Thus, this process can produce high-quality V-bearing molten iron at 1150–1300 °C. However, this molten iron has its own unique properties to be further investigated due to the existence of V-rich carbides. In this paper, V-bearing pig iron with different compositions was prepared. The morphology and structure of V-rich carbides and their precipitation characteristics in the pig iron structures were studied. This study is expected to provide technical support for the industrial practice of low-temperature smelting of vanadium titanomagnetite and the high-value utilization of V-bearing pig iron.

2. Experiment

2.1. Raw Materials and Experimental Procedures

The vanadium titanomagnetite used in the experiment came from a mine in Chaoyang, Liaoning Province, China (composition as shown in Table 1, average particle size 48 μm); the reducing agent anthracite (composition as shown in Table 2, crushed and screened to below 150 μm) and anhydrous Na$_2$CO$_3$ (AR, 99.8 wt%) were purchased from the Sinopharm Chemical Reagents Co., Ltd.(Shanghai, China) The smelting process was carried out in a box-type resistance furnace (Tianjin Zhonghuan Electric Furnace Co., Ltd., Tianjin, China, model: SX-D64163).

Table 1. Chemical composition of vanadium titanomagnetite concentrate (wt.%).

TFe	TiO$_2$	SiO$_2$	CaO	Al$_2$O$_3$	V$_2$O$_5$	MgO	MnO	S	P
47.97	19.32	6.59	2.98	2.29	1.46	0.70	0.41	0.028	0.011

Table 2. Chemical composition of anthracite (wt.%).

FC$_{ad}$	V$_{daf}$	M$_{ad}$	S$_t$	P	A$_d$							
					SiO$_2$	Al$_2$O$_3$	CaO	Fe$_2$O$_3$	TiO$_2$	Na$_2$O	MnO	MgO	Total
81.87	6.32	1.66	0.35	0.028	6.93	2.79	0.95	0.43	0.33	0.21	0.1	0.06	11.81

Firstly, raw ore, anhydrous Na$_2$CO$_3$, and anthracite were mixed in a ratio of 10:6:3 and placed in a graphite clay crucible. When the temperature of the furnace rose to the setting value, it was held on for 10 min, and then the graphite clay crucible was quickly moved into the furnace chamber. The reaction timing began until the temperature stabilized again to the setting value. After the reaction, the crucible was taken out and cooled in the air. After cooling to room temperature, pig iron samples were separated from the slag for analysis. The flow sheet of the smelting process of vanadium titanomagnetite is shown in Figure 1.

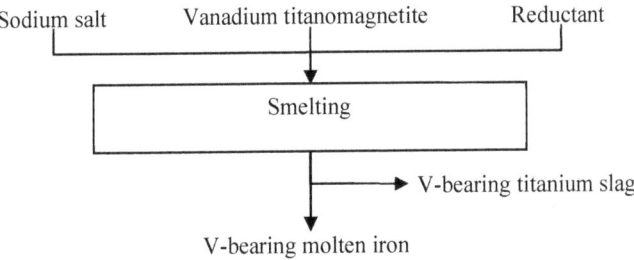

Figure 1. Flow sheet of the smelting process of vanadium titanomagnetite.

By adding auxiliaries such as Na_2CO_3 or NaOH in the smelting process, high-quality V-bearing molten iron with a low content of Ti (<0.03 wt.%), Si (<0.03 wt.%), S (<0.003 wt.%), and P (<0.005 wt.%) was produced. In view of the fact that the contents of C (4–5 wt.%) and V (0.3–2 wt.%) elements in the pig iron, the low-temperature property of the molten iron was mainly affected by these two elements due to the formation of V-rich carbides with a high melting point.

2.2. Theory Background

The Fe–C phase diagram has been studied quite thoroughly [32,33]. However, it is still controversial in some aspects, such as the specific temperature point and C content of the eutectic and eutectoid transformation lines. In general [34–36], the highest content of C in the F phase is 0.0218 wt.%, in the A phase is 2.11 wt.%, and in cementite (Fe_3C) is 6.69 wt.%.

The natural cooling rate is very fast in laboratory conditions. For hypereutectic white cast iron (wt.% (C) = 4.3–6.69%), the phase transition follows the diagram of metastable equilibrium Fe–Fe_3C [33,36]. Firstly, primary cementate (Fe_3C_I) is crystallized from the liquid phase. Then, the isothermal eutectic transformation occurs at the eutectic point (1148 °C, wt.% (C) = 4.3%). Eutectic A phase is precipitated with eutectic Fe_3C; these two phase form ledeburite cast (Ld), and its form at room temperature is L'd. After the liquid phase disappears, the secondary cementate (Fe_3C_{II}) is precipitated from the A phase. Upon approaching the eutectoid point (723 °C, wt.% (C) = 0.77%), the eutectoid F phase is precipitated with eutectoid Fe_3C. These two phases form pearlite (P). Lastly, the third cementate (Fe_3C_{III}) is precipitated from the F phase. Therefore, the equilibrium microstructure of cast iron at room temperature is as follows: primary cementate (Fe_3C_I) + L'd (eutectic Fe_3C + Fe_3C_{II} + P (eutectoid Fe_3C + Fe_3C_{III} + F)).

Fe_3C is an interstitial solid solution with the broad homogeneity range [35,36]. There are still some aspects under be explored, especially the carbide section of the diagram. The situation becomes even more confusing when the V element is present in the system.

2.3. Test Method

The carbon content in pig iron was determined using a CS-2800 carbon sulfur analyzer (NCS Testing Technology CO., Ltd., Beijing, China). The content of vanadium in pig iron was determined by high-sensitivity XRF (PHECDA-PRO, Beijing Ancoren Technology Co., Ltd., Beijing, China). The morphology and composition of the various phases in the pig iron samples were analyzed using a mineral dissociation analyzer (MLA250, FEI Company, Hillsboro, Oregon, United States). Furthermore, the pig iron was thinned by focused ion beam scanning electron microscopy (Helios G4 PFIB, Thermo Scientific™, Waltham, MA, USA), and the fine structure of the V-rich carbides and pig iron matrix was characterized by high-resolution field-emission transmission electron microscopy (JEM-F200, JEOL, Tokyo, Japan).

3. Results and Discussion

3.1. C and V Content of the Pig Iron

Several samples of V-bearing pig iron were prepared under different smelting conditions. The effects of different melting temperature and reaction time on the contents of C and V elements in the samples were investigated. The results are shown in Figure 2.

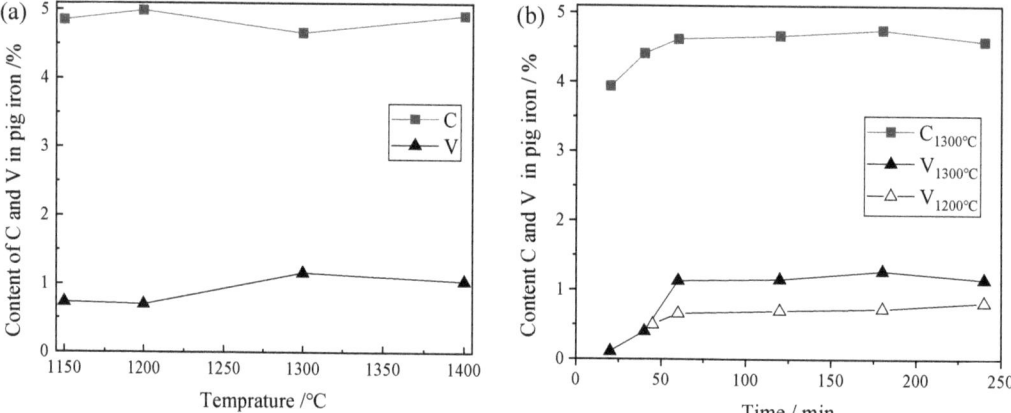

Figure 2. Content of C and V elements in pig iron at different smelting parameters: (**a**) effect of smelting temperature; (**b**) effect of smelting time.

As shown in Figure 2a, the C content in pig iron at different melting temperatures had little change with an average content of about 4.85 wt.%. However, the V content in pig iron increased with the increase in temperature. The V content was 0.70 wt.% at 1200 °C and 1.16 wt.% at 1300 °C.

The variation trend of V content in pig iron was almost consistent with that of C in Figure 2b. When the smelting temperature was 1300 °C, the iron was separated with the slag at 20 min, the content of C was 3.94 wt.%, and the content of V was only 0.11 wt.%. Then, the content of C and V increased rapidly. After 60 min of reaction, the content of C and V tended to change gently, and the content of C and V remained at about 4.8 wt.% and 1.1 wt.%, respectively. When the smelting temperature was 1200 °C, it took about 45 min for the separation of iron and slag to be completed. At this time, the content of V was 0.5 wt.%, and then it slowly increased to 0.7–0.9%.

It could be inferred that the solubility of V element in pig iron was greatly affected by the C content. The significant difference in V content in pig iron at 1200 °C and 1300 °C may be due to the transformation of the V-rich carbides in pig iron.

3.2. Morphology and Composition

The content of C was different in various phases; thus, it could be inferred that the content of V in different pig iron structures was also changing. For hypereutectic white cast iron, it depended on its solubility in the Fe_3C phase and the F phase, as well as the proportion of the two phases in the structure. The microstructure of V-bearing pig iron can be further analyzed in Figure 3.

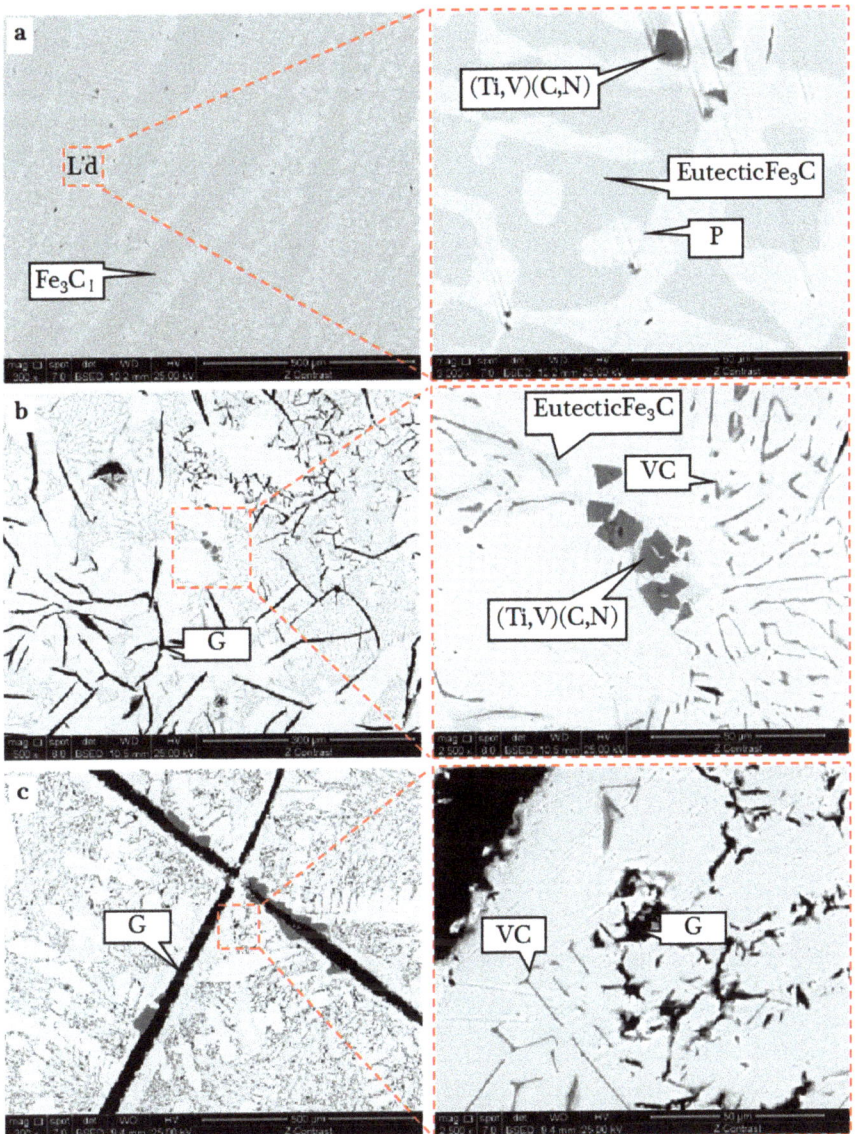

Figure 3. SEM images of the V-bearing pig iron at different smelting temperature: (**a**) 1200 °C, 120 min; (**b**) 1300 °C, 120 min; (**c**) 1400 °C, 120 min.

As shown in Figure 3a, V-bearing pig iron at 1200 °C was mainly composed of the dark-gray long-strip Fe3CI phase and light and dark L′d phase. The bright color phase was the P phase, and the gray phase was a mixture of eutectic Fe_3C and Fe_3C_{II}. There were dark-gray particles in different L′d phases. The L′d structure almost disappeared at 1300 °C (Figure 3b), and there appeared a blacker fine flake graphite phase (G) which separated the iron matrix, while the P phase region expanded substantially. The original eutectic Fe_3C region was transformed into a large number of gray worm-like and rod-shaped particles with centripetal converging distribution, while the dark-gray cube particles were in the core. Figure 3c showed the appearance of large flake G-phase and large cube particles over 100 μm at 1400 °C. The number of gray worm-like and rod-shaped particles increased, and the distribution was more regular, indicating that the phase precipitation had better crystallinity with the increase in the smelting temperature. The components of different phases in pig iron at 1200 °C were analyzed by EDS point scanning as shown in Figure 4 and Table 3.

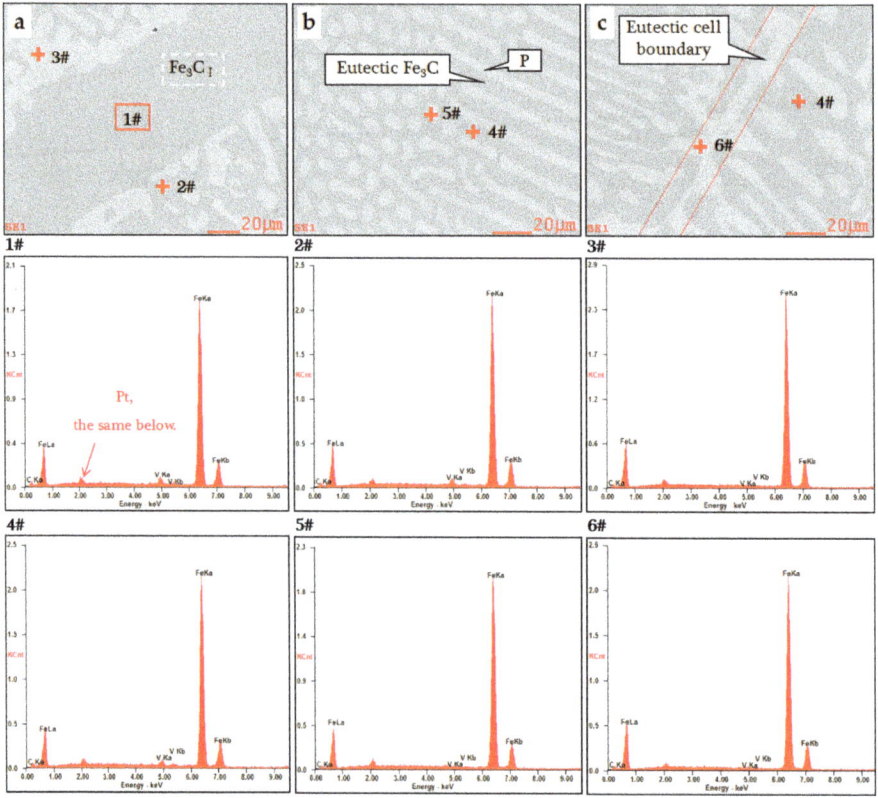

Figure 4. SEM-EDS images of the V-bearing pig iron at 1200 °C: (**a**) Fe_3C_I and adjacent L'd phases; (**b**) L'd phase consisting of eutectic Fe_3C and pearlite phases; (**c**) eutectic cell boundary region.

Table 3. EDS results of the V-bearing pig iron at 1200 °C.

Element (wt.%)	1#	2#	3#	4#	5#	6#
Fe	85.82	86.94	91.68	88.78	91.38	92.26
C	12.00	11.16	8.00	9.54	8.11	7.44
V	2.18	1.92	0.33	1.68	0.51	0.31

The results showed that the highest content of V was 2.18 wt.% (Figure 4a), which existed in the Fe_3C_I, while the lowest V content in P was 0.31 wt.% (Figure 4c). Although EDS results had a large deviation, it was obvious that the Fe_3C phase could enrich V, and the solubility of V in the A phase was limited. Assuming that all Fe_3C had the same V content, it could be inferred that the V content in F phase was very low; the value reported in the literature [22] was only 0.09 wt.%.

According to the relevant theories in Section 2.2, Fe_3C_I was the first phase precipitated from the V-bearing molten iron; accordingly, it had the highest content of V. Then the V content in the subsequent precipitated eutectic Fe_3C gradually decreased due to the insufficient content of V. According to the lever rule, the content of Fe_3C_I in the V-bearing pig iron is presented in Equation (1).

$$w_{Fe_3C_I} = \frac{4.85 - 4.3}{6.69 - 4.3} = 23\%. \tag{1}$$

Then, the LD phase accounted for 87%. The theoretical content of eutectic Fe_3C precipitated from LD is presented in Equation (2).

$$w_{Fe_3C} = \frac{4.3}{6.69} \times 87\% = 55.9\%. \tag{2}$$

Theoretically, for this hypereutectic white cast iron product, the content of Fe_3C phase was 78.9% and that of F phase was 21.1%. The chemical formula of this V-rich Fe_3C could be written as $(Fe_{1-x},V_x)_3C$, $(0 \leq x \leq 1)$ (for short, $(Fe,V)_3C$).

The components of different phases in pig iron at 1300 °C were analyzed by EDS point scanning and mapping scanning, as shown in Figure 5 and Table 4.

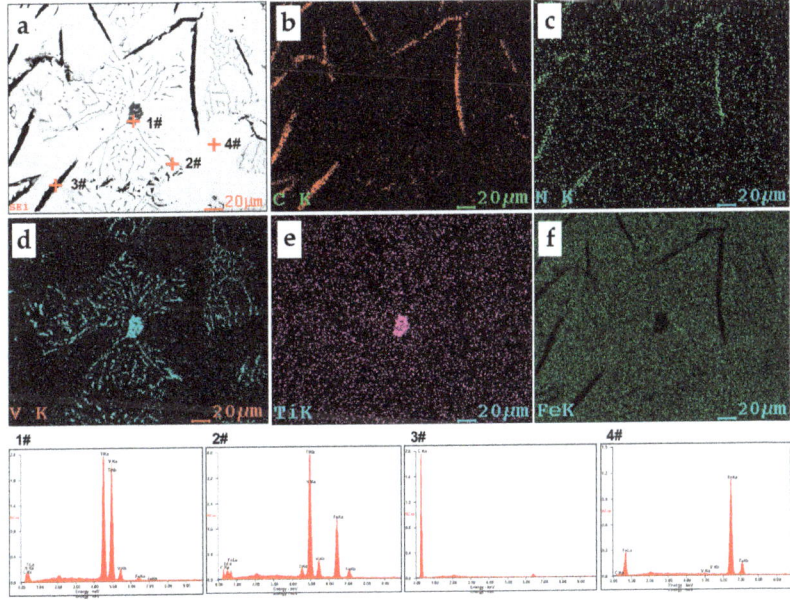

Figure 5. SEM-EDS images of the V-bearing pig iron at 1300 °C: (**a**) (Ti,V)(C,N),VC, graphite, and austenite phases in pig iron; (**b–f**) mapping scanning measurement of C, N, V, Ti, and Fe elements.

Table 4. EDS results of the V-bearing pig iron at 1300 °C.

Element (wt.%)	1#	2#	3#	4#
Fe	2.03	32.94	/	91.26
C	13.83	21	100	8.04
N	6.92	/	/	/
V	35.70	43.09	/	0.70
Ti	41.51	2.97	/	/

In Figure 5a, the content of V and Ti in regular cube particles was 35.70 wt.% and 41.51 wt.%, respectively. These secondary phase particles were (Ti,V)(C,N). The worm-like and rod-like gray particles were mainly composed of VC, with a small amount of Ti. Due to the small size of the particle, EDS scanning could inevitably identify a relatively high content of matrix Fe. After removing Fe, the content of V in VC phase reached 65 wt.%. Considering that there was no V in the G phase and (Ti,V)(C,N) particles were few, the precipitation of VC almost enriched the majority of V from the molten iron.

By further comparison of Figure 3a,c, it can be seen that the eutectic (Fe,V)$_3$C region originally in L'd was greatly reduced, while a large amount of VC phase appeared in the gap of the G phase, indicating that (Fe,V)$_3$C solid solution was in a kind of metastable phase, which would decompose into the G phase and generate VC under certain conditions. The morphology of VC was consistent with the literature [15–18]. It also showed that excessive element V, a strongly graphitized element, could reduce the stability of Fe$_3$C.

To sum up, the V content in molten iron was low below 1300 °C. The liquid iron had good fluidity, and the impurity content was very low. The obtained pig iron was mainly composed of (Fe,V)$_3$C and P phase. Above 1300 °C, the V content in molten iron increased by more than 1 wt.%; (Fe,V)$_3$C would lost its stability and decomposed along with the precipitation of graphite and VC during the solidification process.

3.3. Interface Structures and Phase Transformation Process

High-resolution structures of pig iron were detected by FIB-TEM. Figure 6a shows the FIB section of the pig iron sample. In Figure 6b, it can be seen that the V-bearing pig iron structure contained three phases: (Fe,V)$_3$C, VC, and F.

Figure 6c shows the α-Fe matrix and eutectic Fe$_3$C in the P phase. According to the above analysis, the eutectoid Fe$_3$C was also enriched with V. Surprisingly, the atomic structure of the (Fe,V)$_3$C solid solution was perfectly preserved due to the eutectoid transition from A to P being a solid phase transition process. We can clearly see the obvious interface undulating transition layer in the Figure 6d,e.

Figure 6f shows the high-resolution structure of VC phases, which were full of various microdefects, microtwins, semi-colattice nanostructures, and colattice nanoparticles. Figure 6g,h shows the interface structure of VC and F phases. It can be seen that there were small fluctuations of about 3–4 atomic layers thickness at the interface, from which the mismatches of semi-coherent interface micro-regions and noncoherent dislocations can be seem. The VC and A phases were eutectic precipitated from the (Fe,V)$_3$C matrix. When the temperature was above 723 °C, both phases had an FCC structure. Therefore, the interface should be a coherent or semi-coherent interface. When the temperature was below 723 °C, the A phase transformed into the F phase. Its structure changed from FCC to BCC, during which the atomic rearrangement led to the mismatches of the interfacial lattice. As can be seen from the Figure 6h, there was a difference of 0.005 nm in the crystal plane spacing between VC and F phase; thus, it showed periodic mismatches, forming a stepwise dislocation.

As shown in Figure 6I, there were many vacancies and defects in the complex orthogonal lattice structure of Fe$_3$C. When the V atoms were solidly dissolved into the Fe$_3$C lattice, they would partially displace the Fe atom and formed (Fe,V)$_3$C solid solution. In Figure 6II, we can see the VC cluster structure formed by more than six V atoms converging together. The VC precursor crystal nucleus is shown in Figure 6III. Figure 6IV shows the local structure of VC particles. The process of VC precipitation from (Fe,V)$_3$C is clearly displayed at the atomic level through the TEM images.

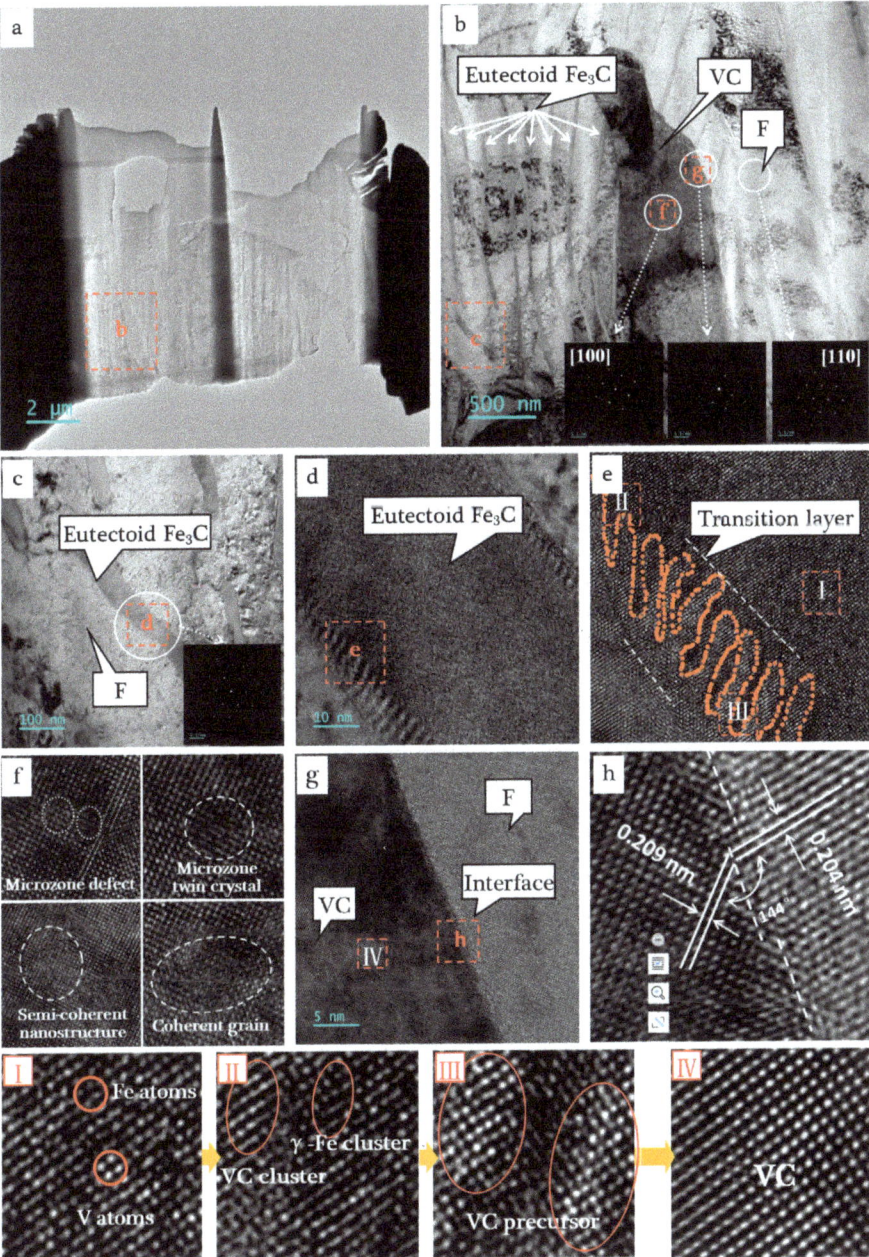

Figure 6. TEM image and electron diffraction pattern of vanadium-bearing pig iron. (**a**) FIB section; (**b**) VC and pearlite matrix; (**c**) eutectic cementite and ferrite matrix; (**d**) (Fe,V)$_3$C; (**e**) interface fluctuation zone; (**f**) nanoparticles in VC phase; (**g**) interface between VC and ferrite matrix; (**h**) interface structure; (**I**) (Fe,V)$_3$C solid solution structure; (**II**) VC cluster; (**III**) VC grain; (**IV**) VC.

The clear images of the transition process provided by TEM helped in explaining the whole process of VC phase precipitation from the $(Fe,V)_3C$ phase during the solidification of the V-bearing molten iron.

(a) The nucleation induction stage. With the decrease in the temperature, the first crystallized $(Fe,V)_3C_I$ phase was unstable. It decomposed into large G_I flakes and released V element into the liquid phase. As a result, C content decreased while V content increased in the adjacent liquid phase. Then, the A phase was precipitated out of the G_I phase as the heterogeneous nucleation interface [37]. C and V elements were released into the adjacent liquid phase due to their low solubility in A phase, which then induced the formation of $(Fe,V)_3C_I$ and further decomposed into G_I.

(b) The nucleation pregnant stage. With the continuous progress of the process (1), the content of C in the residual liquid phase dropped to the eutectic component, while the content of V element increased and dissolved into the eutectic $(Fe,V)_3C$ phase. The size effect caused lattice distortion with the increase in solid solubility, leading to a gradual decrease in the stability of the $(Fe,V)_3C$ phase. Then, many VC cluster structures were formed due to the fluctuation of V concentration and energy.

(c) The nucleus formation stage. When the total distortion energy reached the critical value, these clusters first adsorbed to the A phase matrix interface under induction of the FCC structure. Then, they further absorbed the adjacent VC clusters. Finally, they grew into a VC precursor nucleus which started from the A phase matrix and protruded into the $(Fe,V)_3C$ phase. For the eutectoid process, due to the difficulty of diffusion in the solid phase transition process and the absence of redundant V atoms, the VC precursor crystal nucleus could not be further grown; finally, the two-phase undulating interface structure was formed. Wang [27] made a similar discovery in V-microalloyed steels.

(d) The precipitation and growth stage. For the eutectic process, VC precursor crystal nuclei further converged and grew, forming the VC primary crystal nuclei. With the progress of the reaction, the content of C and V elements adjacent to the crystal nuclei decreased, while the content outside was higher, thus forming a concentration gradient of C and V elements decreased to the grain center. Therefore, the growth direction of VC deviated from the center of the sphere. Under the restriction of the separation of the A phase and G phase, the VC phase was precipitated continuously and finally formed a radiating eutectic cell [13]. If there were (Ti,V)(C,N) second-phase particles or other impurity particles precipitated earlier in the molten iron, they would become the core of VC eutectic cell as the heterogeneous nucleated crystal species.

In this process, the large precipitation of $(Fe,V)_3C_I$ and VC with a high melting point led to poor fluidity of the V-bearing molten iron. On the surface of molten iron, $(Fe,V)_3C_I$ decomposed and a large amount of G_I precipitated due to the rapid temperature drop. The eutectic precipitation temperature of the G phase was 1154 °C [32], and the solidification temperature of the molten iron actually increased. Furthermore, G burned and released a large amount of CO_2 after contact with air, resulting in huge pores on the surface of cast iron and large number of small pores in the interior. This seriously affected the composition and quality of the pig iron.

4. Conclusions

V-bearing molten iron was obtained by adding Na_2CO_3 in the smelting process of vanadium titanomagnetite at low temperature. The morphology and structure of V-rich carbides in V-bearing pig iron were studied, and the precipitation characteristics of V-rich carbides in molten iron were discussed.

(1) Vanadium trended to form carbide in iron and steel structure. There were two main forms of V-rich carbides in low-temperature V-bearing pig iron: $(Fe,V)_3C$ solid solution and VC solid solution.

(2) The microstructure of pig iron and the existence form of V-rich carbides changed at different melting temperatures. In the temperature range of 1150 °C to 1250 °C, the metallographic morphology were P and $(Fe,V)_3C$ phase. When the temperature exceeded 1300 °C, the P, G, and VC phases were mainly present.

(3) When V atoms were solidly dissolved into the Fe_3C lattice, they partially displaced the Fe atoms and formed $(Fe,V)_3C$ solid solution. In the eutectic transition process, the supersaturated V element destroyed the stability of $(Fe,V)_3C$ and promoted the precipitation of G and VC. The existence of VC clusters and VC precursor nuclei were observed by TEM, which confirmed the possibility of this transition process at the atomic perspective.

Above 1300 °C, the rapid crystallization of $(Fe,V)_3C$ and its transition to the G and VC phases led to an increase in viscosity and poor fluidity of the V-bearing molten iron, resulting in a large number of porosity defects. The process seriously affects the composition and quality of the pig iron product. Therefore, it is recommended that the melting temperature should be controlled at no more than 1300 °C.

Author Contributions: Conceptualization, L.C. and D.C.; funding acquisition, D.C., H.Z., Y.Z., L.W. and T.Q.; investigation, L.C. and X.S.; methodology, L.C., H.Z. and Y.Z.; project administration, D.C., H.Z., Y.Z. and L.W.; resources, D.C., L.W. and T.Q.; supervision, D.C., H.Z., L.W. and T.Q.; validation, L.C., D.C., X.S. and L.W.; visualization, L.C., Y.Z., Y.L. and F.M.; writing—original draft, L.C., Y.L. and F.M.; writing—review and editing, L.C., D.C., H.Z. and L.W. All authors have read and agreed to the published version of the manuscript.

Funding: Financial support for this research was provided by Strategic Priority Research Program of the Chinese Academy of Sciences (Funder: Desheng Chen, Funding number: XDC04010102, and Funder: Lina Wang, Funding number: XDC04010100), the Special Project for Transformation of Major Technological Achievements in HeBei province (Funder: Desheng Chen, Funding number: 19044012Z), the science and technology program of Hengshui (Funder: Hongxin Zhao, Funding number: 2020016004B), the Province Key R&D Program of Hebei (Funder: Yulan Zhen, Funding number: 20374105D), and the National Natural Science Foundation of China (Funder: Yahui Liu, Funding number: 22078343).

Data Availability Statement: Data are contained within the article.

Conflicts of Interest: The authors declare no conflict of interest.

References

1. Wu, Y.; Chen, D.H.; Liu, W.H.; Sun, Z.H.; Zhang, B.X.; He, R. Global vanadium industry development report 2021. *Iron Steel Va-Nadium Titan.* **2022**, *43*, 1–9. [CrossRef]
2. Xie, H.E.; Hu, P.; Zheng, K.; Zhu, F.X. Study on phase and chemical composition of V-Ti sinter during softening, melting and dripping process. *Iron Steel Vanadium Titan.* **2022**, *43*, 107–117. [CrossRef]
3. Husslage, W.M.; Bakker, T.; Steeghs, A.G.S.; Reuter, M.; Heerema, R.H. Flow of molten slag and iron at 1500 °C to 1600 °C through packed coke beds. *Met. Mater. Trans. B* **2005**, *36*, 765–776. [CrossRef]
4. Wen, G.Y.; Yan, Y.Z.; Zhao, S.Z.; Huang, J.J.; Jiang, G.H.; Yang, X.M. Properties of liquid iron containing vanadium and titanium. *Iron Steel* **1996**, *31*, 6–11.
5. Yu, T.; Jiang, T.; Wen, J.; Sun, H.; Li, M.; Peng, Y. Effect of chemical composition on the element distribution, phase composition and calcification roasting process of vanadium slag. *Int. J. Miner. Met. Mater.* **2022**, *29*, 2144–2151. [CrossRef]
6. Zhang, J.; Zhang, W.; Xue, Z. An Environment-Friendly Process Featuring Calcified Roasting and Precipitation Purification to Prepare Vanadium Pentoxide from the Converter Vanadium Slag. *Metals* **2019**, *9*, 21. [CrossRef]
7. Wang, Y.-N.; Song, W.-C.; Li, H. Vanadium extraction and dephosphorization from V-bearing hot metal with fluxes containing CaO. *J. Central South Univ.* **2015**, *22*, 2887–2893. [CrossRef]
8. Wang, H.C.; Chen, E.B.; Dong, Y.C.; Li, W.C. Analysis of the thermodynamic Property of Fe-C-V melt. *J. Univ. Sci. Technol. Beijing* **2000**, *22*, 312–315.
9. He, Y.Y.; Liu, Q.C.; Yang, J.; Chen, Q.X.; Ao, W.Z. Experimental Investigation on Fluidity of Hot Metal Bearing Titanium. *Iron Steel Vanadium Titan.* **2010**, *31*, 10–14.
10. Zhang, J.L.; Wei, M.F.; Guo, H.W.; Mao, R.; Hu, Z.W.; Zhao, Y.B. Effect of Ti and Si on the viscosity and solidification properties of molten iron. *J. Univ. Sci. Technol. Beijing* **2013**, *35*, 994–999. [CrossRef]
11. Hou, P.; Yu, W.Z.; Bai, C.G.; Pan, C.; Yuan, W.N.; Li, T. Viscous flow properties and influencing factors of vanadium-titanium magnetite smelting iron. *Iron Steel* **2022**, *57*, 57–65. [CrossRef]

12. Hwang, K.C.; Lee, S.; Lee, H.C. Effects of alloying elements on microstructure and fracture properties of cast high speed steel rolls: Part I: Microstructural analysis. *Mater. Sci. Eng. A* **1998**, *254*, 282–295. [CrossRef]
13. Wei, S.Z.; Ni, F.; Zhu, J.H.; Long, R.; Xu, L.J. Solidification process of Fe-C alloy containing rich vanadium. *J. Iron Steel Res.* **2005**, *17*, 56–64.
14. Xu, L.J.; Li, Z.; Wei, S.Z. VC precipitation and retained austenite transformation of high-vanadium high-speed steel during tem-pering. *Heat Treat. Met.* **2016**, *41*, 6–11. [CrossRef]
15. Kesri, R.; Durand-Charre, M. Metallurgical structure and phase diagram of Fe–C–V system: Comparison with other systems forming MC carbides. *Mater. Sci. Technol.* **1988**, *4*, 692–699. [CrossRef]
16. Kawalec, M.; Fraś, E. Structure, Mechanical Properties and Wear Resistance of High-vanadium Cast Iron. *ISIJ Int.* **2008**, *48*, 518–524. [CrossRef]
17. Fras, E.; Kawalec, M.; Lopez, H. Solidification microstructures and mechanical properties of high-vanadium Fe–C–V and Fe–C–V–Si alloys. *Mater. Sci. Eng. A* **2009**, *524*, 193–203. [CrossRef]
18. Tian, Y.; Yu, H.; Zhou, T.; Wang, K.; Zhu, Z. Revealing morphology rules of MX precipitates in Ti-V-Nb multi-microalloyed steels. *Mater. Charact.* **2022**, *188*, 111919. [CrossRef]
19. Li, X.; Li, H.; Liu, L.; Deng, X.; Wang, Z. The formation mechanism of complex carbides in Nb-V microalloyed steel. *Mater. Lett.* **2022**, *311*, 131544. [CrossRef]
20. Hu, B.-H.; Cai, Q.-W.; Wu, H.-B. Kinetics of Precipitation Behavior of Second Phase Particles in Ferritic Ti-Mo Microalloyed Steel. *J. Iron Steel Res. Int.* **2013**, *20*, 69–77. [CrossRef]
21. Wang, K.; Wang, L.J.; Wang, Q.; He, J.C.; Liu, C.M. Grain growth of deformation induced ferrite in V microalloyed low carbon steel during controlled cooling process. *J. Iron Steel Res.* **2011**, *18*, 377–382. [CrossRef]
22. Liu, T.M.; Zhou, S.Z.; Zuo, R.L.; Mei, D.S.; Deng, J.H. Effect of vanadium content on VC precipitation in PD3 steel. *J. Chongqing Univ.* **2001**, *6*, 78–81. [CrossRef]
23. Pan, X.-L.; Umemoto, M. Precipitation Characteristics and Mechanism of Vanadium Carbides in a V-Microalloyed Medium-Carbon Steel. *Acta Met. Sin. English Lett.* **2018**, *31*, 1197–1206. [CrossRef]
24. Hashemi, S.G.; Eghbali, B. Analysis of the formation conditions and characteristics of interphase and random vanadium precipitation in a low-carbon steel during isothermal heat treatment. *Int. J. Miner. Met. Mater.* **2018**, *25*, 339–349. [CrossRef]
25. Aldazabal, J.; Garcia-Mateo, C.; Capdevila, C. Simulation of V(CN) Precipitation in Steels Allowing for Local Concentration Fluctuations. *Mater. Trans.* **2006**, *47*, 2732–2736. [CrossRef]
26. Li, X.; Zhao, L.; Wang, X.; Zhao, Y. Precipitation and hetero-nucleation effect of V(C, N) in V-microalloyed steel. *J. Wuhan Univ. Technol. Sci. Ed.* **2008**, *23*, 844–849. [CrossRef]
27. Wang, H.; Li, Y.; Detemple, E.; Eggeler, G. Revealing the two-step nucleation and growth mechanism of vanadium carbonitrides in microalloyed steels. *Scr. Mater.* **2020**, *187*, 350–354. [CrossRef]
28. Zhang, Y.-M.; Yi, L.-Y.; Wang, L.-N.; Chen, D.-S.; Wang, W.-J.; Liu, Y.-H.; Zhao, H.-X.; Qi, T. A novel process for the recovery of iron, titanium, and vanadium from vanadium-bearing titanomagnetite: Sodium modification–direct reduction coupled process. *Int. J. Miner. Met. Mater.* **2017**, *24*, 504–511. [CrossRef]
29. Shi, L.-Y.; Zhen, Y.-L.; Chen, D.-S.; Wang, L.-N.; Qi, T. Carbothermic Reduction of Vanadium-Titanium Magnetite in Molten NaOH. *ISIJ Int.* **2018**, *58*, 627–632. [CrossRef]
30. Chen, L.; Zhen, Y.; Zhang, G.; Chen, D.; Wang, L.; Zhao, H.; Meng, F.; Qi, T. Carbothermic reduction of vanadium titanomagnetite with the assistance of sodium carbonate. *Int. J. Miner. Met. Mater.* **2022**, *29*, 239–247. [CrossRef]
31. Chen, L.-M.; Zhen, Y.-L.; Zhang, G.-H.; Chen, D.; Wang, L.; Zhao, H.; Liu, Y.; Meng, F.; Wang, M.; Qi, T. Mechanism of Sodium Carbonate-Assisted Carbothermic Reduction of Titanomagnetite Concentrate. *Met. Mater. Trans. B* **2022**, *53*, 2272–2292. [CrossRef]
32. Zhukov, A.A. Phase diagram of alloys of the system Fe-C. *Met. Sci. Heat Treat.* **1988**, *30*, 249–255. [CrossRef]
33. Okamoto, H. The C-Fe (carbon-iron) system. *J. Phase Equilibria Diffus.* **1992**, *13*, 543–565. [CrossRef]
34. Raghavan, V. C-Fe-V (carbon-iron-vanadium). *J. Phase Equilibria Diffus.* **1993**, *14*, 622–623. [CrossRef]
35. Davydov, S.V. Phase Equilibria in the Carbide Region of Iron–Carbon Phase Diagram. *Steel Transl.* **2020**, *50*, 888–896. [CrossRef]
36. Leineweber, A.; Shang, S.; Liu, Z. C-vacancy concentration in cementite, Fe_3C1^-, in equilibrium with α-Fe[C] and γ-Fe[C]. *Acta Mater.* **2015**, *86*, 374–384. [CrossRef]
37. Zhai, Q.Y.; Xu, J.F.; Yuan, S. Formation mechanism of the austenite shell around nodular graphite in the hypereutectic nodular iron. *Foundry* **2001**, *1*, 18–21.

Disclaimer/Publisher's Note: The statements, opinions and data contained in all publications are solely those of the individual author(s) and contributor(s) and not of MDPI and/or the editor(s). MDPI and/or the editor(s) disclaim responsibility for any injury to people or property resulting from any ideas, methods, instructions or products referred to in the content.

Article

Ginzburg–Landau Analysis on the Physical Properties of the Kagome Superconductor CsV_3Sb_5

Tianyi Han [1], Jiantao Che [1], Chenxiao Ye [2] and Hai Huang [1,*]

1 Department of Mathematics and Physics, North China Electric Power University, Beijing 102206, China
2 School of Nuclear Science and Engineering, North China Electric Power University, Beijing 102206, China
* Correspondence: huanghai@ncepu.edu.cn

Abstract: The kagome lattice consisting of corner-sharing triangles has been studied in the context of quantum physics for more than seventy years. For the novel discovered kagome superconductor CsV_3Sb_5, identifying the pairing symmetry of order parameter remained an elusive problem until now. Based on the two-band Ginzburg–Landau theory, we study the temperature dependence of upper critical field and magnetic penetration depth for this compound. All theoretical results are consistent with the experimental data, which strongly indicates the existence of two-gap s-wave superconductivity in this system. In addition, it is worth noting that the anisotropy of effective masses in the band with large (or small) gap is about 70 (or 2.4). With the calculation of the Kadowaki–Woods ratio as 0.58×10^{-5} μΩ cm mol^2 K^2 mJ^{-2}, the semi-heavy-fermion feature is suggested in the compound CsV_3Sb_5.

Keywords: Ginzburg–Landau theory; two-gap s-wave superconductivity; CsV_3Sb_5; heavy fermion system

1. Introduction

The recently discovered kagome metal series AV_3Sb_5 (A = K, Rb, Cs) exhibit topologically nontrivial band structures, chiral charge order, charge density wave (CDW) and superconductivity, presenting a unique platform for realizing exotic electronic states [1–6]. These materials crystallize in the $P6/mmm$ space group with ideal kagome nets of V atoms which are coordinated by Sb atoms. The kagome layers of CsV_3Sb_5 are sandwiched by extra antimonene layers and Cs layers, as shown in the inset of Figure 1. With the decrease in temperature, the CDW phase transition takes place at about 94 K for CsV_3Sb_5 [7–11]. Based on the density functional theory, the first principle calculations show that the CDW observed in this family of compounds is a consequence of the atomic displacement from the high-symmetry positions of the kagome network [12,13]. Meanwhile, the experimental measurements with Raman spectroscopy have also confirmed the dynamical lattice distortions in the CDW phase [14–16]. Then, below about 2.5 K superconductivity is observed and coexists with the CDW order without further structural transitions [17–23].

For a series of hexagonal symmetric layered materials, the electronic band structure and topological properties have already been carried out based on the numerical ab initio calculations [24,25]. Furthermore, it is also well known that the two-dimensional kagome lattice hosts a pair of Dirac bands protected from the lattice symmetry and will trigger the correlated topological states of matter. For the three-dimensional crystal CsV_3Sb_5, V $3d$ and Sb $5p$ orbitals play dominant contributions to the density of states near the Fermi level, and the nontrivial band crossing is extended along a one-dimensional line in the Brillouin zone. Such band structure features are associated with a $\mathbb{Z}_2$ topological index [2,26], and the topological surface states can be easily observed if a direct gap exists for every momentum in this system. Additionally, electronic transport and heat capacity measurements reveal a large Kadowaki–Woods (KW) ratio. It indicates the V-based kagome prototype structure may be of potential interest as a host of correlated electron phenomenon, particularly as a heavy-fermion material.

Up to now, several theoretical and experimental investigations have already been performed on the pairing symmetry of the kagome compound CsV_3Sb_5. Multiband structure of this compound was predicted by previous theoretical calculations and then confirmed by angle-resolved photoemission spectroscopy studies [27–31]. Meanwhile, the measurement of nuclear magnetic resonance on this kagome metal showed a Hebel–Slichter coherence peak just below T_c, indicating that CsV_3Sb_5 is an s-wave superconductor [11]. Magnetic penetration depth of this system measured by tunneling diode oscillator displayed a clear exponential behavior at low temperatures, which also provides evidence for the nodeless structure in this compound. Furthermore, experimental data on temperature dependence of the superfluid density and electronic specific heat can be well described by two-gap superconductivity scenario [32], which is consistent with the presence of multiple Fermi surfaces in this system. Obviously, to date there is still no general consensus on the form of superconducting order parameter in CsV_3Sb_5 and further explorations to elucidate this issue are necessary.

The main motivation of the present paper is to identify the form of order parameter in this kagome superconductor. Based on the two-band Ginzburg–Landau (GL) theory, we study the temperature dependence of upper critical field and magnetic penetration depth for this compound. Our results can fit the experimental data well in a broad temperature range, which thus strongly suggests CsV_3Sb_5 as a two-gap s-wave superconductor. We can also obtain the effective mass of the electron in the c-axis for the first band as $38m_e$ and only $0.31m_e$ for the other band. With this semi-heavy-fermion feature, we can qualitatively understand the experimental value of the KW ratio in CsV_3Sb_5.

The paper is organized as follows: In the next section, we discuss the two-band GL theory. We derive the formula for the critical temperature and discuss how to properly choose the parameters in the GL theory. In Section 3, we calculate the upper critical field H_{c2} for the kagome superconductor CsV_3Sb_5. Then in Section 4, we work out the magnetic penetration depth for this compound. In Section 5, we discuss the KW ratio and semi-heavy-fermion feature in this material. Finally, Section 6 contains the conclusion of the paper.

2. Two-Band Anisotropic Ginzburg–Landau Theory

Taking into account the multi-gap characteristics of V-based superconductors, we can note the two-band GL free energy functional as [33–35].

$$F = \int d^3\mathbf{r}(f_1 + f_2 + f_{12} + \mathbf{H}^2/8\pi), \tag{1}$$

with

$$f_i = \frac{\hbar^2}{2m_i}\left|\left(\partial_x - \frac{2ieA_x}{\hbar c}\right)\Psi_i\right|^2 + \frac{\hbar^2}{2m_i}\left|\left(\partial_y - \frac{2ieA_y}{\hbar c}\right)\Psi_i\right|^2 + \frac{\hbar^2}{2m_{iz}}\left|\left(\partial_z - \frac{2ieA_z}{\hbar c}\right)\Psi_i\right|^2 \\ - \alpha_i(T)|\Psi_i|^2 + \frac{\beta_i}{2}|\Psi_i|^4 \tag{2}$$

and

$$f_{12} = \eta_{12}(\Psi_1^*\Psi_2 + \text{c.c.}). \tag{3}$$

Here, f_i ($i = 1, 2$) is the free energy density for each band and f_{12} is the interaction-free energy density. $\Psi_i \propto \sqrt{N_i}\Delta_i$ with N_i the density of states at the Fermi level is the superconducting order parameter and $N_1/N_2 = 0.79/0.21$ from the specific heat data in CsV_3Sb_5 [32]. m_i and m_{iz} denote the effective masses in the ab-plane and in the c-direction for band i. From the measurement of Shubnikov–de Haas oscillations with the magnetic field parallel to the c-direction, we have $m_1 = 0.55m_e$ and $m_2 = 0.13m_e$ [36]. η_{12} is the Josephson coupling constant. The coefficient α_i is a function of temperature, while β_i is independent of temperature. If the interband interaction is neglected, the functional can be reduced to two independent single-band problems with the corresponding critical temperatures T_{c1} and T_{c2}, respectively. Thus, the parameters α_1 and α_2 can be approximately expressed as $\alpha_i = \alpha_{i0}(1 - T/T_{ci})$ with α_{i0} the proportionality constant [37]. $\mathbf{H} = \nabla \times \mathbf{A}$ is magnetic field and $\mathbf{A} = (A_x, A_y, A_z)$ is the vector potential.

By minimizing the free energy F with Ψ_i^*, we can obtain the GL equations for the description of the two-band superconductivity

$$\hat{M}_{11}\Psi_1 + \hat{M}_{12}\Psi_2 = 0 \tag{4}$$

and

$$\hat{M}_{21}\Psi_1 + \hat{M}_{22}\Psi_2 = 0, \tag{5}$$

with

$$\hat{M}_{ii} = -\frac{\hbar^2}{2m_i}\left(\partial_x - \frac{2ieA_x}{\hbar c}\right)^2 - \frac{\hbar^2}{2m_i}\left(\partial_y - \frac{2ieA_y}{\hbar c}\right)^2 - \frac{\hbar^2}{2m_{iz}}\left(\partial_z - \frac{2ieA_z}{\hbar c}\right)^2 \\ - \alpha_i + \beta_i|\Psi_i|^2 \tag{6}$$

and

$$\hat{M}_{12} = \hat{M}_{21} = \eta_{12}. \tag{7}$$

By minimizing the free energy F with the vector potential $\mathbf{A}$, we then obtain the equation for the current $\mathbf{j} = (j_x, j_y, j_z)$ as

$$\nabla \times \mathbf{H} = 4\pi \mathbf{j} \tag{8}$$

with

$$j_x = \frac{e}{ic}\sum_i \left(\frac{\hbar}{m_i}\Psi_i^* \partial_x \Psi_i - \frac{\hbar}{m_i}\Psi_i \partial_x \Psi_i^* - \frac{4ie}{m_i c}\Psi_i^* \Psi_i A_x\right), \tag{9}$$

$$j_y = \frac{e}{ic}\sum_i \left(\frac{\hbar}{m_i}\Psi_i^* \partial_y \Psi_i - \frac{\hbar}{m_i}\Psi_i \partial_y \Psi_i^* - \frac{4ie}{m_i c}\Psi_i^* \Psi_i A_y\right) \tag{10}$$

and

$$j_z = \frac{e}{ic}\sum_i \left(\frac{\hbar}{m_{iz}}\Psi_i^* \partial_z \Psi_i - \frac{\hbar}{m_{iz}}\Psi_i \partial_z \Psi_i^* - \frac{4ie}{m_{iz} c}\Psi_i^* \Psi_i A_z\right). \tag{11}$$

Equations (4), (5) and (8) are the fundamental GL equations for the two-gap superconductors. In the absence of fields and gradients, Equations (4) and (5) give

$$\left(-\alpha_1 + \beta_1|\Psi_1|^2\right)\Psi_1 + \eta_{12}\Psi_2 = 0 \tag{12}$$

and

$$\eta_{12}\Psi_1 + \left(-\alpha_2 + \beta_2|\Psi_2|^2\right)\Psi_2 = 0. \tag{13}$$

At $T \to T_c$, we get

$$\alpha_1(T_c)\alpha_2(T_c) = \eta_{12}^2. \tag{14}$$

With $T_c = 2.5$ K for CsV$_3$Sb$_5$, we can easily get η_{12} from the equation above.

In principle, we can derive the parameters in our GL theory from microscopic two-band BCS theory. Following Ref. [35], if we compare the microscopic forms of α_i and β_i, we can obtain two useful relations $\alpha_{10}/\alpha_{20} = T_{c1}/T_{c2}$ and $\beta_1 = \beta_2$. Since two superconducting gaps appear at $1.6k_B T_c$ and $0.63k_B T_c$ [32], we can approximate T_{c1}/T_{c2} as $1.6/0.63 \approx 2.5$. In addition, it has been proven that the ratio of energy gaps at zero temperature is equal to that at critical temperature [38], and according to Equations (12)–(14) the ratio of energy gaps at T_c can be written as $|(\Psi_1/\sqrt{N_1})/(\Psi_2/\sqrt{N_2})| = \sqrt{(N_2/N_1)[\alpha_2(T_c)/\alpha_1(T_c)]}$. Then with simple algebra, we can obtain $T_{c1} = 2.4$ K from this condition.

3. Calculation on the Upper Critical Field of CsV_3Sb_5

3.1. The Upper Critical Field Parallel to the c-Axis

Now let us solve the problem of the nucleation of superconductivity in the presence of a field **H**. With the magnetic field along the c-axis, the vector potential **A** can be chosen as **A** = $(0, Hx, 0)$. Since the vector potential depends only on x, similar to the single-band case, we can look for solution with the form

$$\Psi_i = e^{ik_y y} e^{ik_z z} f_i(x). \tag{15}$$

Near the upper critical field, the quartic terms in Equation (2) can be ignored, so the linearized two-band GL equations take the form

$$\hat{M}_{11} f_1(x) + \hat{M}_{12} f_2(x) = 0 \tag{16}$$

and

$$\hat{M}_{21} f_1(x) + \hat{M}_{22} f_2(x) = 0 \tag{17}$$

with

$$\hat{M}_{ii} = -\frac{\hbar^2}{2m_i} \frac{d^2}{dx^2} + \frac{1}{2} m_i \omega_i^2 (x - x_0)^2 - \alpha_i + \frac{\hbar^2}{2m_{iz}} k_z^2. \tag{18}$$

Here $\omega_i = 2eH/m_i c$ and $x_0 = \hbar c k_y / 2eH$. Thus, inclusion of the factor $e^{ik_y y}$ only shifts the location of the minimum of the effective potential. This is unimportant for the present, but it will become important when we deal with superconductivity near surfaces of finite samples [39,40]. We can also set $k_z = 0$ if we only consider the upper critical field [39].

At $\eta_{12} = 0$, we can obtain the solutions to Equations (16) and (17) immediately by noting that, for each band, it is the Schrödinger equation for a particle bound in a harmonic oscillator potential. The resulting harmonic oscillator eigenvalues are

$$E_{i,n} = (n + 1/2) \hbar \omega_i - \alpha_i. \quad (n = 0, 1, 2, \ldots) \tag{19}$$

If $\eta_{12} \neq 0$, Equations (16) and (17) describe a system of two coupled oscillators. We can set the form of the solutions as

$$f_1(x) = c_1 \left(\frac{b}{\pi}\right)^{1/4} e^{-bx^2/2} \tag{20}$$

and

$$f_2(x) = c_2 \left(\frac{b}{\pi}\right)^{1/4} e^{-bx^2/2}. \tag{21}$$

Here c_1, c_2 are constants, and $b = 2\pi H / \Phi_0$ with the magnetic flux quantum $\Phi_0 = \pi \hbar c / e$. Thus, for $n = 0$, we can transform Equations (16) and (17) into

$$E_{1,0} c_1 + \eta_{12} c_2 = \varepsilon c_1 = 0 \tag{22}$$

and

$$\eta_{12} c_1 + E_{2,0} c_2 = \varepsilon c_2 = 0. \tag{23}$$

with ε the eigenvalue of the matrix.

Then we can obtain the upper critical field parallel to the c-axis from the minimum energy eigenvalue

$$\varepsilon_{min} = \frac{1}{2} \left[(E_{1,0} + E_{2,0}) - \sqrt{(E_{1,0} - E_{2,0})^2 + 4\eta_{12}^2} \right] = 0, \tag{24}$$

which can be simplified as

$$E_{1,0} E_{2,0} = \eta_{12}^2. \tag{25}$$

With $E_{i,0} = \hbar\omega_i/2 - \alpha_i$ and $\omega_i = 2eH_{c2}^{\parallel c}/m_i c$, we get from Equation (25)

$$\left(\frac{\hbar e}{m_1 c}H_{c2}^{\parallel c} - \alpha_1\right)\left(\frac{\hbar e}{m_2 c}H_{c2}^{\parallel c} - \alpha_2\right) = \eta_{12}^2. \tag{26}$$

Simple algebra shows that the upper critical field can be expressed as

$$H_{c2}^{\parallel c} = \frac{\Phi_0}{2\pi\hbar^2}\left[(m_1\alpha_1 + m_2\alpha_2) + \sqrt{(m_1\alpha_1 - m_2\alpha_2)^2 + 4m_1 m_2 \eta_{12}^2}\right]. \tag{27}$$

Single crystals of CsV$_3$Sb$_5$ can be synthesized via a self-flux growth method [41–43]. In order to prevent the reaction of Cs with air and water, all the preparation processes are performed in an argon glovebox. After high temperature reaction in the furnace, the excess flux is removed by water and a millimeter-sized single crystal can be obtained. The as-grown CsV$_3$Sb$_5$ single crystals are stable in the air. Then electrical transport measurements can be carried out in a Quantum Design physical property measurement system (PPMS-14T), and magnetization measurements can be performed in a SQUID magnetometer (MPMS-5T). For CsV$_3$Sb$_5$, the experimental data of the upper critical field can be measured following these steps and then shown in Figure 1.

To fit the experimental measurement, we choose the GL parameter $\alpha_{10} = 0.11$ meV. According to Equation (27), we plot the theoretical result of $H_{c2}^{\parallel c}$ as the solid line in Figure 1. Note that the experimental data are almost linear and our calculation fits the experimental measurement well.

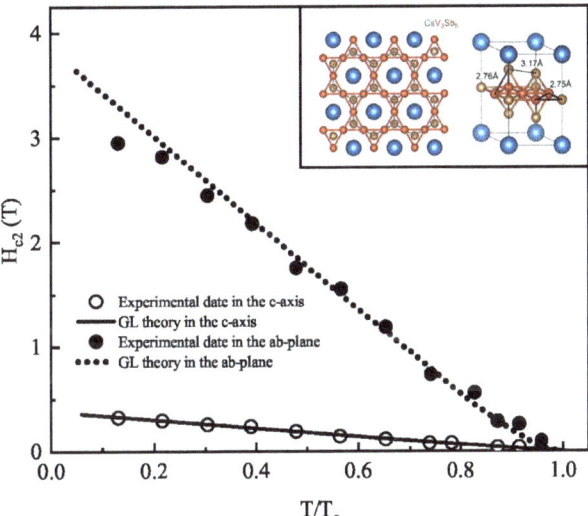

Figure 1. Upper critical field $H_{c2}^{\parallel c}$ (solid line) and $H_{c2}^{\parallel ab}$ (dotted line) as function of temperature. The experimental data are from Ref. [43]. The inset shows the schematic crystal structure of CsV$_3$Sb$_5$.

3.2. The Upper Critical Field Parallel to the ab-Plane

In this subsection, we will study the nucleation of superconductivity with the magnetic field **H** applied in the *ab*-plane. We set $\mathbf{H} = (0, H, 0)$ and take $\mathbf{A} = (0, 0, -Hx)$. Similarly, we look for a solution with the form (15). Close to the upper critical field we can also obtain the linearized GL Equations (16) and (17), but the diagonal element of the $\hat{M}$-matrix changes into

$$\hat{M}_{ii} = -\frac{\hbar^2}{2m_i}\frac{d^2}{dx^2} + \frac{1}{2}m_i\omega_i'^2(x + x_0')^2 - \alpha_i, \tag{28}$$

where $\omega'_i = 2eH/c\sqrt{m_i m_{iz}}$ and $x'_0 = \hbar c k_z/2eH$.

If $\eta_{12} = 0$, analogous to the analysis of the last subsection, the harmonic oscillator eigenvalues are

$$E'_{i,n} = (n+1/2)\hbar\omega'_i - \alpha_i. \quad (n = 0, 1, 2, \ldots) \tag{29}$$

If $\eta_{12} \neq 0$, we cannot get an exact result due to the mixing between the minimum and higher-level eigenfunctions. We thus follow a variational approach. We look for a solution in the form

$$f_1(x) = c_1 g_1(x) = c_1 \left(\frac{b_1}{\pi}\right)^{1/4} e^{-b_1 x^2/2} \tag{30}$$

and

$$f_2(x) = c_2 g_2(x) = c_2 \left(\frac{b_2}{\pi}\right)^{1/4} e^{-b_2 x^2/2}, \tag{31}$$

with b_1 and b_2 the variational parameters. Introducing $D_{ij} = \langle g_i | \hat{M}_{ij} | g_j \rangle$, detailed calculations give

$$D_{11} = \frac{\hbar^2 b_1}{4m_1} + \frac{e^2 H^2}{m_{1z} c^2 b_1} - \alpha_1, \tag{32}$$

$$D_{22} = \frac{\hbar^2 b_2}{4m_2} + \frac{e^2 H^2}{m_{2z} c^2 b_2} - \alpha_2 \tag{33}$$

and

$$D_{12} = D_{21} = \eta_{12} (b_1 b_2)^{1/4} \left(\frac{2}{b_1 + b_2}\right)^{1/2}. \tag{34}$$

Then we can transform Equations (16), (17) and (28) into

$$D_{11} c_1 + D_{12} c_2 = \varepsilon' c_1 = 0 \tag{35}$$

and

$$D_{21} c_1 + D_{22} c_2 = \varepsilon' c_2 = 0. \tag{36}$$

Let ε' denote the eigenvalue of the D-matrix. The upper critical field corresponds to the minimum eigenvalue $\varepsilon'_{min} = 0$, and it is available from Equations (35) and (36) as

$$\varepsilon'_{min} = \frac{1}{2}\left[(D_{11} + D_{22}) - \sqrt{(D_{11} - D_{22})^2 + 4D_{12}D_{21}}\right]. \tag{37}$$

Minimizing ε'_{min} with respect to b_1 and b_2

$$\frac{\partial \varepsilon'_{min}}{\partial b_1} = 0 \quad \text{and} \quad \frac{\partial \varepsilon'_{min}}{\partial b_2} = 0, \tag{38}$$

and combining with

$$\varepsilon'_{min}\left(H_{c2}^{\|ab}, b_1, b_2\right) = 0, \tag{39}$$

we can obtain the upper critical field $H_{c2}^{\|ab}$ at an arbitrary temperature.

We choose the GL parameters $m_{1z} = 38 m_e$ and $m_{2z} = 0.31 m_e$ to fit the experimental data. By numerically solving three nonlinear Equations (38) and (39), we plot the theoretical result of $H_{c2}^{\|ab}$ as the dotted line in Figure 1. Note that our calculation is in agreement with the experimental measurement of $H_{c2}^{\|ab}$ in temperature down to $0.2 T_c$.

4. Calculation on the Magnetic Penetration Depth of CsV_3Sb_5

Now we begin to calculate the magnetic penetration depth for this two-band superconductor. In the presence of the weak fields, the solution takes the form [39]

$$\Psi_i(\mathbf{r}) = |\Psi_i|e^{i\varphi_i(\mathbf{r})}, \tag{40}$$

where $|\Psi_i|$ is constant. If the external field is applied parallel to the c-axis, we can set $\mathbf{A} = (A_x, A_y, 0)$, and without loss of generality we consider the phase factor φ_i as a function of x and y. From Equation (8), we get

$$\frac{\nabla \times \mathbf{H}}{4\pi} = \frac{2e\hbar}{m_1 c}|\Psi_1|^2\left(\nabla\varphi_1 - \frac{2e}{\hbar c}\mathbf{A}\right) + \frac{2e\hbar}{m_2 c}|\Psi_2|^2\left(\nabla\varphi_2 - \frac{2e}{\hbar c}\mathbf{A}\right). \tag{41}$$

Then following the standard procedure in Ref. [39], we can rewrite Equation (41) as

$$\nabla^2\mathbf{H} - \left(\frac{16\pi e^2}{m_1 c^2}|\Psi_1|^2 + \frac{16\pi e^2}{m_2 c^2}|\Psi_2|^2\right)\mathbf{H} = 0. \tag{42}$$

Therefore, we can obtain the magnetic penetration depth in the ab-plane as

$$\lambda_{ab} = \left(\frac{m_1 m_2 c^2}{16\pi m_2 e^2|\Psi_1|^2 + 16\pi m_1 e^2|\Psi_2|^2}\right)^{1/2}. \tag{43}$$

Similarly, the magnetic penetration depth along the c-axis is given by

$$\lambda_c = \left(\frac{m_{1z} m_{2z} c^2}{16\pi m_{2z} e^2|\Psi_1|^2 + 16\pi m_{1z} e^2|\Psi_2|^2}\right)^{1/2}. \tag{44}$$

We take $\beta_1 = 1.3 \times 10^{-2}$ meV·μm^3 to fit the experimental data on the magnetic penetration depth. First of all, $|\Psi_1|$ and $|\Psi_2|$ as function of temperature can be numerically obtained from Equations (12) and (13). Then from Equations (43) and (44), we plot λ_{ab} and λ_c as function of temperature in Figure 2. From Figure 2, we can see that our theoretical calculation can fit the experimental data well almost in the whole temperature range.

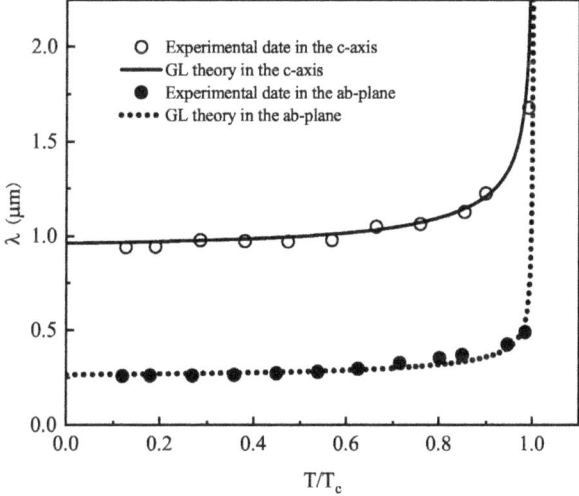

Figure 2. Magnetic penetration depth along the c-axis (solid line) and in the ab-plane (dotted line) as function of temperature. The experimental data are from Ref. [43].

At this stage, we would also like to point out that Gupta et al. also tried to fit the experimental data of the magnetic penetration depth with the d-wave model [43]. However, compared with the two-gap s-wave model, the d-wave model does not describe the data well, which provides further evidence for the nodeless structure in this compound.

5. KW Ratio and the Semi-Heavy–Fermion System

In this section, we would like to discuss the KW ratio and semi-heavy-fermion feature in the compound CsV$_3$Sb$_5$. Since the discovery by Steglich et al. of superconductivity in the high-effective-mass ($\sim$100m_e) electrons in CeCu$_2$Si$_2$, the search for and characterization of such heavy-fermion systems has been a rapidly growing field of study [44]. In a Fermi liquid, the electronic contribution to the heat capacity has a linear temperature dependence $C_{el}(T) = \gamma T$, and at low temperatures the resistivity varies as $\rho(T) = \rho_0 + AT^2$. This is observed experimentally when electron–electron scattering, which gives rise to the quadratic term, dominates over electron–phonon scattering in the process. In a number of typical transition metals, we have $A/\gamma^2 \approx 0.09 \times 10^{-5}$ $\mu\Omega$ cm mol^2 K^2 mJ^{-2} even though γ^2 varies by an order of magnitude across the materials studied. Meanwhile, it was found in many heavy-fermion compounds A/γ^2 reaches 1.0×10^{-5} $\mu\Omega$ cm mol^2 K^2 mJ^{-2} despite the large mass renormalization. Because of this remarkable behavior A/γ^2 has become known as the KW ratio, and large value of this ratio is treated as a robust signature of heavy-fermion systems [45,46].

With the effective masses in the ab-plane ($m_1 = 0.55m_e$, $m_2 = 0.13m_e$) [36] and those along the c-direction ($m_{1z} = 38m_e$, $m_{2z} = 0.31m_e$) from the two-band GL theory, we can expect that the first band in CsV$_3$Sb$_5$ will show the heavy-fermion properties, while the other band can be treated as the normal metal. Meanwhile, for this kagome crystal we have resistivity coefficient $A = 2.3 \times 10^{-3}$ $\mu\Omega$ cm K^{-2} from the electronic transport measurement and Sommerfeld factor $\gamma = 20$ mJ mol^{-1} K^{-2} from the specific heat data [32]. It is thus reasonable that CsV$_3$Sb$_5$, as a semi-heavy-fermion compound, possesses a medium KW ratio $A/\gamma^2 \approx 0.58 \times 10^{-5}$ $\mu\Omega$ cm mol^2 K^2 mJ^{-2} between 0.09×10^{-5} and 1.0×10^{-5} $\mu\Omega$ cm mol^2 K^2 mJ^{-2}.

6. Conclusions

In summary, based on the two-band anisotropic GL theory, we studied the temperature dependence of upper critical field and magnetic penetration depth for the kagome superconductor CsV$_3$Sb$_5$. Our theoretical results fit the experimental data in a broad temperature range, pointing to the existence of two-gap s-wave superconductivity in this system. From the large anisotropy of effective masses in the first band, we also suggest that CsV$_3$Sb$_5$ is a semi-heavy-fermion compound. The possible mechanism of the semi-heavy-fermion state and other problems for these kinds of materials are reserved for further investigations.

Author Contributions: Conceptualization, T.H., J.C., C.Y. and H.H.; methodology, T.H., J.C., C.Y. and H.H.; software, T.H.; validation, T.H., J.C., C.Y. and H.H.; formal analysis, T.H., J.C., C.Y. and H.H.; investigation, T.H., J.C., C.Y. and H.H.; resources, H.H.; data curation, T.H., J.C., C.Y. and H.H.; writing—original draft preparation, T.H.; writing—review and editing, T.H., J.C., C.Y. and H.H.; visualization, T.H.; supervision, J.C., C.Y. and H.H.; project administration, H.H. All authors have read and agreed to the published version of the manuscript.

Funding: This research received no external funding.

Data Availability Statement: Not applicable.

Conflicts of Interest: The authors declare no conflict of interest.

References

1. Ortiz, B.R.; Gomes, L.C.; Morey, J.R.; Winiarski, M.; Oswald, I.W.H.; Mangum, J.S.; Neilson, J.R.; Ertekin, E.; McQueen, T.M.; Toberer, E.S. New kagome prototype materials: Discovery of KV$_3$Sb$_5$, RbV$_3$Sb$_5$ and CsV$_3$Sb$_5$. *Phys. Rev. Mater.* **2019**, *3*, 094407. [CrossRef]
2. Ortiz, B.R.; Teicher, S.M.L.; Hu, Y.; Zuo, J.L.; Sarte, P.M.; Schueller, E.C.; Abeykoon, A.M.M.; Krogstad, M.J.; Rosenkranz, S.; Osborn, R.; et al. CsV$_3$Sb$_5$: A $\mathbb{Z}_2$ topological kagome metal with a superconducting ground state. *Phys. Rev. Lett.* **2020**, *125*, 247002. [CrossRef] [PubMed]
3. Ortiz, B.R.; Sarte, P.M.; Kenney, E.M.; Graf, M.J.; Teicher, S.M.L.; Seshadri, R.; Wilson, S.D. Superconductivity in the $\mathbb{Z}_2$ kagome metal KV$_3$Sb$_5$. *Phys. Rev. Mater.* **2021**, *5*, 034801. [CrossRef]
4. Yin, Q.W.; Tu, Z.J.; Gong, C.S.; Fu, Y.; Yan, S.H.; Lei, H.C. Superconductivity and normal-state properties of kagome metal RbV$_3$Sb$_5$ single crystals. *Chin. Phys. Lett.* **2021**, *38*, 037403. [CrossRef]
5. Shumiya, N.; Hossain, M.S.; Yin, J.X.; Jiang, Y.X.; Ortiz, B.R.; Liu, H.X.; Shi, Y.G.; Yin, Q.W.; Lei, H.C.; Zhang, S.T.S.; et al. Intrinsic nature of chiral charge order in the kagome superconductor RbV$_3$Sb$_5$. *Phys. Rev. B* **2021**, *104*, 035131. [CrossRef]
6. Guguchia, Z.; Mielke, C., III; Das, D.; Gupta, R.; Yin, J.X.; Liu, H.; Yin, Q.; Christensen, M.H.; Tu, Z.; Gong, C.; et al. Tunable unconventional kagome superconductivity in charge ordered RbV$_3$Sb$_5$ and KV$_3$Sb$_5$. *Nat. Commun.* **2023**, *14*, 153. [CrossRef]
7. Li, H.X.; Zhang, T.T.; Yilmaz, T.; Pai, Y.Y.; Marvinney, C.E.; Said, A.; Yin, Q.W.; Gong, C.S.; Tu, Z.J.; Vescovo, E.; et al. Observation of unconventional charge density wave without acoustic phonon anomaly in kagome superconductors AV$_3$Sb$_5$(A = Rb, Cs). *Phys. Rev. X* **2021**, *11*, 031050. [CrossRef]
8. Liang, Z.W.; Hou, X.Y.; Zhang, F.; Ma, W.R.; Wu, P.; Zhang, Z.Y.; Yu, F.H.; Ying, J.J.; Jiang, K.; Shan, L.; et al. Three-dimensional charge density wave and surface-dependent vortex-core states in a kagome superconductor CsV$_3$Sb$_5$. *Phys. Rev. X* **2021**, *11*, 031026. [CrossRef]
9. Zhao, H.; Li, H.; Ortiz, B.R.; Teicher, S.M.L.; Park, T.; Ye, M.X.; Wang, Z.Q.; Balents, L.; Wilson, S.D.; Zeljkovic, I. Cascade of correlated electron states in the kagome superconductor CsV$_3$Sb$_5$. *Nature* **2021**, *599*, 216. [CrossRef]
10. Ortiz, B.R.; Teicher, S.M.L.; Kautzsch, L.; Sarte, P.M.; Ratcliff, N.; Harter, J.; Ruff, J.P.C.; Seshadri, R.; Wilson, S.D. Fermi surface mapping and the nature of charge-density-wave order in the kagome superconductor CsV$_3$Sb$_5$. *Phys. Rev. X* **2021**, *11*, 041030. [CrossRef]
11. Mu, C.; Yin, Q.W.; Tu, Z.J.; Gong, C.S.; Lei, H.C.; Li, Z.; Luo, J.L. S-wave superconductivity in kagome metal CsV$_3$Sb$_5$ revealed by $^{121/123}$Sb NQR and ^{51}V NMR measurements. *Chin. Phys. Lett.* **2021**, *38*, 077402. [CrossRef]
12. Ptok, A.; Kobialka, A.; Sternik, M.; Lazewski, J.; Jochym, P.T.; Oles, M.A.; Piekarz, P. Dynamical study of the origin of the charge density wave in AV$_3$Sb$_5$ (A = K, Rb, Cs) compounds. *Phys. Rev. B* **2022**, *105*, 235134. [CrossRef]
13. Subedi, A. Hexagonal-to-base-centered-orthorhombic 4Q charge density wave order in kagome metals KV$_3$Sb$_5$, RbV$_3$Sb$_5$, and CsV$_3$Sb$_5$. *Phys. Rev. Mater.* **2022**, *6*, 015001. [CrossRef]
14. Wang, Z.X.; Wu, Q.; Yin, Q.W.; Gong, C.S.; Tu, Z.J.; Lin, T.; Liu, Q.M.; Shi, L.Y.; Zhang, S.J.; Wu, D.; et al. Unconventional charge density wave and photoinduced lattice symmetry change in the kagome metal CsV$_3$Sb$_5$ probed by time-resolved spectroscopy. *Phys. Rev. B* **2021**, *104*, 165110. [CrossRef]
15. Ratcliff, N.; Hallett, L.; Ortiz, B.R.; Wilson, S.D.; Harter, J.W. Coherent phonon spectroscopy and interlayer modulation of charge density wave order in the kagome metal CsV$_3$Sb$_5$. *Phys. Rev. Mater.* **2021**, *5*, L111801. [CrossRef]
16. Wulferding, D.; Lee, S.; Choi, Y.; Yin, Q.W.; Tu, Z.J.; Gong, C.S.; Lei, H.C.; Yousuf, S.; Song, J.; Lee, H.; et al. Emergent nematicity and intrinsic versus extrinsic electronic scattering processes in the kagome metal CsV$_3$Sb$_5$. *Phys. Rev. Res.* **2022**, *4*, 023215. [CrossRef]
17. Ni, S.L.; Ma, S.; Zhang, Y.H.; Yuan, J.; Yang, H.T.; Lu, Z.Y.W.; Wang, N.N.; Sun, J.P.; Zhao, Z.; Li, D.; et al. Anisotropic superconducting properties of kagome metal CsV$_3$Sb$_5$. *Chin. Phys. Lett.* **2021**, *38*, 057403. [CrossRef]
18. Tan, H.X.; Liu, Y.Z.; Wang, Z.Q.; Yan, B.H. Charge density waves and electronic properties of superconducting kagome metals. *Phys. Rev. Lett.* **2021**, *127*, 046401. [CrossRef]
19. Chen, H.; Yang, H.T.; Hu, B.; Zhao, Z.; Yuan, J.; Xing, Y.Q.; Qian, G.J.; Huang, Z.H.; Li, G.; Ye, Y.H.; et al. Roton pair density wave in a strong-coupling kagome superconductor. *Nature* **2021**, *599*, 222. [CrossRef]
20. Jiang, Y.X.; Yin, J.X.; Denner, M.M.; Shumiya, N.; Ortiz, B.R.; Xu, G.; Guguchia, Z.; He, Y.Y.; Hossain, M.S.; Liu, X.X.; et al. Unconventional chiral charge order in kagome superconductor KV$_3$Sb$_5$. *Nat. Mater.* **2021**, *20*, 1353. [CrossRef]
21. Du, F.; Luo, S.S.; Ortiz, B.R.; Chen, Y.; Duan, W.Y.; Zhang, D.T.; Lu, X.; Wilson, S.D.; Song, Y.; Yuan, H.Q. Pressure-induced double superconducting domes and charge instability in the kagome metal KV$_3$Sb$_5$. *Phys. Rev. B* **2021**, *103*, L220504. [CrossRef]
22. Uykur, E.; Ortiz, B.R.; Wilson, S.D.; Dressel, M.; Tsirlin, A.A. Optical detection of the density-wave instability in the kagome metal KV$_3$Sb$_5$. *NPJ Quantum Mater.* **2022**, *7*, 16. [CrossRef]
23. Zhao, C.C.; Wang, L.S.; Xia, W.; Yin, Q.W.; Ni, J.M.; Huang, Y.Y.; Tu, C.P.; Tao, Z.C.; Tu, Z.J.; Gong, C.S.; et al. Nodal superconductivity and superconducting domes in the topological kagome metal CsV$_3$Sb$_5$. *arXiv* **2021**, arXiv:abs/2102.08356.
24. Freitas, R.R.Q.; Rivelino, R.; de Brito Mota, F.; de Castilho, C.M.C.; Kakanakova-Georgieva, A.; Gueorguiev, G.K. Topological insulating phases in two-dimensional bismuth-containing single layers preserved by hydrogenation. *J. Phys. Chem. C* **2015**, *119*, 23599. [CrossRef]

25. Freitas, R.R.Q.; de Brito Mota, F.; Rivelino, R.; de Castilho, C.M.C.; Kakanakova-Georgieva, A.; Gueorguiev, G.K. Spin-orbit-induced gap modification in buckled honeycomb XBi and XBi$_3$ (X = B, Al, Ga, In) sheets. *J. Phys. Condens. Matter* **2015**, *27*, 485306. [CrossRef] [PubMed]
26. Neupert, T.; Denner, M.M.; Yin, J.X.; Thomale, R.; Hasan, M.Z. Charge order and superconductivity in kagome materials. *Nat. Phys.* **2022**, *18*, 137. [CrossRef]
27. Liu, Z.H.; Zhao, N.N.; Yin, Q.W.; Gong, C.S.; Tu, Z.J.; Li, M.; Song, W.H.; Liu, Z.T.; Shen, D.W.; Huang, Y.B.; et al. Charge-density-wave-induced bands renormalization and energy gaps in a kagome superconductor RbV$_3$Sb$_5$. *Phys. Rev. X* **2021**, *11*, 041010. [CrossRef]
28. Chen, X.; Zhan, X.H.; Wang, X.J.; Deng, J.; Liu, X.B.; Chen, X.; Guo, J.G.; Chen, X.L. Highly robust reentrant superconductivity in CsV$_3$Sb$_5$ under pressure. *Chin. Phys. Lett.* **2021**, *38*, 057402. [CrossRef]
29. Ye, L.D.; Kang, M.G.; Liu, J.W.; Von Cube, F.; Wicker, C.R.; Suzuki, T.; Jozwiak, C.; Bostwick, A.; Rotenberg, E.; Bell, D.C.; et al. Massive Dirac fermions in a ferromagnetic kagome metal. *Nature* **2018**, *555*, 638. [CrossRef]
30. Guo, H.M.; Franz, M. Topological insulator on the kagome lattice. *Phys. Rev. B* **2009**, *80*, 113102. [CrossRef]
31. Mazin, I.I.; Jeschke, H.O.; Lechermann, F.; Lee, H.; Fink, M.; Thomale, R.; Valentí, R. Theoretical prediction of a strongly correlated Dirac metal. *Nat. Commun.* **2014**, *5*, 4261. [CrossRef] [PubMed]
32. Duan, W.Y.; Nie, Z.Y.; Luo, S.S.; Yu, F.H.; Ortiz, B.R.; Yin, L.C.; Su, H.; Du, F.; Wang, A.; Chen, Y.; et al. Nodeless superconductivity in the kagome metal CsV$_3$Sb$_5$. *Sci. China Phys. Mech. Astron.* **2021**, *64*, 107462. [CrossRef]
33. Doh, H.; Sigrist, M.; Cho, B.K.; Lee, S.I. Phenomenological theory of superconductivity and magnetism in Ho$_{1-x}$Dy$_x$Ni$_2$B$_2$C. *Phys. Rev. Lett.* **1999**, *83*, 5350. [CrossRef]
34. Askerzade, I.N.; Gencer, A.; Güclü, N. On the Ginzburg-Landau analysis of the upper critical field H_{c2} in MgB$_2$. *Supercond. Sci. Technol.* **2002**, *15*, L13. [CrossRef]
35. Zhitomirsky, M.E.; Dao, V.H. Ginzburg-Landau theory of vortices in a multigap superconductor. *Phys. Rev. B* **2004**, *69*, 054508. [CrossRef]
36. Fu, Y.; Zhao, N.N.; Chen, Z.; Yin, Q.W.; Tu, Z.J.; Gong, C.S.; Xi, C.Y.; Zhu, X.D.; Sun, Y.P.; Liu, K.; et al. Quantum transport evidence of topological band structures of kagome superconductor CsV$_3$Sb$_5$. *Phys. Rev. Lett.* **2021**, *12*, 207002. [CrossRef] [PubMed]
37. Kong, Y.; Dolgov, O.V.; Jepsen, O.; Andersen, O.K. Electron-phonon interaction in the normal and superconducting states of MgB$_2$. *Phys. Rev. B* **2001**, *64*, 020501(R). [CrossRef]
38. Kresin, V.Z. Transport properties and determination of the basic parameters of superconductors with overlapping bands. *J. Low Temp. Phys.* **1973**, *11*, 519. [CrossRef]
39. Tinkham, M. *Introduction to Superconductivity*; McGraw-Hill Inc: New York, NY, USA, 1996.
40. de Gennes, P.G. *Superconductivity of Metals and Alloys*; Westview Press: New York, NY, USA, 1966.
41. Yu, F.H.; Wu, T.; Wang, Z.Y.; Lei, B.; Zhuo, W.Z.; Ying, J.J.; Chen, X.H. Concurrence of anomalous Hall effect and charge density wave in a superconducting topological kagome metal. *Phys. Rev. B* **2021**, *104*, L041103. [CrossRef]
42. Yang, S.Y.; Wang, Y.J.; Ortiz, B.R.; Liu, D.; Gayles, J.; Derunova, E.; Gonzalez-Hernandez, R.; Šmejkal, L.; Chen, Y.L.; Parkin, S.S.P.; et al. Giant, unconventional anomalous Hall effect in the metallic frustrated magnet candidate, KV$_3$Sb$_5$. *Sci. Adv.* **2020**, *6*, eabb6003. [CrossRef] [PubMed]
43. Gupta, R.; Das, D.; Mielke, C.H., III; Guguchia, Z.; Shiroka, T.; Baines, C.; Bartkowiak, M.; Luetkens, H.; Khasanov, R.; Yin, Q.; et al. Microscopic evidence for anisotropic multigap superconductivity in the CsV$_3$Sb$_5$ kagome superconductor. *NPJ Quantum Mater.* **2022**, *7*, 49. [CrossRef]
44. Steglich, F.; Aarts, J.; Bredl, C.D.; Lieke, W.; Meschede, D.; Franz, W.; Schäfer, H. Superconductivity in the presence of strong pauli paramagnetism: CeCu$_2$Si$_2$. *Phys. Rev. Lett.* **1979**, *43*, 1892. [CrossRef]
45. Kadowaki, K.; Woods, S.B. Universal relationship of the resistivity and specific heat in heavy-fermion compounds. *Solid State Commun.* **1986**, *58*, 507. [CrossRef]
46. Jacko, A.C.; Fjærestad, J.O.; Powell, B.J. A unified explanation of the Kadowaki-Woods ratio in strongly correlated metals. *Nat. Phys.* **2009**, *5*, 422. [CrossRef]

Disclaimer/Publisher's Note: The statements, opinions and data contained in all publications are solely those of the individual author(s) and contributor(s) and not of MDPI and/or the editor(s). MDPI and/or the editor(s) disclaim responsibility for any injury to people or property resulting from any ideas, methods, instructions or products referred to in the content.

Article

The Effect of Cr Additive on the Mechanical Properties of Ti-Al Intermetallics by First-Principles Calculations

Hui Wang, Fuyong Su * and Zhi Wen

School of Energy and Environmental Engineering, University of Science & Technology Beijing, Beijing 100083, China
* Correspondence: sfyong@ustb.edu.cn

Abstract: The structure, elastic properties and electronic structure of Ti-Al intermetallics including Ti_3Al (space group P63/mmc), TiAl (space group I4/mmm) and $TiAl_3$ (space group P4/mmm) are systematically studied by first-principles calculations. The results show that Ti-Al intermetallics can exist stably whether Cr replaces Ti or Al. The ductility of the alloy cannot be improved when Ti is replaced in Cr-doped TiAl and $TiAl_3$. However, when it replaces Al, the alloy has better ductility. In Ti_3Al, the ductility can be improved regardless of whether Cr replaces Ti or Al, and the effect is better when it replaces Al. The bond in Ti-Al intermetallics is mainly a Ti-Ti metal bond. The metal bond between Ti-Ti is strengthened and a solid metal bond is formed between Cr and Ti, inducing a better ductility of the material, after Cr replaces Al in Ti-Al intermetallics.

Keywords: Ti-Al intermetallics; first-principles calculations; Cr-doped; ductility

Citation: Wang, H.; Su, F.; Wen, Z. The Effect of Cr Additive on the Mechanical Properties of Ti-Al Intermetallics by First-Principles Calculations. *Crystals* **2023**, *13*, 488. https://doi.org/10.3390/cryst13030488

Academic Editor: Jacek Ćwik

Received: 20 February 2023
Revised: 5 March 2023
Accepted: 8 March 2023
Published: 11 March 2023

Copyright: © 2023 by the authors. Licensee MDPI, Basel, Switzerland. This article is an open access article distributed under the terms and conditions of the Creative Commons Attribution (CC BY) license (https://creativecommons.org/licenses/by/4.0/).

1. Introduction

As with Ni-based superalloys, Ti-Al intermetallics have good elastic modulus, excellent oxidation resistance and high specific strength. However, their density is only half that of Ni-based superalloys. Therefore, Ti-Al alloys are the most ideal substitutes for Ni-based superalloys. In addition, Ti-Al intermetallics have good flame retardancy and can replace expensive Ti-based components. Ti-Al intermetallics have become a new generation of key materials in aerospace, automotive and other industries [1]. After extensive theoretical and experimental research, researchers have found many ways to improve most mechanical properties of Ti-Al intermetallics. However, their room temperature brittleness still greatly limits their wide application. Song et al. [2] highlighted that $TiAl_3$ has the highest hardness and brittleness among the three alloys. Jian et al. [3] recently evaluated the effect of Al concentration in Ti-Al-based alloys on the mechanical properties. They found that as Al increased, the hardness of the materials increased and their ductility decreased. To improve the ductility of TiAl-based alloys, researchers found that ternary or more complex alloy additives can have a good effect. By changing the electronic structure of alloy materials and improving the type and strength of electronic bonding bonds, the improvement in the ductility and toughness of materials can be achieved. These additives exist in interstitial or alternative forms [4,5], and the most commonly used alloying elements are Cr, V, Nb, La, Ta and Cd.

In order to improve the properties of Ti-Al intermetallics, researchers have conducted a terrific amount of theoretical and experimental research. The position of alloying elements on different sublattices in Ti-Al-based alloys has a great influence on their mechanical properties. Therefore, it is necessary to discuss the occupation of alloying atoms in Ti-Al-based alloys. Zhang et al. [6] observed that Mn tends to replace Al in TiAl. Song et al. [7] pointed out that in $TiAl_3$, Y, Zr, Nb, Mo and Sb preferentially replace Ti atoms, while Ga and In preferentially replace Al atoms. The occupying position of V, Mn, Co and Ge depends on the concentration ratio of Ti and Al in Ti-Al intermetallics.

In recent years, the method of improving material properties by doping alloying elements has attracted more attention [8–11]. Fan et al. [8] found that La, Ce, Pr, Nd and Sm can be used as strengthening phases to improve the rigidity and tensile strength of Al alloys, which is significantly helpful as a way of enhancing the mechanical properties and thermodynamic stability of Al alloys. Qi et al. [10] detected that the increase in Zr content will result in the aggravation of element segregation in a CoCrFeNiZrx alloy, forming a Zr-rich intermetallic compound. The compressive yield strength and fracture strain of the alloy are also improved. Trong et al. [11] found that the concentration of Au in a NiAu alloy can change the microstructure of materials. Related studies have reported that W and Y can improve the oxidation resistance [12,13] and C and Si can enhance the creep property of Ti-Al alloys [14,15]. Liu et al. [16] considered the influence of Ru on the microstructure change and mechanical properties of TiAl through thermal compression and the three point bend test. The experimental results displayed that Ru can increase the strength of TiAl and improve its ductility. Tetsui [17] studied the effect of Nb concentration in the Ti-50Al-xNb alloy on the malleability and high temperature strength of the material, and found that there was a suitable composition near Ti-46Al-7.5Nb (at. %) in the casting material. Liu et al. [18] discussed the influence of Nb content on the microstructure and yield strength of a high-Nb-containing TiAl-based alloy, and found that the Nb content remarkably improved the high-temperature strength. Duan et al. [19] investigated the mechanical properties of Ti-16Al-8Nb and Ti-16Al-8Nb-1Sn through experiments, and pointed out that the high temperature ductility of Ti-16Al-8Nb was significantly improved after adding 1% Sn. Music et al. [20] investigated the influence of some transition metals on the elastic constants of γ-TiAl and $\alpha 2$-Ti_3Al using ab initio calculations and found that alloying could improve their bulk modulus B. Yuan et al. [21] found that Ta particles can enhance the thermal deformation ability of TiAl through experimental research. Therefore, adding alloys is significant as a way of reducing the brittleness of Ti-Al intermetallics.

As a commonly used alloying element, Cr has a good effect on improving the properties of Ti-Al alloys. Xiong [22] and Ye [23] found that adding Cr into a TiAl alloy can make the β phase more stable and effectively inhibit ω_0 phase precipitation. Tetsui [24] found that a TiAl alloy has better impact resistance after adding Cr. Moreover, it was found that the Ti-Al-Cr ternary alloy showed higher oxidation resistance [25,26]. However, there are not many researches on the influence of Cr on the mechanical properties of Ti-Al intermetallics. If Cr can significantly improve the ductility of Ti-Al intermetallics, it will supply terrific assistance for the design and manufacture of Ti-Al intermetallics in the future. Therefore, it is necessary to explore the mechanical properties of Ti-Al-Cr ternary alloys. In this study, we calculated the total energy, formation energy, elastic constant and electronic structure of Ti-Al intermetallics containing Cr by using the first-principles method, and the position of Cr in a Ti-Al-based alloy and its effect on the ductility of a Ti-Al-based alloy were discussed.

2. Calculation Method

Our calculations were based on the framework of density functional theory (DFT) [27,28] that is implemented in the Vienna ab initio Simulation Package (VASP) [29,30]. The VASP can calculate the formation enthalpy, binding energy, dissolution enthalpy and elasticity of the crystal to judge its structural stability. It can also calculate the electronic structure in the crystal, predict the bonding situation and explain the macroscopic properties. The projection enhancement wave (PAW) is used to describe the interaction between atomic nucleus and valence electron. All calculations were performed using the projector augmented wave (PAW) [31,32] potentials and the generalized gradient approximation (GGA) in the exchange correlation function of Perdew–Burke–Ernzerhof (PBE) [33] was used for all calculations. The plane wave cutoff energy was 450 eV, which has been found to be sufficient for all the elements considered in this work to obtain precise energetics. The supercells of Ti_3Al (2 × 1 × 1), TiAl (4 × 2 × 2) and $TiAl_3$ (2 × 2 × 1) were commonly arranged with 32 atoms. The k-point meshes of these three structures were selected as 4 × 4 × 10, 4 × 8 × 6 and 6 × 6 × 5, respectively. In order to make the results more accurate, the conjugate gradient minimization method was used for structural optimization to allow the relaxation of crystal cells and internal atoms. The convergence criteria for energy and force were 1×10^{-5} eV and 0.02 eV/Å, respectively. The PAW pseudopotentials of Ni, Ti_sv and Cr_sv were used to simulate the real pseudopotentials. The calculation of elastic constant adopted energy-strain method, which is realized in VASPKIT code. After obtaining the stable structure of each system, the total energy, elastic constant, density of states and other physical properties of each system were calculated.

3. Results and Discussion

It is well known that Ti-Al binary compounds usually have several different crystal structures. There are three main kinds: Ti_3Al, TiAl and $TiAl_3$. The structures of Ti_3Al, TiAl and $TiAl_3$ belong to space groups of P63/mmc (Ti_3Al), P4/mmm (TiAl) and I4/mmm ($TiAl_3$), respectively. The lattice structures of these crystals are shown in Figure 1, where the purple balls represent Al, and the gray balls represent Ti.

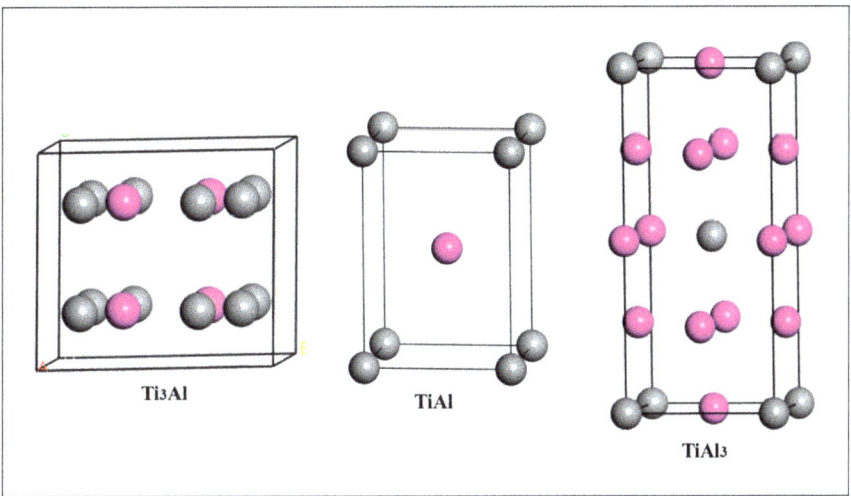

Figure 1. Crystal structures of three kinds of Ti-Al intermetallics.

In this paper, lattice constants of $Ti_{24}Al_8$, $Ti_{16}Al_{16}$ and Ti_8Al_{24} are optimized, and the equilibrium lattice constants at the minimum energy of the crystal are obtained. Table 1 summarizes the equilibrium lattice parameters calculated in this paper and the data calculated by other researchers. The difference between the calculated results and the data available from previous studies is within 1%, which indicates the good accuracy of this work. Therefore, it can be considered that the equilibrium lattice constant is suitable for the subsequent calculation.

Alloying elements occupy different sublattice positions in Ti-Al intermetallics, which seriously affects the lattice constants, electronic structures and elastic constants of their solid solutions. Studies have shown that the occupancy of some alloy atoms depends on the ratio of Ti and Al. Thus, it is necessary to identify the site preference of Cr first. In this paper, the formation energies of Cr occupying Ti and Al sites in Ti-Al intermetallics are calculated, and the occupation preference of alloy atoms is described according to the formation energy.

Table 1. Equilibrium lattice constants of Ti-Al intermetallics.

Phase	Space Group	a/Å	b/Å	c/Å
Ti$_3$Al	P6$_3$/mmc	5.736	5.736	4.638
	P6$_3$/mmc	5.70 [a]	5.70 [a]	4.616 [a]
TiAl	P4/mmc	2.815	2.815	4.071
	P4/mmc	2.842 [b]	2.842 [b]	4.076 [b]
		2.832 [c]	2.832 [c]	4.07 [c]
TiAl$_3$	I4/mmm	3.842	3.842	8.609
	I4/mmm	3.849 [d]	3.849 [d]	8.632 [d]
		3.856 [e]	3.856 [e]	8.622 [e]

[a] Calculation results in Reference [34]. [b] Calculation results in Reference [2]. [c] Calculation results in Reference [35]. [d] Calculation results in Reference [3]. [e] Calculation results in Reference [36].

The structural stability of the crystal is closely related to the formation energy. The smaller the negative value of crystal formation energy, the more stable the structure of the formed compound is. For Cr-doped Ti-Al-based alloys, the formation energy can be expressed as:

$$E^f = E_{total} - N_{Al}E_{solid}^{Al} - N_{Ti}E_{solid}^{Ti} - N_{Cr}E_{solid}^{Cr} \quad (1)$$

where E_{total} is the total energy of the crystal; N_{Al}, N_{Ti} and N_{Cr} are the number of Al, Ti and Cr atoms in the crystal, respectively; and E_{solid}^{Al}, E_{solid}^{Ti} and E_{solid}^{Cr} are the average energy per atom in elemental Al, Ti and Cr, respectively, which are -3.742, -7.762 and -9.486 eV, respectively. The calculation results of the total energy and formation energy of each intermetallic compound are listed in Table 2.

Table 2. The E_{total} and E^f of each system.

System	Site Occupancy	E_{total}/eV	E^f/eV
Ti$_7$Al$_{24}$Cr	Ti	−165.315	−11.687
Ti$_8$Al$_{23}$Cr	Al	−168.919	−11.271
Ti$_{15}$Al$_{16}$Cr	Ti	−197.844	−12.056
Ti$_{16}$Al$_{15}$Cr	Al	−201.726	−11.918
Ti$_{23}$Al$_8$Cr	Ti	−226.189	−8.241
Ti$_{24}$Al$_7$Cr	Al	−228.869	−6.901

As shown in Table 2, the formation energies of Ti$_7$Al$_{24}$Cr, Ti$_{15}$Al$_{16}$Cr and Ti$_{23}$Al$_8$Cr are lower than those of Ti$_8$Al$_{23}$Cr, Ti$_{16}$Al$_{15}$Cr and Ti$_{23}$Al$_8$Cr, respectively, indicating that the Cr atom is more likely to replace the Ti atom in Ti-Al intermetallics. Nevertheless, the formation energy of all alloys is negative, indicating that these ternary alloys can exist stably. In other words, Cr can exist stably in Ti$_3$Al, TiAl and TiAl$_3$ alloys regardless of whether Cr replaces Ti or Al.

The low ductility and toughness of Ti-Al intermetallics at room temperature limit their applications. In order to improve their ductility, adding alloying elements to Ti-Al intermetallics is an effective way. Elastic constant is one of the important parameters that characterizes the mechanical properties of materials, and can calculate the mechanical stability and stress resistance of materials. The periodic arrangement of crystal atoms makes its structure symmetrical. The Ti$_3$Al, TiAl and TiAl$_3$ intermetallics studied in this paper are tetragonal systems. Doped alloy atoms cause appropriate lattice distortion, making a tetragonal system into an orthorhombic system. The orthogonal system has nine independent elastic constants (C_{11}, C_{22}, C_{33}, C_{12}, C_{13}, C_{23}, C_{44}, C_{55} and C_{66}). We calculated the elastic constants of Cr substituting Ti and Al in the intermetallics to determine the mechanical stability and mechanical properties of ternary alloys, and the calculation results are shown in Table 3.

Table 3. The elastic constants of doped Ti-Al intermetallics.

Phase	C_{11}	C_{22}	C_{33}	C_{44}	C_{55}	C_{66}	C_{12}	C_{13}	C_{23}
Ti_8Al_{24}	196.79	196.79	214.96	94.16	94.16	124.87	83.98	48.93	48.93
$Ti_7Al_{24}Cr$	189.26	189.26	214.50	90.20	90.20	124.50	83.76	41.14	41.14
$Ti_8Al_{23}Cr$	191.45	191.45	209.58	94.63	94.63	124.73	90.59	56.19	56.19
$Ti_{16}Al_{16}$	199.26	199.26	177.29	113.50	113.50	42.11	62.74	89.03	89.03
$Ti_{15}Al_{16}Cr$	204.02	204.40	175.40	110.73	109.79	39.27	55.32	84.88	84.96
$Ti_{16}Al_{15}Cr$	183.85	186.74	169.14	106.97	106.86	31.68	78.39	96.39	91.24
$Ti_{24}Al_8$	195.27	194.46	232.09	56.91	56.92	51.64	85.68	69.98	69.98
$Ti_{23}Al_8Cr$	187.07	178.15	236.73	48.28	52.07	32.93	104.07	68.97	72.96
$Ti_{24}Al_7Cr$	171.39	169.70	221.66	49.40	49.11	27.03	113.28	74.20	76.90

According to Born's criterion of elastic stability, the elastic constant of a tetragonal crystal must simultaneously satisfy Equations (2)–(4) so that the crystal can exist stably:

$$(C_{11} - C_{12}) > 0, (C_{11} + C_{33} - 2C_{13}) > 0 \tag{2}$$

$$C_{11} > 0, C_{33} > 0, C_{44} > 0, C_{66} > 0, \tag{3}$$

$$(2C_{11} + C_{33} + 2C_{12} + 4C_{13}) > 0 \tag{4}$$

By substituting the calculated elastic constants of $Ti_7Al_{24}Cr$, $Ti_8Al_{23}Cr$, $Ti_{15}Al_{16}Cr$, $Ti_{16}Al_{15}Cr$, $Ti_{23}Al_8Cr$ and $Ti_{24}Al_7Cr$ into Equations (2)–(4), it is found that they all meet these necessary conditions, indicating that they can exist stably.

Based on the elastic constants, the bulk modulus B, Young's modulus E, shear modulus G and Poisson's ratio v can be further calculated. Since the compression resistance of materials is positively correlated with the bulk modulus B, the bulk modulus B is usually used to describe the compression resistance of materials. Shear modulus G and Young's modulus E can be used to characterize the strength and wear resistance of materials. Large shear modulus G and Young's modulus E indicate high strength and wear resistance of the material. The Pugh's B/G ratio can be used to evaluate the ductility of materials, and the critical value of the B/G ratio is 1.75 [37]. If the material's B/G ratio is greater than 1.75, the material is considered to have ductility, and the larger the ratio, the better the ductility of the material. If the material B/G ratio is less than 1.75, the material is considered brittle, and the smaller the ratio, the more brittle the material. In general, shear modulus and Young's modulus E can be used to characterize the strength and wear resistance of materials. A large shear modulus and Young's modulus E indicates high strength and wear resistance of materials. The B/G ratio can be used to evaluate the ductility of materials. A small B/G ratio denotes a good ductility of materials. Poisson's ratio v refers to the ratio of transverse normal strain to axial normal strain of a material under uniaxial tension or compression, which reflects the transverse deformation of the material. The shear stability of the material can be quantified by calculating the Poisson's ratio v of the material. Generally speaking, when Poisson's ratio v is in the range of 0.1~0.3, the material is hard. When Poisson's ratio v is between 0.3 and 0.4, the material has proper plasticity and low hardness. Bulk modulus B, shear modulus G and Poisson's ratio v are calculated using Equations (5), (6) and (7), respectively:

$$B = \frac{1}{9}(C_{11} + C_{22} + C_{33}) + \frac{2}{9}(C_{12} + C_{13} + C_{23}) \tag{5}$$

$$G = \frac{1}{15}(C_{11} + C_{22} + C_{33}) - \frac{1}{15}(C_{12} + C_{13} + C_{23}) + \frac{1}{5}(C_{44} + C_{55} + C_{66}) \tag{6}$$

$$v = \frac{3B - 2G}{6B + 2G} \tag{7}$$

The calculated values of bulk modulus B, shear modulus G and elastic modulus ratio G/B of Ti_8Al_{24}, $Ti_7Al_{24}Cr$, $Ti_8Al_{23}Cr$, $Ti_{16}Al_{16}$, $Ti_{15}Al_{16}Cr$, $Ti_{16}Al_{15}Cr$, $Ti_{24}Al_8$, $Ti_{23}Al_8Cr$ and $Ti_{24}Al_7Cr$ alloys are summarized in Table 4.

Table 4. The bulk modulus B (GPa), shear modulus G (GPa), the B/G ratio and Poisson's ratio v of doped Ti-Al intermetallics.

Phase	B	G	B/G	v
Ti_8Al_{24}	108.022	91.083	1.186	0.171
$Ti_7Al_{24}Cr$	102.786	89.445	1.149	0.163
$Ti_8Al_{23}Cr$	110.935	88.764	1.250	0.184
$Ti_{16}Al_{16}$	117.490	76.154	1.543	0.233
$Ti_{15}Al_{16}Cr$	114.905	75.868	1.515	0.229
$Ti_{16}Al_{15}Cr$	119.083	67.350	1.768	0.262
$Ti_{24}Al_8$	119.233	59.506	2.004	0.286
$Ti_{23}Al_8Cr$	121.552	50.386	2.412	0.318
$Ti_{24}Al_7Cr$	121.278	45.000	2.695	0.335

Comparing the elastic constants of Ti_8Al_{24}, $Ti_{16}Al_{16}$ and $Ti_{24}Al_8$, it can be seen that the B/G ratio of Ti_8Al_{24} and $Ti_{16}Al_{16}$ are less than 1.75, which indicates that they are brittle materials. The B/G ratio of $Ti_{24}Al_8$ is more than 1.75, which indicates that it is a ductile material. With the increase in Ti content, the bulk modulus B, B/G ratio and Poisson's ratio v of the material increase gradually, and the compression resistance and ductility of the three Ti-Al-based alloys are gradually improved. Comparing the B/G ratio of Ti_8Al_{24}, $Ti_7Al_{24}Cr$ and $Ti_8Al_{23}Cr$, it can be noted that the B/G ratio of $Ti_7Al_{24}Cr$ is 1.149, which is smaller than that of Ti_8Al_{24}. This indicates that when Cr replaces the Ti atom in Ti_8Al_{24}, its ductility cannot be improved. In addition, the bulk modulus B decreases, and the anti-compression performance of the material becomes worse. However, when Cr replaces the Al atom in Ti_8Al_{24}, the B/G ratio of $Ti_8Al_{23}Cr$ is 1.250. Although it is still a brittle material, it shows that this substitution method improved the ductility. The v of $Ti_7Al_{24}Cr$ is 0.163 (which is 0.008 smaller than that of Ti_8Al_{24}), while the v of $Ti_8Al_{23}Cr$ is 0.184 (which is 0.013 larger than that of Ti_8Al_{24}). It shows that Cr can increase the anisotropy by replacing the Al atom in Ti_8Al_{24}, rather than replacing the Ti atoms.

Comparing the B/G ratios of $Ti_{16}Al_{16}$, $Ti_{15}Al_{16}Cr$ and $Ti_{16}Al_{15}Cr$, it is found that the B/G ratio of $Ti_{15}Al_{16}Cr$ is the smallest, which indicates that when Cr replaces Ti atoms in $Ti_{16}Al_{16}$, it cannot improve the compression and ductility properties of the material. When Cr replaces the Al atom in $Ti_{16}Al_{16}$, the bulk modulus B and B/G ratio of $Ti_{16}Al_{15}Cr$ increase. Moreover, the B/G ratio of $Ti_{16}Al_{15}Cr$ is 1.768; this value is larger than 1.75, indicating that the material exhibits ductility. It shows that when Cr replaces the Al atom, the ductility and compression resistance of the material significantly improves. The Poisson's ratio v of $Ti_{15}Al_{16}Cr$ is 0.229, which is 0.004 less than that of Ti_8Al_{24}. It can be predicted that the plasticity of $Ti_{15}Al_{16}Cr$ is basically unchanged compared with Ti_8Al_{24}. However, the Poisson's ratio v of $Ti_8Al_{23}Cr$ is 0.262, which is 0.029 higher than that of Ti_8Al_{24}, indicating that the substitution of Cr for Al in Ti_8Al_{24} not only preserves the high hardness of the material but also significantly increases its plasticity.

Comparing the bulk modulus B and B/G ratios of $Ti_{24}Al_8$, $Ti_{23}Al_8Cr$ and $Ti_{24}Al_7Cr$, it is found that their B/G ratios are more than 1.75 and they are ductile materials. The bulk modulus B and B/G ratios of $Ti_{23}Al_8Cr$ and $Ti_{24}Al_7Cr$ are higher than those of $Ti_{24}Al_8$, which indicates that Cr can improve the ductility and compression resistance of $Ti_{24}Al_8$ whether it replaces Ti or Al atoms. However, the bulk modulus B and B/G ratio of $Ti_{24}Al_7Cr$ are the largest. The replacement of Al by Cr in $Ti_{24}Al_8$ can better improve its mechanical properties. In conclusion, the substitution of Al atoms with Cr in Ti_8Al_{24}, $Ti_{16}Al_{16}$ and $Ti_{24}Al_8$ alloys improves the ductility of the material. The Poisson ratio of $Ti_{24}Al_8$ is 0.286, which belongs to a high hardness alloy, and compared with $Ti_{24}Al_8$, the Poisson's ratios v of $Ti_{23}Al_8Cr$ and $Ti_{24}Al_7Cr$ increased by 0.032 and 0.049, respectively. It shows that, regardless of whether Cr replaces the Al atom or Ti atom in Ti_8Al_{24}, it will increase the plasticity of the material. However, the Poisson's ratios v of $Ti_{23}Al_8Cr$ and $Ti_{24}Al_7Cr$ are between 0.3–0.4, and the hardness of the materials is low.

The properties of materials are closely related to their internal chemical bonds. The density of states indicates the number of electrons in the unit energy range. It can be used as an intuitive result of the energy band structure, which can reflect the distribution of electrons in various orbits, thus reflecting the interaction between atoms. In order to further clarify the reason why the elastic property of Ti-Al intermetallics is affected by the substitution of Ti and Al by Cr, we discuss the electronic structure of these Ti-Al intermetallics.

Figures 2–4 show the partial density of states data of elements in Ti_8Al_{24}, $Ti_7Al_{24}Cr$, $Ti_8Al_{23}Cr$, $Ti_{16}Al_{16}$, $Ti_{15}Al_{16}Cr$, $Ti_{16}Al_{15}Cr$, $Ti_{24}Al_8$, $Ti_{23}Al_8Cr$ and $Ti_{24}Al_7Cr$. It can be seen from these figures that the Ti-d-Ti-d interaction is dominant in the Ti-Al intermetallics, and the bonding characteristics are metallic. From Figure 2c, it can be noted that for Ti_8Al_{24}, there is an overlap between Al s, Al p and Ti d orbitals from about −2 eV to 5 eV, and there is a low energy weak spd hybrid band. However, the energy is so small that the covalent bonds formed by hybridization between Al and Ti elements are almost negligible. When Cr replaces Ti, the peaks of state density of Ti and Al decrease, and the interactions between Ti-Ti, Ti-Al and Al-Al become weaker. When Cr replaces Al, the peak of the density of states of Ti increases, while Al decreases. The interaction between Ti-Ti is strengthened, and the interaction between Al-Al is weakened.

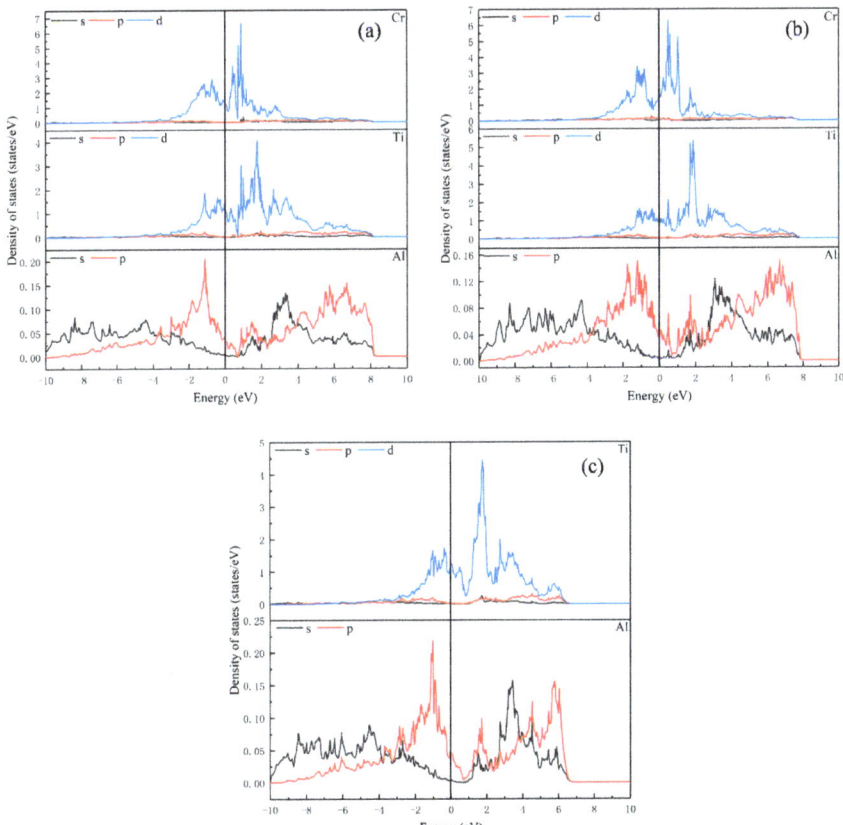

Figure 2. Partial density of states for (**a**) $Ti_7Al_{24}Cr$; (**b**) $Ti_8Al_{23}Cr$; (**c**) Ti_8Al_{24}.

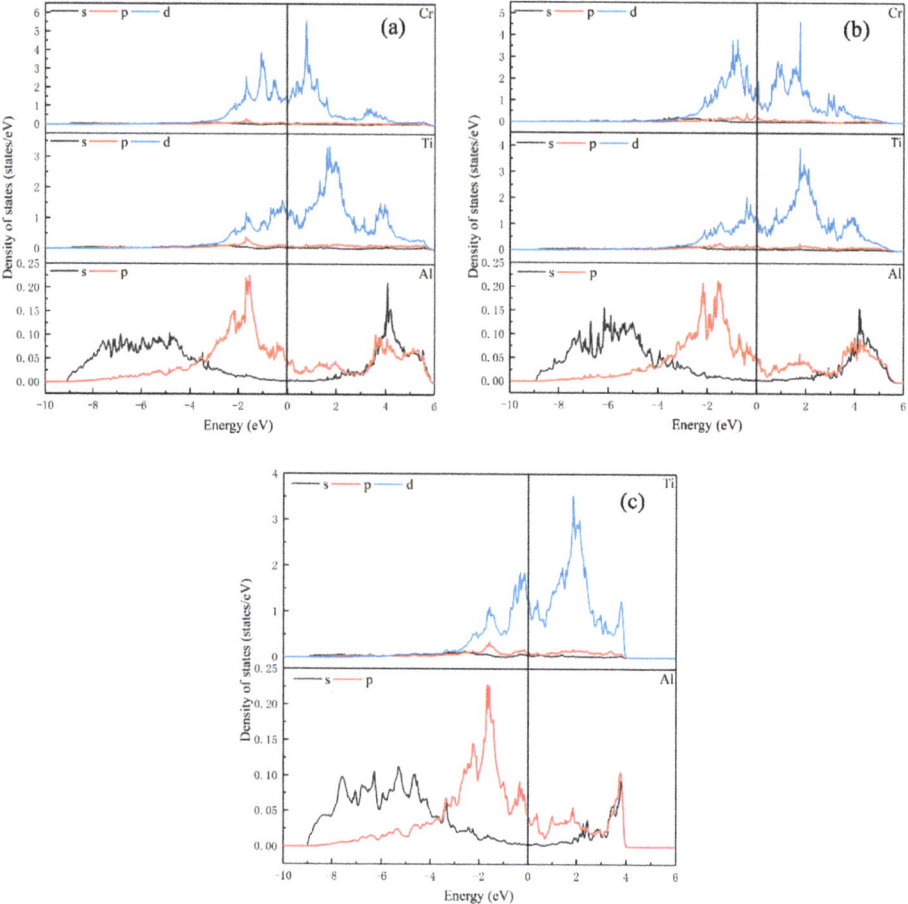

Figure 3. Partial density of states for (**a**) $Ti_{15}Al_{16}Cr$; (**b**) $Ti_{16}Al_{15}Cr$; (**c**) $Ti_{16}Al_{16}$.

As can be noted from Figure 3c, for $Ti_{16}Al_{16}$, the orbital overlap between atoms is mainly concentrated at −3 eV to 4 eV. When Cr replaces Ti or Al, the overlap between the atoms gets a little wider. In Figure 3a, when Cr replaces Ti, the peak of density of states of Ti and Al decreases compared with $Ti_{16}Al_{16}$, and the interaction between Ti-Ti, Ti-Al and Al-Al becomes weaker. In Figure 3b, when Cr replaces Al, the peak of density of states of Cr is somewhat lower than that of $Ti_{15}Al_{16}Cr$, but it is still larger than that of Ti. There is strong d orbital hybridization with Ti atoms from approximately −3 eV to 4 eV. The Ti-Ti interaction is enhanced by increasing the peak of density of states of Ti.

As shown in Figure 4c, for $Ti_{24}Al_8$, the orbital overlap between atoms is mainly between −3 eV and 3 eV. In Figure 4a, unlike Ti_8Al_{24} and $Ti_{16}Al_{16}$, the peak of density of states of Ti in $Ti_{23}Al_8Cr$ is not significantly reduced after replacing Ti with Cr. However, the peak of density of states of the Cr atom is larger than that of Ti, and there are strong metal bonds between the Cr atom and the surrounding Ti atoms between −3 eV and 3 eV. Thus, the ductility of the alloy is also enhanced after Cr replaces the Al in $Ti_{24}Al_8$.

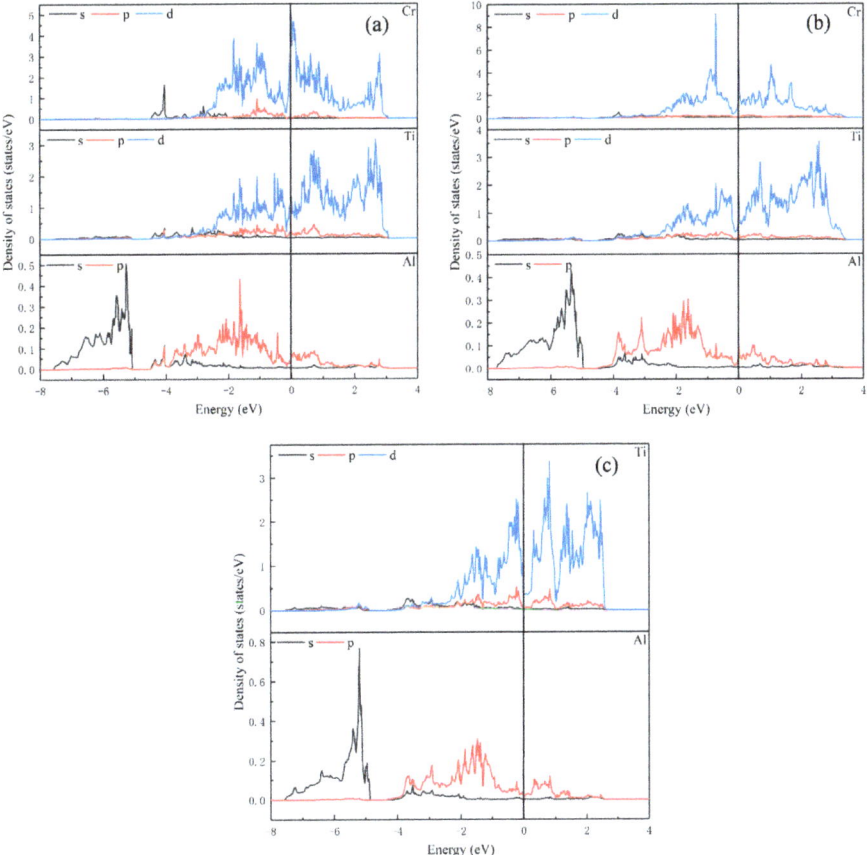

Figure 4. Partial density of states for (**a**) $Ti_{23}Al_8Cr$; (**b**) $Ti_{24}Al_7Cr$; (**c**) $Ti_{24}Al_8$.

By analyzing the partial density of states data of elements in nine alloys, including Ti_8Al_{24}, $Ti_7Al_{24}Cr$, $Ti_8Al_{23}Cr$, $Ti_{16}Al_{16}$, $Ti_{15}Al_{16}Cr$, $Ti_{16}Al_{15}Cr$, $Ti_{24}Al_8$, $Ti_{23}Al_8Cr$ and $Ti_{24}Al_7Cr$, it can be seen that the main reason for the ductility reduction in Ti_8Al_{24} and $Ti_{16}Al_{16}$ is the weakening of the Ti-Ti metal bond's strength after replacing Ti with Cr. After replacing Ti atoms with Cr in $Ti_{24}Al_8$, the main reason for the increase in the ductility of the material is that the peak of the density of states of Ti is not significantly reduced, and Cr forms a solid metal bond with the surrounding Ti atoms. After Cr replaces Al in Ti-Al intermetallics, the reason for the increase in material ductility is that Cr and Ti form a solid metal bond, which improves the metal bond strength of the system.

4. Conclusions

(1) The Cr atom prefers to replace Ti atoms in Ti-Al intermetallics. However, according to the formation energy and Born–Huang criterion, whether the Cr atom replaces Ti or Al, the system has energy stability and mechanical stability;

(2) The ductility of TiAl and $TiAl_3$ compounds cannot be improved by the Cr atom replacing the Ti atom but can be improved by the Cr atom replacing the Al atom. In the Ti_3Al compound, the Cr atom can improve the ductility no matter whether it replaces Ti or Al, and the effect is better with the latter;

(3) After replacing Al with Cr, the metal bond between Ti-Ti is strengthened, and a strong metal bond is formed between Cr and Ti, thus improving the ductility of Ti-Al intermetallics.

Author Contributions: Methodology, H.W. and F.S.; validation, Z.W.; investigation, H.W.; resources, F.S.; data curation, F.S. and Z.W.; writing—original draft, H.W.; writing—review & editing, H.W. and F.S.; visualization, F.S. and Z.W.; project administration, F.S. and Z.W. All authors have read and agreed to the published version of the manuscript.

Funding: This work was supported by the National Natural Science Foundation of China (Grant NO. 51706017).

Data Availability Statement: The original data can be obtained from the first author of this article.

Conflicts of Interest: The authors declare that they have no known competing financial interest or personal relationship that could have appeared to influence the work reported in this paper.

References

1. Zhao, K.; Feng, N.; Wang, Y. Fabrication of Ti-Al intermetallics by a two-stage aluminothermic reduction process using Na2TiF6. *Intermetallics* **2017**, *85*, 156–162. [CrossRef]
2. Song, Y.; Dou, Z.; Zhang, T.; Liu, Y.; Wang, G.C. First-principles calculation on the structural, elastic and thermodynamic properties of Ti-Al intermetallics. *Mater. Res. Express* **2019**, *6*, 1065. [CrossRef]
3. Jian, Y.; Huang, Z.; Xing, J.; Sun, L.; Liu, Y.; Gao, P. Phase stability, mechanical properties and electronic structures of Ti Al binary compounds by first principles calculations. *Mater. Chem. Phys.* **2019**, *221*, 311–321. [CrossRef]
4. Jiang, C. First-principles study of site occupancy of dilute 3d, 4d and 5d transition metal solutes in L10 TiAl. *Acta. Mater.* **2008**, *56*, 6224–6231. [CrossRef]
5. Tang, P.; Tang, B.; Su, X. First-principles studies of typical long-period superstructures Al5Ti3, h-Al2Ti and r-Al2Ti in Al-rich TiAl alloys. *Comput. Mater. Sci.* **2011**, *50*, 1467–1476. [CrossRef]
6. Zhang, L.G.Z.A. Ab initio pseudopotential calculations on the effect of Mn doped on lattice parameters of L10 TiAl. *Intermetallics* **2000**, *8*, 637–641.
7. Song, Y.; Yang, R.; Li, D.; Hu, Z.Q.; Guo, Z.X. A first principles study of the influence of alloying elements on TiAl: Site preference. *Intermetallics* **2000**, *8*, 563–568. [CrossRef]
8. Fan, T.; Lin, L.; Liang, H.; Ma, Y.; Tang, Y.; Hu, T.; Ruan, Z.; Chen, D.; Wu, Y. First-principles study of the structural, mechanical and thermodynamic properties of Al11RE3 in aluminum alloys. *Crystals* **2023**, *13*, 347. [CrossRef]
9. Saraç, U.; Trong, D.N.; Baykul, M.C.; Long, V.C.; Ţălu, Ş. Tuning structural properties, morphology and magnetic characteristics of nanostructured Ni-Co-Fe/ITO ternary alloys by galvanostatic pretreatment process. *Microsc. Res. Tech.* **2022**, *85*, 3945–3954. [CrossRef]
10. Qi, W.; Wang, W.; Yang, X.; Xie, L.; Zhang, J.; Li, D.; Zhang, Y. Effect of Zr on phase separation, mechanical and corrosion behavior of heterogeneous CoCrFeNiZrx high-entropy alloy. *J. Mater. Sci. Technol.* **2022**, *109*, 76–85. [CrossRef]
11. Trong, D.N.; Long, V.C.; Ţălu, Ş. The Structure and Crystallizing Process of NiAu Alloy: A Molecular Dynamics Simulation Method. *J. Compos. Sci.* **2021**, *5*, 18. [CrossRef]
12. Couret, A.; Voisin, T.; Thomas, M.; Monchoux, J.P. Development of a TiAl Alloy by Spark Plasma Sintering. *JOM* **2017**, *69*, 2576–2582. [CrossRef]
13. Tang, S.Q.; Qu, S.J.; Feng, A.H.; Shen, J.; Chen, D.L. Core-multishell globular oxidation in a new TiAlNbCr alloy at high temperatures. *Sci. Rep.* **2017**, *7*, 3483. [CrossRef]
14. Kastenhuber, M.; Klein, T.; Clemens, H.; Mayer, S. Tailoring microstructure and chemical composition of advanced γ-TiAl based alloys for improved creep resistance. *Intermetallics* **2018**, *97*, 27–33. [CrossRef]
15. Zhou, C.; Zeng, F.P.; Liu, B.; Zhao, K.; Lu, J.; Qiu, C.; Li, J.; He, Y. Effects of Si on Microstructures and High Temperature Properties of Beta Stabilized TiAl. *Alloy. Mater. Trans.* **2016**, *57*, 461–465. [CrossRef]
16. Liu, Q.; Nash, P. The effect of Ruthenium addition on the microstructure and mechanical properties of TiAl alloys. *Intermetallics* **2011**, *19*, 1282–1290. [CrossRef]
17. Tetsui, T. Effects of high niobium addition on the mechanical properties and high-temperature deformability of gamma TiAl alloy. *Intermetallics* **2002**, *10*, 239–245. [CrossRef]
18. Liu, Z.C.; Lin, J.P.; Li, S.J.; Chen, G.L. Effects of Nb and Al on the microstructures and mechanical properties of high Nb containing TiAl base alloys. *Intermetallics* **2002**, *10*, 653–659. [CrossRef]
19. Duan, Q.; Luan, Q.; Liu, J.; Peng, L. Microstructure and mechanical properties of directionally solidified high-Nb containing Ti–Al alloys. *Mater. Des.* **2010**, *31*, 3499–3503. [CrossRef]
20. Music, D.; Schneider, J. Effect of transition metal additives on electronic structure and elastic properties of TiAl and Ti3Al. *Phys. Rev. B* **2006**, *74*, 174110. [CrossRef]

21. Yuan, C.; Liu, B.; Liu, Y.X.; Yong, L.I.U. Processing map and hot deformation behavior of Ta-particle reinforced TiAl composite. *T. Nonferr. Met. Soc.* **2020**, *30*, 657–667. [CrossRef]
22. Xiong, J.; Song, L.; Guo, X.; Liu, X.; Zhang, W.; Zhang, T. Inhibition of ω_0 phase precipitation in TNM-based TiAl alloys by Cr and Mn. *Intermetallics* **2023**, *153*, 107774. [CrossRef]
23. Ye, L.; Wang, H.; Zhou, G.; Hu, Q.M.; Yang, R. Phase stability of TiAl-X (X=V, Nb, Ta, Cr, Mo, W, and Mn) alloys. *J. Alloys Compd.* **2020**, *819*, 153291. [CrossRef]
24. Tetsui, T. Impact resistance of commercially applied TiAl alloys and simple-composition TiAl alloys at various temperatures. *Metals* **2022**, *12*, 2003. [CrossRef]
25. Fox-Rabinovich, G.S.; Weatherly, G.C.; Wilkinson, D.S.; Kovalev, A.I.; Wainstein, D.L. The role of chromium in protective alumina scale formation during the oxidation of ternary TiAlCr alloys in air. *Intermetallics* **2004**, *12*, 165–180. [CrossRef]
26. Brady, M.P.; Wright, I.G.; Gleeson, B. Alloy design strategies for promoting protective oxide-scale formation. *JOM* **2000**, *52*, 16–21. [CrossRef]
27. Kohn, W.; Sham, L.J. Self-Consistent Equations Including Exchange and Correlation Effects. *Phys. Rev.* **1965**, *140*, 1133–1141. [CrossRef]
28. Kohn, W.; Hohenberg, P. Inhomogeneous Electron Gas. *Phys. Rev.* **1964**, *136*, 864–871.
29. Kresse, G. Efficient iterative schemes for ab initio total-energy calculations using a plane-wave basis set. *Phys. Rev. B* **1996**, *54*, 11169–11186. [CrossRef]
30. Kresse, G.; Furthmüller, J. Efficiency of ab-initio total energy calculations for metals and semiconductors using a plane-wave basis set. *Comput. Mater. Sci.* **1996**, *6*, 15–50. [CrossRef]
31. Kresse, G.; Joubert, D. From ultrasoft pseudopotentials to the projector augmented-wave method. *Phys. Rev. B* **1999**, *59*, 1758–1775. [CrossRef]
32. Blochl, P.E. Projector augmented-wave method. *Phys. Rev. B* **1994**, *50*, 17953–17979. [CrossRef]
33. Perdew, J.P.; Burke, K.; Ernzerhof, M. Generalized Gradient Approximation Made Simple. *Phys. Rev. Lett.* **1996**, *77*, 3865–3868. [CrossRef] [PubMed]
34. Tan, J.H.; Zhu, K.J.; Peng, J.H. First-principles simulation on structure property of Ti-Al intermetallics. *J. Comput. Phys.* **2017**, *34*, 365–373.
35. Hultgren, R.; Desai, P.D.; Hawkins, D.T.; Gleiser, M.; Kelley, K.K. Selected values of the thermodynamic properties of binary alloys. *Natl. Stand. Ref. Data Syst.* **1973**, *58*, 1432.
36. Xie, Y.; Tao, H.; Peng, H.; Li, X.; Liu, X.; Peng, K. Atomic states, potential energies, volumes, stability, and brittleness of ordered FCC TiAl3-type alloys. *Phys. B Condens. Matter* **2005**, *366*, 17–37. [CrossRef]
37. Liu, Y.; Cui, X.; Niu, R.; Zhang, S.; Liao, X.; Moss, S.D.; Finkel, P.; Garbrecht, M.; Ringer, S.P.; Cairney, J.M.; et al. Giant room temperature compression and bending in ferroelectric oxide pillars. *Nat. Commun.* **2022**, *13*, 335. [CrossRef]

Disclaimer/Publisher's Note: The statements, opinions and data contained in all publications are solely those of the individual author(s) and contributor(s) and not of MDPI and/or the editor(s). MDPI and/or the editor(s) disclaim responsibility for any injury to people or property resulting from any ideas, methods, instructions or products referred to in the content.

Article

Room and Elevated Temperature Sliding Friction and Wear Behavior of $Al_{0.3}CoFeCrNi$ and $Al_{0.3}CuFeCrNi_2$ High Entropy Alloys

Dheyaa F. Kadhim [1,2], Manindra V. Koricherla [2] and Thomas W. Scharf [2,*]

1 Department of Mechanical Engineering, University of Thi-Qar, Nasiriyah 64001, Iraq
2 Department of Materials Science and Engineering, The University of North Texas, Denton, TX 76207, USA
* Correspondence: thomas.scharf@unt.edu

Abstract: In this study, processing–structure–property relations were systematically investigated at room and elevated temperatures for two FCC $Al_{0.3}CoFeCrNi$ and $Al_{0.3}CuFeCrNi_2$ high-entropy alloys (HEAs), also known as complex concentrated alloys (CCAs), prepared by conventional arc-melting. It was determined that both alloys exhibit FCC single-phase solid solution structure. Micro-indentation and sliding wear tests were performed to study the hardness and tribological behavior and mechanisms at room and elevated temperatures. During room-temperature sliding, both alloys exhibit similar friction behavior, with an average steady-state coefficient of friction (COF) of ~0.8. Upon increasing sliding temperatures to 300 °C, the average COF decreased to a lowest value of ~0.3 for $Al_{0.3}CuFeCrNi_2$. Mechanistic wear studies showed this was due to the low interfacial shear strength tribofilms formed inside the wear tracks. Raman spectroscopy and energy dispersive spectroscopy determined the tribofilms were predominantly composed of binary oxides and multi-element solid solution oxides. While the tribofilms at elevated temperatures lowered the COF values, the respective wear rates in both alloys were higher compared to room-temperature sliding, due to thermal softening during 300 °C sliding. Thus, these single FCC-phase HEAs provide no further benefit in wear resistance at elevated temperatures, and likely will have similar implications for other single FCC-phase HEAs.

Keywords: complex concentrated alloys; high-entropy alloys; sliding friction and wear; tribofilm

Citation: Kadhim, D.F.; Koricherla, M.V.; Scharf, T.W. Room and Elevated Temperature Sliding Friction and Wear Behavior of $Al_{0.3}CoFeCrNi$ and $Al_{0.3}CuFeCrNi_2$ High Entropy Alloys. *Crystals* **2023**, *13*, 609. https://doi.org/10.3390/cryst13040609

Academic Editor: Jacek Ćwik

Received: 17 February 2023
Revised: 26 March 2023
Accepted: 30 March 2023
Published: 2 April 2023

Copyright: © 2023 by the authors. Licensee MDPI, Basel, Switzerland. This article is an open access article distributed under the terms and conditions of the Creative Commons Attribution (CC BY) license (https://creativecommons.org/licenses/by/4.0/).

1. Introduction

In recent years, numerous metallic alloys have been studied that contain multiple principal and equimolar elements, typically with five or more major constituents in the range of 5–35 at. %. This new approach to designing such alloys with multiple principal elements has led to the emergence of high-entropy alloys (HEAs), which were proposed by Yeh et al. [1] and Cantor et al. [2], and have more recently been denoted as complex concentrated alloys (CCAs) [3,4]. Unlike conventional alloys, core effects are present, such as high entropy, lattice distortion, sluggish diffusion, and cocktail effects [1–4]. The high-entropy effect results from the higher configurational entropy in these alloys compared to conventional alloys. Due to the high entropy of mixing, these alloys may favor the formation of simple solid solutions and prevent the generation of hard but brittle intermetallic compounds. As a result, these alloys can exhibit superior wear, oxidation and corrosion resistance, as well as high-temperature strength [3–6]. Moreover, recent studies have indicated that $Al_{0.5}CoCrCuFeNiTi_x$, $Al_xCoCrCuFeNi$ and $AlCoCrFeNiTi_x$ HEAs exhibit a variety of microstructures and mechanical properties, with face-centered cubic (FCC), body-centered cubic (BCC) structures, or a mixture of both [7,8].

According to studies on the tribological performance of HEAs, composition is a key factor in wear resistance. The effect of Al addition on the $FeCoCrNiAl_x$ HEA was examined by Liu et al. [9], who concluded that the wear mechanisms changed from previously mixed wear modes of abrasive, adhesive, and oxidative wear to a mixture of abrasive and oxidative wear. Such a change shows that the FeCoCrNiAl$_x$ HEA coating's wear resistance was significantly enhanced by the addition of Al. In a tribological study of Al$_{0.65}$CoCrFeNi, Miao et al. [10] found that only abrasive wear features could be seen on the remelted layer, whereas adhesive wear predominated on the substrate. As a result, the average friction coefficient and wear rate of the remelted layer were reduced by 23% and 80%, respectively, compared to the substrate. Wu et al. [11] reported on the adhesive wear behavior of Al$_x$CoCrCuFeNi alloys, showing that increasing the aluminum content resulted in the HEA structure transforming from FCC to BCC phases, subsequently raising the hardness value and lowering the wear rate. In addition, a low Al percentage has been found to favor a single-phase FCC lattice in the well-studied Al$_x$CoCrFeNi HEA, whereas a greater Al fraction results in a BCC phase [12].

The well-studied Al$_x$CoCrFeNi HEA system microstructures have been shown to change from having a single FCC phase to a single BCC phase with increasing Al content, with the Al$_{0.3}$CoCrFeNi HEA being the only one to form a solid solution with an FCC structure. This particular HEA will serve as a baseline alloy in the present study. According to several studies, this HEA exhibits good mechanical properties such as high plasticity, work-hardening capacity, and a balance between cryogenic strength and ductility [13–16]. Due to the inherent property of the FCC structure, Al$_{0.3}$CoCrFeNi HEA has a relatively low strength at ambient temperature, with a room-temperature yield strength between 150 and 350 MPa [13]; its melting point is at 1870 K [14]. The mechanical properties of Al$_{0.3}$CoCrFeNi HEAs have been significantly improved through the application of thermo-mechanical processing techniques, such as cold rolling and subsequent annealing. For example, with 90% cold rolling and 550 °C annealing, Gwalani et al. [15] produced a more complex microstructure with hierarchical features of ultra-fine grains, fine-scale B2 and σ precipitates, and nanoclusters that resulted in the tensile yield strength significantly increasing from 160 to 1800 MPa. The fully recrystallized states along with grain refinement strengthening and precipitation strengthening were the focus of earlier investigations into strengthening FCC Al$_{0.3}$CoCrFeNi HEAs. Jiao et al. [16] used instrumented nanoindentation to examine the mechanical characteristics of Al$_{0.3}$CoCrFeNi and AlCoCrFeNi HEAs over a wide range of loading rates to determine an excellent combination of strength and ductility.

Based on the above studies, Al$_{0.3}$CoFeCrNi HEAs exhibit a good balance of mechanical properties, so their effects on tribological properties are of interest. However, there have been limited systematic investigations on the room-temperature friction and wear behavior and mechanisms of single FCC-phase Al$_{0.3}$CoFeCrNi, and no studies on Al$_{0.3}$CuFeCrNi$_2$ HEA. Additionally, the tribological behavior and mechanistic studies are unknown for these alloys at elevated temperatures, at which alternative structural alloys are of interest to replace bearing steels, such 440C and 52100, which oxidize and form iron oxide deleterious phases. It is at elevated temperatures such as 300 °C, used in this study, that mechanical and tribological processes begin to occur in metallic alloys, such as thermal softening and tribochemical (oxidation) reactions, respectively. Therefore, the objective of the present study was to investigate the influence and mechanisms of Ni content on the tribological properties and corresponding tribofilm evolution of Al$_{0.3}$CoFeCrNi and Al$_{0.3}$CuFeCrNi$_2$ HEAs, both at room temperature and during 300 °C sliding.

2. Experimental Methods

Two alloys with different Ni content and containing Co or Cu were studied: Al$_{0.3}$CoFeCrNi (A1) and Al$_{0.3}$CuFeCrNi$_2$ (A2). Alloy ingots were prepared by arc-melting and casting. The mixtures of the alloying elements with purities higher than 99.5 wt.% were melted in an argon atmosphere several times to improve the chemical homogeneity of the ingots in the liquid state. Both alloys were cold rolled to a 20% reduction in thickness with ten passes and solutionized at 1150 °C for 30 min, followed by water-quenching for homogenization. For microstructural and property investigation, the samples were ground and polished by standard metallographic techniques to dimensions of approximately

19 mm (length) × 14 mm (width) × 7 mm (thick) with mass of ~14 g. Surface microstructural characterization was carried out in a field emission gun scanning electron microscope (SEM) using a FEI Nova 200 dual-beam SEM. The SEM is also equipped with energy dispersive x-ray spectroscopy (EDS) for chemical analysis of the alloys before and after sliding (wear track maps). Room and elevated temperature (300 °C after the sliding tests) microhardness measurements were performed using a Shimadzu Vickers hardness indenter with a normal load of 9.807 N at a hold time of 10 s. A total of ten measurements were recorded for both alloys at spacings of ~1 mm apart.

The friction behavior of both alloys was Investigated using a Falex ISC-200 pin-on-disk tribometer, following the ASTM G99 standard. The sliding coefficient of friction (COF) was measured at room and elevated (300 °C) temperature in lab air (40% relative humidity). The current study was limited to 300 °C sliding since elevated temperature studies require a significant amount of time. The HEAs were tested in unidirectional sliding against a Si_3N_4 ball counterface (3.175 mm diameter) with hardness of 22 GPa to avoid ball wear and transfer to the wear tracks. The sliding speed was 8.5 mm/s for all tests, with a normal load of 0.25 N. Based on these values, the initial maximum Hertzian contact stress is ~0.6 GPa, which was chosen to be below the yield strength of these alloys. The total sliding distance was 200 m for all tests that took about 6 h. At least three measurements were made for each HEA for repeatability purposes. After each test, an optical microscope was used to image the worn surfaces of the HEAs and the Si_3N_4 counterfaces. A stylus surface profilometer (Veeco Dektak 150 Profilometer) was used to measure wear track depths. At least eight profilometry traces were taken across each wear scar to obtain the cross sectional worn area. The wear factor/rate was calculated as the removed volume loss divided by the applied load and the total sliding distance. The volume loss can be calculated by multiplying the area of the worn surface by the circumference of the circular wear track, assuming uniform wear. Crystal structures were identified with an X-ray diffractometer (Rigaku Ultima III) under radiation conditions of 30 kV, 20 mA, a CuKα anode, and a scanning speed of 2 degrees/minute. Representative wear surfaces were analyzed using SEM and EDS to acquire both secondary electron (SE) and backscatter electron (BSE) images as well as elemental wear maps, respectively. In addition, a Raman spectrometer (Thermo Electron Almega XR) was used to determine tribo-chemical phases on the wear surfaces using a 532 nm laser wavelength.

3. Results and Discussion

3.1. Microstructure and Phase Analysis

Figure 1 shows the XRD scans of the $Al_{0.3}CoFeCrNi$ (A1) and $Al_{0.3}CuFeCrNi_2$ (A2) HEAs. It is evident there is one set of fundamental FCC reflections that verifies an FCC solid solution crystal structure without the presence of secondary phases. Using the indexed (111) reflections, the lattice constants (a) for A1 and A2 alloys are calculated to be 3.591 Å and 3.589 Å, respectively. Based on the resolution of the x-ray diffractometer, these values are within experimental error, and thus there are no lattice parameter differences between the two single FCC-phase HEAs. These HEAs retain a single phase at higher temperatures when cooled rapidly due to the entropy of too many alloying elements, along with sluggish diffusion kinetics.

Figure 2 and Table 1 show representative SEM images and corresponding EDS chemical analysis, respectively, of the A1 and A2 alloys. The alloys exhibit microcrystalline structures with elongated grains (sizes in the range of 300–600 μm) and have no compositional segregation. Wang et al. [8] also reported that as-cast $Al_{0.3}CoCrFeNi$ HEA has an FCC single phase with a columnar microstructure. Similarly, Kao et al. [17] determined that $Al_{0.37}CoCrFeNi$ has an FCC single-phase crystal structure, also shown by Guo et al. [18] for $Al_xCrCuFeNi_2$ HEAs. The latter authors also determined that $Al_{0.5}CrCuFeNi_2$ alloy exhibits a dendritic microstructure. The SEM images in Figure 2 show evidence of possible interdendritic and dendritic regions retaining an FCC phase with high closeness in the lattice constants, also reported in [19,20]. Ng et al. [21] showed that a $Al_{0.5}CrCuFcNi_2$ alloy exhibits an FCC single phase, but due to XRD peak overlaps, they surmised there were two disordered FCC phases with very close lattice parameters at ~3.59 Å, which is equivalent to the above calculated lattice parameters for A1 and A2 alloys. Gwalani et al. [22] determined with transmission electron microscopy and selected area diffraction patterns that $Al_{0.3}CuFeCrNi_2$ (with the same composition as A2 alloy) has an FCC-type solid solution at ambient temperature. They further stated the high mixing entropy and sluggish effects that decrease the Gibbs energy develop a solid solution rather than intermetallic compounds, and as a result were surmised to be the reasons for the formation of a single solid solution instead of intermetallic compounds [22]. From a thermodynamic point of view, the mixing enthalpy overcomes and leads to decomposition in the matrix to form two phases.

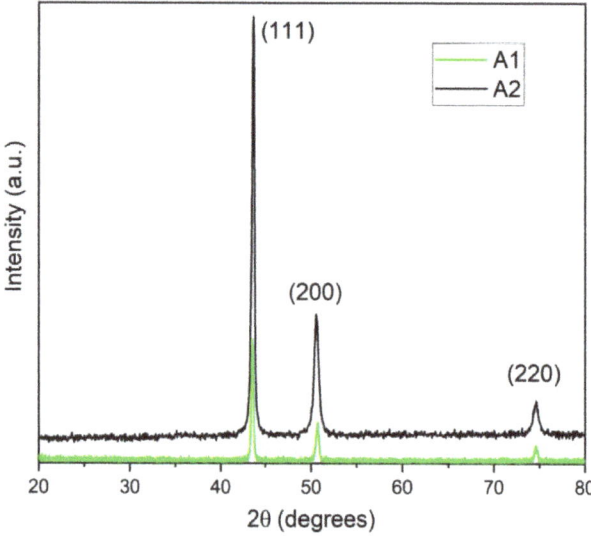

Figure 1. XRD patterns of A1 and A2 alloys showing FCC reflections.

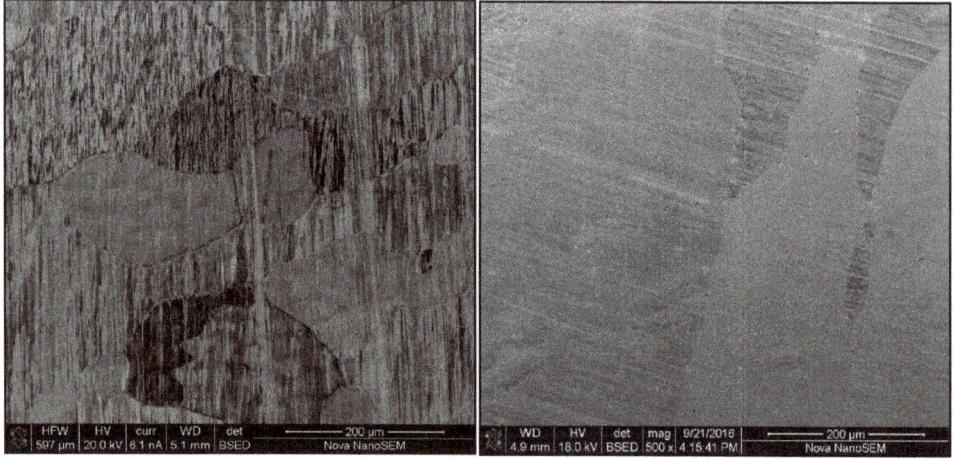

Figure 2. SEM images of the A1 (**left**) and A2 (**right**) alloys.

Table 1. Chemical composition (in at%) of A1 and A2 alloys.

	Al	Co/Cu	Fe	Cr	Ni
A1	6.64	22.40	22.52	22.97	25.47
A2	5.69	19.47	17.67	18.52	38.65

3.2. Microhardness

It is crucial to assess the hardness of the material in order to understand the likely underlying wear process, since material hardness is often correlated to wear resistance [23]. Table 2 lists the averaged microhardness values for the A1 and A2 alloys at room temperature (RT) and 300 °C (acquired after the sliding wear tests and outside the wear tracks). Both alloys exhibit similar RT hardness values. However, at 300 °C, there is thermal softening in A1 and hardening in A2 that could be due to Ni content differences, resulting in the likely formation of a harder NiO scale on the A2 alloy. In addition, Qiu et al. [24] determined that adding more Ni to $Al_2CrFeCoCuTiNi_x$ HEAs resulted in increasing RT microhardness and strength. They attributed this increase to the Ni content that increased the content of the BCC crystal structure in the alloys. In contrast, López Ríos et al. [25] reported that because of Cr and Fe precipitates dissolving in the nickel-rich matrix and forming a stable solid solution, the hardness values for AlCrFeCoNi decreased to 562 HV, 455 HV, and 316 HV, with reduction in nickel concentration. Kuo et al. [26] determined that a $CuFeTiZrNi_{0.1}$ alloy exhibits a microhardness of 935 HV, wherein increasing the FCC phase in the alloy is correlated to an increase in the Ni content. Therefore, they concluded that the low hardness of the FCC phase causes the alloy's hardness to gradually decrease with increasing Ni content; however, the hardness was not measured at elevated temperatures, e.g., at 300 °C, where the effects of oxide scales and thermal softening can influence the microhardness values.

Table 2. Average microhardness (HV_1) values and standard deviations of A1 and A2 alloys at RT and 300 °C.

	RT	300 °C
A1	168 ± 3	144 ± 10
A2	164 ± 4	188 ± 18

3.3. Friction and Wear Behavior

Table 3 lists the averaged steady-state COF and wear factor values for A1 and A2 alloys during RT and 300 °C sliding. The RT COF values of ~0.8 for both alloys during steady-state sliding are relatively high friction values that are consistent with Si_3N_4 sliding on Ni and other metallic materials [27,28]. However, these values are considerably higher than other HEAs with hard secondary phases; for example, COF values of ~0.3 were observed in HEA $Al_{0.25}Ti_{0.75}CoCrFeNi$ [29], since harder alloys exhibit a smaller real area of sliding contact, thereby leading to lower frictional forces. The high RT friction behavior of both alloys is due to the single FCC phase being relatively soft, which is corroborated by the relatively low hardness values listed in Table 2. During sliding at 300 °C, the averaged COF values decreased for both alloys, but considerably more in the A2 alloy, which will be further discussed in the next section.

Table 3. Average steady-state COF values and wear factors ($mm^3/N·m$) with standard deviations of A1 and A2 alloys at RT and 300 °C.

	RT	300 °C	RT	300 °C
	COF		Wear factors	
A1	0.84 ± 0.06	0.54 ± 0.04	$6.6 ± 0.8 \times 10^{-5}$	$1.2 ± 0.3 \times 10^{-4}$
A2	0.78 ± 0.06	0.29 ± 0.05	$5.3 ± 1.1 \times 10^{-5}$	$9.8 ± 0.4 \times 10^{-5}$

Representative cross-sectional wear track depth and width profiles are shown in Figure 3 for A1 and A2 alloys acquired after sliding at RT and 300 °C. It is evident that during RT sliding, the alloys exhibit a smaller worn area compared to sliding at 300 °C, despite the higher COF values. The cross sectional worn areas were used to calculate the corresponding wear factors listed in Table 3. The wear factors listed in Table 3 agree well with the RT COF trends, i.e., there are similar wear factors for both alloys at RT. However, during 300 °C sliding, there is a slightly lower wear factor for the A2 alloy, since there was no thermal softening, as evidenced by the increase in hardness. Compared to RT sliding wear factors, both alloys have higher wear factors, suggesting that in addition to hardness, there are other factors active that will be discussed in the next section. Since both HEAs exhibit single FCC-phase structures without the presence of typical secondary phases present in structural alloys, e.g., hard intermetallic or carbide phases, the wear factors are about an order of magnitude higher. For example, the aforementioned $Al_{0.25}Ti_{0.75}CoCrFeNi$ BCC HEA has hard intermetallic $L2_1$ and χ phases responsible for lower sliding wear rates in the order of 1×10^{-5} mm^3/N·m [29]. However, a similar single FCC-phase HEA $Al_{0.1}CoCrFeNi$ to that of A1 alloy, but with lower Al content and slightly softer, exhibits a higher RT sliding wear factor of 1.9×10^{-4} mm^3/N·m [30]. Based on the above, single FCC-phase HEAs do not provide adequate wear resistance compared to HEAs with hard secondary phases, making them similar to other structural alloys such as bearing steels with chromium carbide precipitates. Only when hard secondary phases are present do HEAs such as $Al_{0.25}Ti_{0.75}CoCrFeNi$ have comparable wear rates to those measured for chromium carbide bearing steel 440C, values of ~1×10^{-5} mm^3/N·m [29].

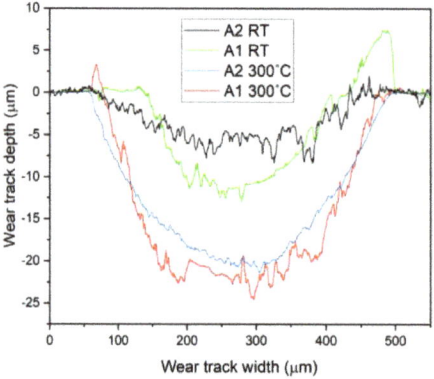

Figure 3. Representative cross-sectional wear track depths and widths of A1 and A2 alloys after RT and 300 °C sliding.

3.4. Friction and Wear Mechanisms

Figure 4 shows representative SEM images and EDS maps of the A1 alloy wear track after RT sliding. The wear track shows signs of abrasive wear, i.e., microabrasion with grooves along the sliding direction. There is also oxidative wear based on the dark features in the BSE images (meaning a higher atomic number contrast) that coincides with the oxygen EDS map in Figure 4. Based on the elemental maps, there is not one particular metal that shows a preference for oxidative wear. This suggests that there is a mixed metal oxide tribofilm on the wear surface. These flattened oxide patches inside the wear track are indicative of a surface fatigue wear mode that results in metallic oxide wear fragments delaminating from the surface. With repeated sliding, the wear fragments can either be ejected from the sliding contact or become entrapped beneath the Si_3N_4 counterface, with the latter pathway contributing to the formation of micro-grooves by a three-body abrasive wear mode.

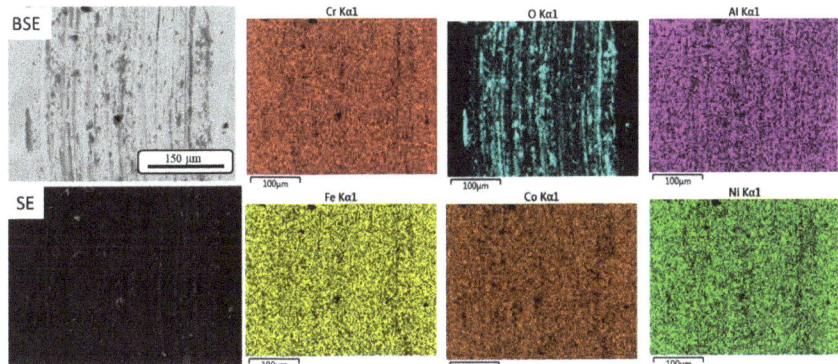

Figure 4. BSE and SE images and corresponding EDS elemental maps of the A1 alloy wear track after RT sliding.

Figure 5 shows representative SEM images and EDS maps of the A2 alloy wear track after RT sliding. Similar to the A1 alloy, there is micro-ploughing/micro-grooving inside the wear track, indicating a micro-abrasive wear mode with striations running parallel to the sliding direction and across the entire wear track length. There are also oxidative/surface fatigue wear modes present, although the wear track width is slightly smaller, and there is slightly less micro-abrasion/metal oxide tribofilm covering the wear track. This accounts for the slightly lower RT COF and wear factor values listed in Table 3 for the A2 alloy.

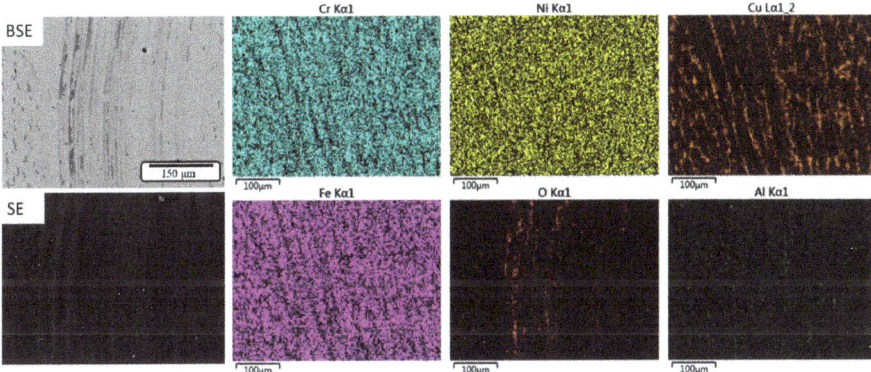

Figure 5. BSE and SE images and corresponding EDS elemental maps of the A2 alloy wear track after RT sliding.

Figure 6 shows SEM images and EDS maps of the A1 alloy wear track after 300 °C sliding. It is evident there is an increased amount of oxidative wear compared to RT sliding for this alloy, based on the BSE image and corresponding oxygen EDS wear map covering almost the entire wear track. Furthermore, it appears that this oxide tribofilm is also a mix of all the metallic elements, based on the EDS maps that do not show a preference for any particular metal oxide phase. Despite the lower COF due to the lower interfacial shear strength oxide tribofilm, the wear track width has increased in size compared to the RT wear track. This is supported by the increased wear factor listed in Table 3, due to thermal softening for this A1 alloy, which is based on the lower hardness value of $HV_1 = 144$. More severe adhesive wear also occurs in the wear track shown by the SEM images in Figure 6, since these wear tracks are covered with a compact tribofilm (often referred to as an oxide glaze layer), indicating a change in wear mechanisms. Abrasive grooves along the sliding direction are still visible on the wear track in areas not covered by the oxidized tribofilm. In addition, there is evidence of fragmented oxide wear debris that further acts similar to abrasive particles to accelerate the wear process, resulting in the high wear factor. Therefore, while the low interfacial shear strength oxide tribofilm provides low COF values, it not protective at this elevated sliding temperature compared to RT sliding.

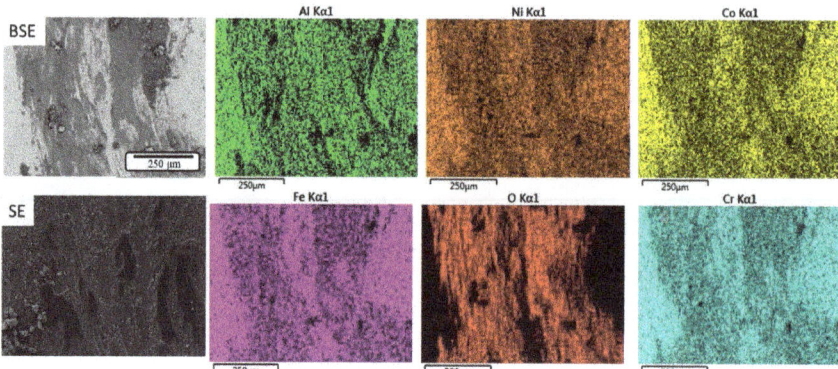

Figure 6. BSE and SE images and corresponding EDS elemental maps of the A1 alloy wear track after 300 °C sliding.

In contrast, the A2 alloy exhibits increased hardness, up to $HV_1 = 188$ at 300 °C, which results in a slightly smaller wear track, as shown in Figure 7, and thus a smaller frictional area of contact during sliding. Hence, the A2 alloy has the lowest COF of ~0.29 and a slightly lower wear factor of 9.8×10^{-5} mm^3/N·m compared to the A1 alloy at 300 °C, but this is still higher compared to RT sliding. Hardening in the A2 alloy is likely due to the increased amount of Ni resulting in the formation of a NiO scale, although verification would be needed with cross-sectional microscopy and micro-indentation. The wear track shown in Figure 7 exhibits slightly less oxidative wear compared to the A1 alloy, based on the images and EDS oxygen map. In addition, the fragmented oxide wear debris act as micro-abrasive particles, resulting in increased wear. Like the A1 alloy, the low interfacial shear strength oxide tribofilm results in low COF values, but overall does not provide wear protection at this elevated sliding temperature compared to RT sliding. Similarl to other alloys such as the CoCr-based alloy (Haynes 25), at temperatures greater than 200 °C, the friction coefficient is reduced due to the formation of a protective oxide layer, which minimizes the adhesion between the two contacting surfaces [31]. When the temperature becomes greater than 150 °C, the sintering rates of the metal oxide films increase, which leads to the formation of glazes (Co and Cr oxides) that can supply prolonged protection against friction and reduce the friction coefficient [32]. In addition, low unidirectional sliding friction coefficients for Co-based alloys at temperatures greater than 200 °C can be attributed to the formation of thermally stable oxide glazes on the pin surface, which cause low friction and wear [28].

The Si_3N_4 counterfaces showed some abrasive wear with extruded wear debris at RT, while at 300 °C, the counterfaces exhibited similar features, along with some adhered metal oxide transfer films from the wear tracks. A similar study revealed there is no accumulation of third bodies or transfer films adhered to the Si_3N_4 balls after RT sliding for HEAs $Al_{0.1}CoCrFeNi$ and CoCrFeMnNi [30]. In order to better determine the tribochemcial oxide phases, Raman spectroscopy was performed inside the four wear tracks shown in Figures 4–7. Figure 8 shows representative Raman spectra for the alloys after RT and 300 °C sliding. It is evident there are several oxide phases, both binary oxides and multi-element solid solution oxides, present in both alloys, including CoO, Co_3O_4, Cr_2O_3, NiO, and multi-element Cr_2O_4 and Fe_2O_4 tribochemical phases. These metal oxides tribochemically form at intermediate and higher temperatures via tribo-sintering the compact oxide tribofilm, which has been shown to lower the friction coefficients [28,33]. These oxide phases are in good agreement with the Raman spectra acquired inside FeCrNi medium-entropy alloy and CoCrFeNi HEA wear tracks, respectively [33,34]. The higher-Ni content A2 alloy at 300 °C exhibited a higher intensity peak of NiO and Cr_2O_3 in the tribofilm, which could be responsible for its lower friction and wear. Future studies will explore sliding temperatures higher than 300 °C to determine if there are more protective high-temperature oxide phases in both alloy tribofilms that result in low interfacial shear strength for friction reduction, while simultaneously providing low wear. It is possible that at higher sliding temperatures than 300 °C, the thermal softening process is counteracted by the formation of more protective tribochemical oxide phases that act as solid lubricants, thereby lowering friction and wear.

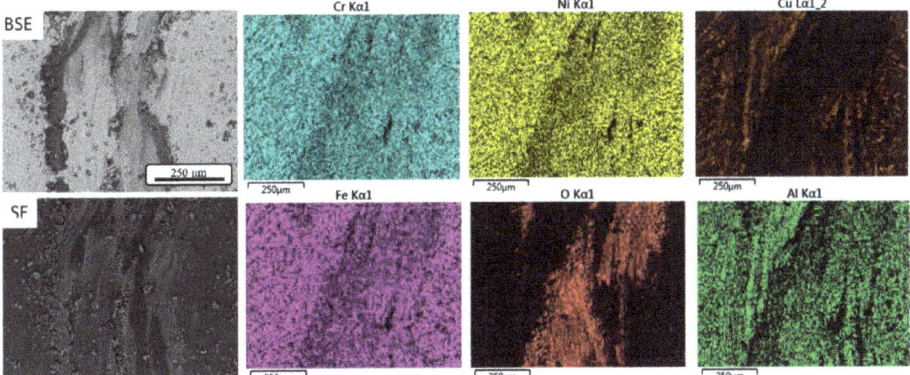

Figure 7. BSE and SE images and corresponding EDS elemental maps of the A2 alloy wear track after 300 °C sliding.

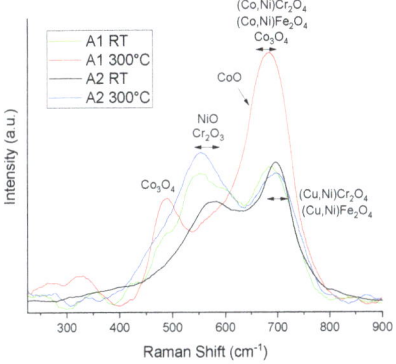

Figure 8. Raman spectra acquired inside A1 alloy and A2 alloy wear tracks after RT and 300 °C sliding.

4. Summary and Conclusions

Two FCC single-phase HEAs alloys with different Ni contents and either Co or Cu were studied: $Al_{0.3}CoFeCrNi$ (A1) and $Al_{0.3}CuFeCrNi_2$ (A2). For the A1 alloy with lower Ni content, micro-indentation and sliding wear tests revealed that the hardness decreased, resulting in thermal softening and a higher wear factor during 300 °C sliding. In contrast, the higher Ni content A2 alloy exhibited increasing hardness and subsequently a slightly lower wear factor. Mechanistic wear studies showed this was due to the oxidative wear, with the formation of low interfacial shear strength tribofilms that covered the wear tracks. Raman spectra determined that the A2 alloy at 300 °C exhibits a higher intensity peak of

NiO and Cr_2O_3 in the oxide tribofilm, which is likely responsible for lowering both the COF and wear factor. However, compared to RT sliding, both alloys provide no wear protection during 300 °C sliding, most likely due to thermal softening in the tribofilms. Thus, these single FCC-phase HEAs provide no further benefit to wear resistance at elevated temperatures, with similar implications likely for other such single FCC-phase HEAs. Lastly, these and other single-phase HEAs without hard secondary phases are no better than current bearing steels.

Author Contributions: Conceptualization, D.F.K. and T.W.S.; methodology, D.F.K., M.V.K. and T.W.S.; formal analysis, D.F.K., M.V.K. and T.W.S.; investigation, D.F.K., M.V.K. and T.W.S.; data curation, D.F.K. and M.V.K.; writing—original draft preparation, D.F.K. and T.W.S.; writing—review and editing, D.F.K. and T.W.S.; supervision, T.W.S. All authors have read and agreed to the published version of the manuscript.

Funding: This research received no external funding.

Data Availability Statement: All data generated or analyzed during this study are included in this published article.

Acknowledgments: We thank Bharat Gwalani for providing the two alloys, and the UNT Materials Research Facility (MRF) for access to SEM/EDS and XRD.

Conflicts of Interest: The authors declare no conflict of interest.

References

1. Yeh, J.W.; Chen, S.K.; Lin, S.J.; Gan, J.Y.; Chin, T.S.; Shun, T.T.; Tsau, C.H.; Chang, S.Y. Nanostructured High-Entropy Alloys with Multiple Principal Elements: Novel Alloy Design Concepts and Outcomes. *Adv. Eng. Mater.* **2004**, *6*, 299–303+274. [CrossRef]
2. Cantor, B.; Chang, I.T.H.; Knight, P.; Vincent, A.J.B. Microstructural Development in Equiatomic Multicomponent Alloys. *Mater. Sci. Eng. A* **2004**, *375–377*, 213–218. [CrossRef]
3. Klenam, D.; Asumadu, T.; Bodunrin, M.; Vandadi, M.; Bond, T.; Merwe, J.; Van Der Rahbar, N.; Soboyejo, W. Cold Spray Coatings of Complex Concentrated Alloys: Critical Assessment of Milestones, Challenges, and Opportunities. *Coatings* **2023**, *13*, 538. [CrossRef]
4. Miracle, D.B.; Senkov, O.N. A Critical Review of High-entropy alloys and Related Concepts. *Acta Mater.* **2017**, *122*, 448–511. [CrossRef]
5. Lin, Y.C.; Cho, Y.H. Elucidating the Microstructure and Wear Behavior for Multicomponent Alloy Clad Layers by in Situ Synthesis. *Surf. Coatings Technol.* **2008**, *202*, 4666–4672. [CrossRef]
6. Zhang, Y.; Li, R. New Advances in High-Entropy Alloys. *Entropy* **2020**, *22*, 1158. [CrossRef]
7. Chuang, M.; Tsai, M.; Wang, W.; Lin, S.; Yeh, J. Microstructure and Wear Behavior of $Al_xCo1.5CrFeNi1.5Ti_y$ High-Entropy Alloys. *Acta Mater.* **2011**, *59*, 6308–6317. [CrossRef]
8. Wang, W.R.; Wang, W.L.; Yeh, J.W. Phases, Microstructure and Mechanical Properties of $Al_xCoCrFeNi$ High-Entropy Alloys at Elevated Temperatures. *J. Alloys Compd.* **2014**, *589*, 143–152. [CrossRef]
9. Liu, Y.; Xu, Z.; Xu, G.; Chen, H. Influence of Al Addition on the Microstructure and Wear Behavior of Laser Cladding $FeCoCrNiAl_x$ High-Entropy Alloy Coatings. *Coatings* **2023**, *13*, 426. [CrossRef]
10. Miao, J.; Li, T.; Li, Q.; Chen, X.; Ren, Z.; Lu, Y. Enhanced Surface Properties of the $Al0.65CoCrFeNi$ High-Entropy Alloy via Laser Remelting. *Materials* **2023**, *16*, 1085. [CrossRef]
11. Wu, J.M.; Lin, S.J.; Yeh, J.W.; Chen, S.K.; Huang, Y.S.; Chen, H.C. Adhesive Wear Behavior of $Al_xCoCrCuFeNi$ High-Entropy Alloys as a Function of Aluminum Content. *Wear* **2006**, *261*, 513–519. [CrossRef]
12. Jiang, J.; Chen, P.; Qiu, J.; Sun, W.; Saikov, I.; Shcherbakov, V.; Alymov, M. Microstructural Evolution and Mechanical Properties of $Al_xCoCrFeNi$ High-Entropy Alloys under Uniaxial Tension: A Molecular Dynamics Simulations Study. *Mater. Today Commun.* **2021**, *28*, 102525. [CrossRef]
13. Shun, T.T.; Du, Y.C. Microstructure and Tensile Behaviors of FCC $Al0.3CoCrFeNi$ High Entropy Alloy. *J. Alloys Compd.* **2009**, *479*, 157–160. [CrossRef]
14. Wang, X.; Zhang, Z.; Wang, Z.; Ren, X. Microstructural Evolution and Tensile Properties of $Al0.3CoCrFeNi$ High-Entropy Alloy Associated with B2 Precipitates. *Materials* **2022**, *15*, 1215. [CrossRef]
15. Gwalani, B.; Gorsse, S.; Choudhuri, D.; Zheng, Y.; Mishra, R.S.; Banerjee, R. Tensile Yield Strength of a Single Bulk $Al0.3CoCrFeNi$ High Entropy Alloy Can Be Tuned from 160 MPa to 1800 MPa. *Scr. Mater.* **2019**, *162*, 18–23. [CrossRef]
16. Jiao, Z.M.; Ma, S.G.; Yuan, G.Z.; Wang, Z.H.; Yang, H.J.; Qiao, J.W. Plastic Deformation of $Al0.3CoCrFeNi$ and $AlCoCrFeNi$ High-Entropy Alloys Under Nanoindentation. *J. Mater. Eng. Perform.* **2015**, *24*, 3077–3083. [CrossRef]
17. Kao, Y.F.; Chen, T.J.; Chen, S.K.; Yeh, J.W. Microstructure and Mechanical Property of As-Cast, -Homogenized, and -Deformed $Al_xCoCrFeNi$ ($0 < x < 2$) High-Entropy Alloys. *J. Alloys Compd.* **2009**, *488*, 57–64. [CrossRef]

18. Guo, S.; Ng, C.; Lu, J.; Liu, C.T. Effect of Valence Electron Concentration on Stability of Fcc or Bcc Phase in High-entropy alloys. *J. Appl. Phys.* **2011**, *109*. [CrossRef]
19. Guo, S.; Ng, C.; Liu, C.T. Anomalous Solidification Microstructures in Co-Free Al XCrCuFeNi2 High-Entropy Alloys. *J. Alloys Compd.* **2013**, *557*, 77–81. [CrossRef]
20. Shun, T.T.; Hung, C.H.; Lee, C.F. Formation of Ordered/Disordered Nanoparticles in FCC High-entropy alloys. *J. Alloys Compd.* **2010**, *493*, 105–109. [CrossRef]
21. Ng, C.; Guo, S.; Luan, J.; Wang, Q.; Lu, J.; Shi, S.; Liu, C.T. Phase Stability and Tensile Properties of Co-Free Al0.5CrCuFeNi2 High-Entropy Alloys. *J. Alloys Compd.* **2014**, *584*, 530–537. [CrossRef]
22. Gwalani, B.; Soni, V.; Choudhuri, D.; Lee, M.; Hwang, J.Y.; Nam, S.J.; Ryu, H.; Hong, S.H.; Banerjee, R. Stability of Ordered L12 and B2 Precipitates in Face Centered Cubic Based High-entropy alloys–Al0.3CoFeCrNi and Al0.3CuFeCrNi2. *Scr. Mater.* **2016**, *123*, 130–134. [CrossRef]
23. Guardian, R.; Rosales-Cadena, I.; Diaz-Reyes, C.; Ruiz-Ochoa, J.A. Wear Evaluation of Copper-Nickel-Aluminum Alloys under Extreme Conditions. *J. Miner. Mater. Charact. Eng.* **2023**, *11*, 16–26. [CrossRef]
24. Qiu, X.-W.; Liu, C.-G. Microstructure and Properties of Al2CrFeCoCuTiNix High-Entropy Alloys Prepared by Laser Cladding. *J. Alloys Compd.* **2013**, *553*, 216–220. [CrossRef]
25. López Ríos, M.; Socorro Perdomo, P.P.; Voiculescu, I.; Geanta, V.; Crăciun, V.; Boerasu, I.; Mirza Rosca, J.C. Effects of Nickel Content on the Microstructure, Microhardness and Corrosion Behavior of High-Entropy AlCoCrFeNix Alloys. *Sci. Rep.* **2020**, *10*, 1–11. [CrossRef]
26. Kuo, P.-C.; Chen, S.-Y.; Yu, W.; Okumura, R.; Iikubo, S.; Laksono, A.D.; Yen, Y.-W.; Pasana, A.S. The Effect of Increasing Nickel Content on the Microstructure, Hardness, and Corrosion Resistance of the CuFeTiZrNix High-Entropy Alloys. *Materials* **2022**, *15*, 3098. [CrossRef]
27. Torgerson, T.B.; Mantri, S.A.; Banerjee, R.; Scharf, T.W. Room and Elevated Temperature Sliding Wear Behavior and Mechanisms of Additively Manufactured Novel Precipitation Strengthened Metallic Composites. *Wear* **2019**, *426–427*, 942–951. [CrossRef]
28. Scharf, T.W.; Prasad, S.V.; Kotula, P.G.; Michael, J.R.; Robino, C.V. Elevated Temperature Tribology of Cobalt and Tantalum-Based Alloys. *Wear* **2015**, *330–331*, 199–208. [CrossRef]
29. Gwalani, B.; Ayyagari, A.V.; Choudhuri, D.; Scharf, T.; Mukherjee, S.; Gibson, M.; Banerjee, R. Microstructure and Wear Resistance of an Intermetallic-based Al0.25Ti0.75CoCrFeNi High Entropy Alloy. *Mater. Chem. Phys.* **2018**, *210*, 197–206. [CrossRef]
30. Ayyagari, A.; Barthelemy, C.; Gwalani, B.; Banerjee, R.; Scharf, T.W.; Mukherjee, S. Reciprocating Sliding Wear Behavior of High-entropy alloys in Dry and Marine Environments. *Mater. Chem. Phys.* **2018**, *210*, 162–169. [CrossRef]
31. Korashy, A.; Attia, H.; Thomson, V.; Oskooei, S. Characterization of Fretting Wear of Cobalt-Based Superalloys at High Temperature for Aero-Engine Combustor Components. *Wear* **2015**, *330–331*, 327–337. [CrossRef]
32. Pauschitz, A.; Roy, M.; Franek, F. Mechanisms of Sliding Wear of Metals and Alloys at Elevated Temperatures. *Tribol. Int.* **2008**, *41*, 584–602. [CrossRef]
33. Fu, A.; Xie, Z.; He, W.; Cao, Y. Effect of Temperature on Tribological Behavior of FeCrNi Medium Entropy Alloy. *Metals* **2023**, *13*, 282. [CrossRef]
34. Zhang, A.; Han, J.; Su, B.; Li, P.; Meng, J. Microstructure, Mechanical Properties and Tribological Performance of CoCrFeNi High Entropy Alloy Matrix Self-Lubricating Composite. *Mater. Des.* **2017**, *114*, 253–263. [CrossRef]

Disclaimer/Publisher's Note: The statements, opinions and data contained in all publications are solely those of the individual author(s) and contributor(s) and not of MDPI and/or the editor(s). MDPI and/or the editor(s) disclaim responsibility for any injury to people or property resulting from any ideas, methods, instructions or products referred to in the content.

Article

Effect of High-Energy Ball Milling in Ternary Material System of (Mg-Sn-Na)

Halit Sübütay [1] and İlyas Şavklıyıldız [2,*]

[1] Department of Metallurgical and Materials Engineering, Selçuk University, Konya 42075, Turkey; halit.subutay@selcuk.edu.tr
[2] Department of Metallurgical and Materials Engineering, Konya Technical University, Konya 42075, Turkey
* Correspondence: isavkliyildiz@ktun.edu.tr

Abstract: In this study, the nature of the ball-milling mechanism in a ternary materials system (Mg-6Sn-1Na) is investigated for proper mechanical alloying. An identical powder mixture for this material system is exposed to different milling durations for a suitable mixture. First, the platelet structure formation is observed on particles with increasing milling duration, mainly formed in <200> direction of the hexagonal crystal structure of the Mg matrix. Then, the flake structure with texture formation is broken into smaller spherical particles with further ball milling up to 12 h. According to EDS analysis, the secondary phases in the Mg matrix are homogenously distributed with a 12-h milling duration which advises a proper mixture in this material system. The solid solution formation is triggered with an 8-h milling duration according to XRD analysis on 101 reflections. Conventional sintering is performed at 350 °C in 2 h for each sample. In bulk samples, XRD data reveal that secondary phases (Mg_2Sn) with island-like structures are formed on the Mg matrix for a milling duration of up to 8 h. These bigger secondary phases are mainly constituted as Mg_2Sn intermetallic forms, which have a negative effect on physical and mechanical properties due to a mismatch in the grain boundary formation. However, the homogenous distribution of secondary phases with a smaller particle size distribution, acquired with 12 h milling time, provides the highest density, modulus of elasticity, and hardness values for this ternary materials system. The ternary materials produced with the 12-h ball-milling process provide an improvement of about 117% in hardness value compared with the cast form.

Keywords: Mg alloys; ball milling; mechanical alloying; ternary material system

Citation: Sübütay, H.; Şavklıyıldız, İ. Effect of High-Energy Ball Milling in Ternary Material System of (Mg-Sn-Na). *Crystals* **2023**, *13*, 1230. https://doi.org/10.3390/cryst13081230

Academic Editor: Jacek Ćwik

Received: 27 July 2023
Revised: 4 August 2023
Accepted: 8 August 2023
Published: 9 August 2023

Copyright: © 2023 by the authors. Licensee MDPI, Basel, Switzerland. This article is an open access article distributed under the terms and conditions of the Creative Commons Attribution (CC BY) license (https://creativecommons.org/licenses/by/4.0/).

1. Introduction

Magnesium and its alloys are the lowest mass density morphological materials with well-shapeable and high-wear resistance, which makes them potential candidates for several applications such as aircraft, biomaterials, and automobiles [1,2]. The primary downside of Mg is its low formability at decreasing temperatures due to the limited activation of shear systems [3]. In addition, another disadvantage is the very low corrosion resistance of Mg [4]. Alloying process for pure Mg is a proper solution for strengthening, raising formability, and improving the corrosion resistance of magnesium alloys [5–7]. The alloying effect of such elements as Ti [8], Cu [9], Al [10], Ni [11], Sn [12], Zn [13], Ca [14], Zr [15], and Mn [16], or such rare earth elements as Ce [17], Sm [18], Dy [19], Er [20], and Y [21], have been commonly employed in Mg and alloys. Such alloying processes and thermomechanical processing, such as high-energy ball milling, have provided a significant enhancement in specific endurance, sinterability, and creep properties [22].

A high-energy ball-milling process, which provides mechanical alloying (MA), is used as the most common method for alloying powders, in which different material systems can mix homogeneously at the ppm level due to the involvement of the continuous cold-welding and fracturing mechanisms [23]. The most common issue in the high-energy ball-milling procedure is a continuous collision between ball-powder jar, which leads to severe plastic deformations on particles [24] and fracture of the platelet particles, causing the variation of lattice strain, crystallite size, and attaining nano-sized structures and phase transformations [25,26]. After mechanical alloying due to reiterative fracture, cold welding, and refracture of powders, new effective surfaces improve chemically, and supersaturation may occur beyond the equilibrium limit [27]. Furthermore, thermomechanical activation of the particle morphology plays a significant role in the high-energy ball-milling process [28]. In the literature, there are publications reporting that the solubility of Titanium in Mg [29] and Magnesium in Ti [30] in studies with mechanical alloying Ti-Mg powders increases. High-energy ball milling is used in the production of composite materials with high physical and mechanical properties, which are complex and expensive to manufacture with traditional methods such as casting [31].

According to the literature survey [32–37], it is clearly seen that several composite materials are manufactured by changing matrix and reinforcement species with different particle size distribution in conjunction with altering milling conditions, such as milling time and processing agent [38]. Wang et al. [39] investigated Mg-%25 wt. Sn composite alloys manufactured by mechanical alloying. They reported that sintering parameters and the number of secondary phases directly affect the mechanical properties by changing the ball-milling duration.

Son et al. [40] studied the relationship between precipitation nature and mechanical properties of multiphase of Mg-Sn-Al-Zn alloys. The intermetallic phases, such as Mg_2Sn, have a major effect on the development of mechanical properties. The creep endurances of binary Mg-Sn and ternary Mg-Sn-Al material systems were studied by Poddar et al. [41], which suggested that the creep behavior of the materials could be developed with the help of intermetallic phases, including Mg_2Sn, which has chemical stability at higher temperatures [42]. Mendis et al. [43] surveyed the role of Na addition to Mg-Sn alloys alloyed by the casting method. In their study, it was clearly seen that an abnormal increase in hardness occurred because the distribution of Na led to an increment in the number of intermetallic precipitates.

Although different studies on Mg-Sn composite alloys fabricates via casting and powder metallurgy are well tabulated in the literature, there is a limited experimental investigation of the effect of the milling time and addition of trace amount Na element on the powder morphology, precipitate formation, intermetallic transformation, and crystallographic properties of high-energy ball-milled Mg-Sn-Na metal powders and their relationship between density, hardness, elastic module of generated composite alloys [44]. For this reason, the main aim of the study is to do an elaborate characterization of powder morphology, the variation of crystallite size, the distribution of reinforcement species (Sn and Na), and the occurrence of Mg_2Sn intermetallic formation with the addition of Na element within the Mg matrix. Moreover, sintered Mg-Sn-Na composite alloys are employed to examine densities, hardness, and elastic modulus.

2. Materials and Methods

2.1. Materials and Ball-Milling Process

In the present study, Sn (6 wt%) and Na (1 wt%) were used to reinforce elements in the Mg matrix to produce alloys with metal matrix. Pure Mg (<53 μm; 99.99 wt% purity) and Sn (<6 μm; 99.9 wt% purity) elemental powders were obtained by Nanografi Company(Ankara, Turkey). Mg, Sn, and Na powders were high-energy ball-milled with tungsten carbide (WC; 10 mm diameter) balls in a tungsten carbide jar (250 mL) using RETSCH-PM 100 (Konya, Turkey) planetary ball-milling device at a rotation speed of 250 rpm.

The ball-to-powder (BPR) massive 5:1 rate was chosen, and the ball-milling time was selected as 0.5, 2, 4, 8, and 12 h. To prevent clusters, cold working, and severe plastic deformation of the powders' duration milling, 4 wt% stearic acid was used as a process control agent (PCA). To avoid excessive temperature rise throughout the ball-milling time, the device was milled for 5 min, then rested for 5 min. After ball milling, the powders were compacted under 500 MPa pressure for 30 s by a cold-press device. To prevent oxidation, all green compacts (cold-pressed specimens) were sintered for 2 h at 350 °C with a stable heating rate of 5 °C/min under the high-purity argon atmosphere. The argon flow ratio was constant at 3 L/min. In Figure 1, the preparation and production of Mg-Sn-Na alloy is represented schematically.

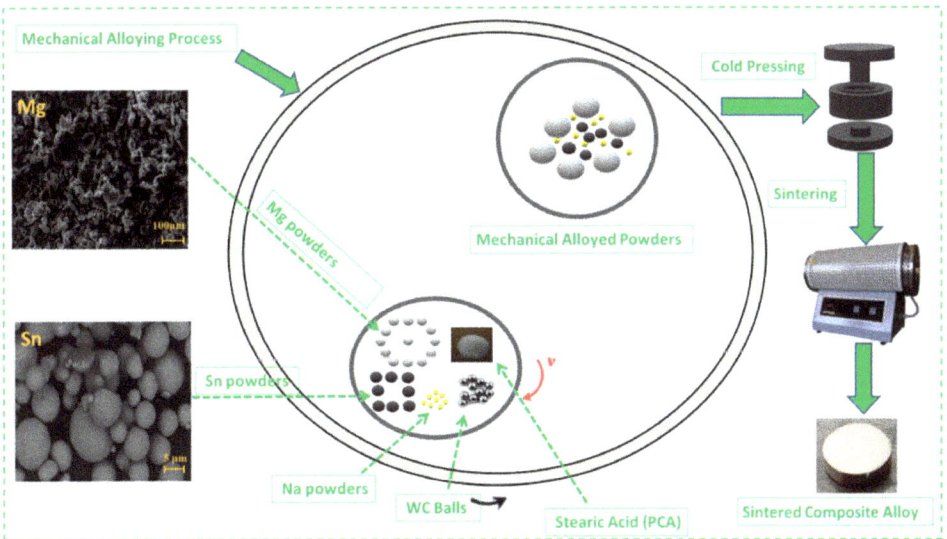

Figure 1. Schematic experimental setup for ball milling and Powder Metallurgy procedure.

2.2. Powder and Sintered Sample Characterization

The morphologies of milled and unmilled powders, along with microstructural progress in sintered composite alloys, were investigated using scanning electron microscopy (SEM, Zeiss EVO LS10, Konya, Turkey). Elemental analysis of the milled powders and dispersion mechanism of sintered composite alloys were investigated with EDS-Mapping and line scan mode. Moreover, all ball-milled powders' particle size distribution (PSD) was examined via a Mastersizer particle size analyzer device. X-ray diffractometer (XRD) was utilized to calculate the crystallite size and define the peak profile parameters. XRD data were gathered by Cu-radiation at wavelength of 1.5406 Å by utilizing a step-scan procedure with 0.15 degrees 2-theta at every step and 4 s at every step dwelling time [45]. Peak profile parameters of all ball-milled powders were detected by the MDI Jade XRD program. For all ball-milled powders, the Debye Scherrer equation was utilized to calculate values of lattice strain and crystallite size (Equation (1)) [46].

$$D_{hkl} = \frac{K \times \lambda}{(\beta_{hkl} \times cos\Phi)} \quad (1)$$

where K (0.94) is the shape coefficient; λ is the wavelength of Cu radiation; β is the FWHM value of the peak, and Φ is the XRD diffraction angle [47,48].

Dislocation density (δ) values were calculated according to Williamson–Smallman formula given in Equation (2) [48,49].

$$\delta = n/D^2 \quad (2)$$

Here, n is nearly 1, and D is the crystallite size. Theoretical density values of sintered composite alloys were calculated utilizing the mixture rule formula, and their experimental density measurements were carried out according to the ASTM B962-17 standard Archimedes' principle [50]. The microhardness values of the sintered composite alloys were surveyed via a micro-Vickers hardness measure procedure. The micro-Vickers hardness test [51] was performed under an applying load of 10 gf and a dwelling time of 20 s. At least five measurements were taken for the average hardness value.

Young's Modulus of sintered composite alloys was used to calculate the ultrasonic wave velocity (UWC) measurement method [48]. UWC measurements were performed in tangency mode to strengthen ASTM E494-20, utilizing the ultrasonic pulse-echo-overlap method (PEOM) [52]. The UWC measurements were carried out utilizing one flaw detector, one-number longitudinal wave tangency transducer (20 MHz), and one-number shear-wave tangency transducer (5 MHz) [53].

3. Results and Discussion

3.1. Powder Morphology

At first glance, the particle morphology of the ternary material system of (Mg-Sn-Na) is preserved after the 0.5 h milling process. In Figure 2, the particles show a spherical form as the initial powder mixture. So, 0.5 h milling duration has almost no effect on the change in both particle morphologies and in the mixing nature of the ternary system, which arouses the need for further milling for proper mixing. Then, the said material system is exposed to a 2-h ball-milling treatment. The flake shape of particles is a sign that the ball-milling procedure in this material system works efficiently, along with extreme plastic deformation. The particles turn into flake structures due to continuous collision between ball–ball and ball–wall with increasing surface area and particle size of up to 70 microns. Furthermore, few cold-welding mechanisms are observed between the small particles and bigger particles. However, the 4-h milling procedure triggers a huge cold-welding mechanism that involves flake particles due to excessive plastic deformation on particles. Eventually, all welded particles create platelet structure formation with an increasing particle size of up to 120 microns, which tripled the initial powder particle size distribution. After this point, the particles start to break into much smaller forms, which is attributed to severe work hardening with further ball milling of up to 8 h in duration. Bigger platelet particles with small broken pieces create a bimodal particle size distribution, as seen in Figure 2. During high-energy ball milling, cold working and fracture mechanism work concomitantly for proper mechanical alloying. However, after a certain milling duration, the fracture mechanism becomes the dominant instrument for the particles, which is due to either extreme work hardening or loss of function of the processing agent. With an 8-h milling process, all particles are exposed to severe plastic deformation and accumulate excessive stress concentration, which eventually affects the sinterability of this material system. Further high-energy ball-milling process of up to 12 h turns the flake morphology into smaller fragmented particle formations. Monomodal particle size distribution is also revealed with 12-h milling time, and particle size distribution of milled powder of less than 10 microns is received, which is well below the initial particle size. The final microstructure formation suggests that a 12-h milling duration is enough to provide sufficient mixing and mechanical alloying for this ternary material system.

The EDS analysis represented in Figure 3 also verifies the homogenous mixture of the ternary system. All elements (Mg-Sn-Na) in this system are uniformly dispersed throughout the sample. However, our previous study revealed that monomodal particle size distribution was acquired via an 8-h milling duration in a binary system of (Mg-9Sn) [48]. So, the presence of Na elements in this ternary system prevents or retards the fracture mechanism, and an extra 4-h milling process is needed to achieve homogenous mixing formation with smaller particle size distribution.

The particle size analyses also show the same tendency regarding particle evolution as a function of milling time. First, the nature of the single peak at higher numbers advises that the particle size is increased with monomodal distribution in this material system up to 4 h of the milling procedure (Figure 4). Then, the bimodal distribution observed within 8 h leads to two separated peaks, according to the particle analyzer. Finally, a single intense peak profile is acquired around the 10-micron range, with 12 milling durations. Overall, the particle size analysis is consistent with SEM micrograph examination, as explained previously.

Figure 2. SEM examination of ball-milled Mg-6Sn-1Na powders at particular milling duration.

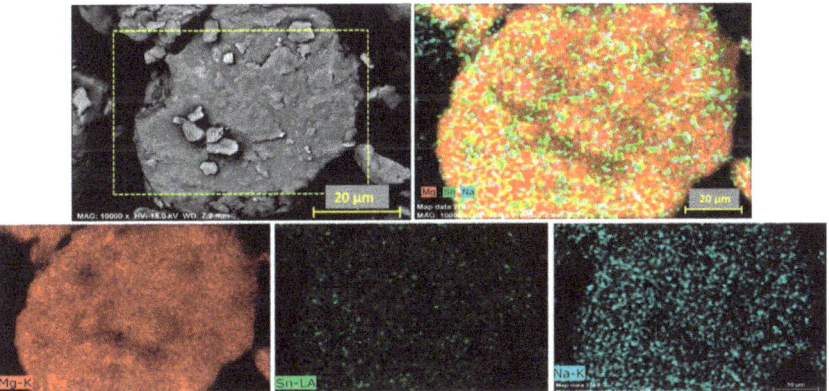

Figure 3. EDS mapping analysis for ball-milled powder at 12-h milling duration.

XRD analysis also helps to understand both phase formation and peak profile evolution for each phase, as represented in Figure 5. Peaks for Mg and Sn phases are detected in up to 2 h of milling conditions. However, the 4-h milling duration makes Sn peaks fade out, and the intermetallic phase of Mg_2Sn starts to appear in the XRD spectrum. This behavior is attributed to the high-energy ball-milling work. So, at least 4-h ball milling needs to trigger intermetallic formation in this system. However, in the Mg-Sn binary system, such behavior was achieved in the 2-h milling procedure, which means that the presence of Na also delays the formation of the intermetallic phase [54,55]. Further ball-milling process maps the Sn out from the XRD spectra, as observed in the 8-h and 12-h conditions. Such observation suggests that the Sn phase goes into both solid solutions in the Mg matrix and chemical reaction for intermetallic phase formation [56]. A closer approach to the major peak (101) of the Mg phase is represented in Figure 5c. The peak position of (101) reflection first has a tendency to shift to higher degrees, which intimates the decrement in the interatomic distance because of the compression stress [57]. A repetitive collision during the milling process creates such compression stress formation on particles as a natural high-energy ball milling. This issue is the case for the milling condition for up to 4-h duration. After that, the 101 peaks start to shift to lower degrees, which means that the interatomic distance for this reflection starts to increase [58]. This singularity is only due to the solid solution formation in the Mg matrix in this case [59]. It is important to state that Na and Sn elements are initiated to dissolve into the Mg matrix due to the high-energy ball milling in the 8-h conditions. Therefore, at least an 8-h milling procedure is needed to trigger solid solution formation in the ternary (Mg-Sn-Na) material system. Nonetheless, in the binary material system of (Mg-9Sn), solid solution formation was observed with a 4-h ball milling [48]. So, the existence of the Na element also postpones the solid solution formation. The peak intensity of (101) reflection is monolithically reduced with increasing milling duration, which advises the reduction in crystallite size or coherent diffraction in domain size. The continuous collision mechanism during ball milling leads crystals to break apart in the (101) direction. Regarding the (002) reflection of the Mg matrix, the (002) reflection shows singularity in terms of peak intensity changes. Peak intensity first starts to increase with the increase in milling duration of up to 4 h and also shows higher relative peak intensity than the major peak (101) reflection of the Mg matrix. Such a result is attributed to the texture formation in a hexagonal crystal structure of Mg. Therefore, we conclude that severe plastic deformation observed in the SEM micrograph is mainly elongated through the (002) direction. Because the interatomic bonding governed in this direction is weak, atoms prefer to be oriented accordingly [60–62]. A further ball-milling process results in a decrease in the peak intensity of (002) reflection, which refers to breaking the coherent diffraction domains for the (002) reflection.

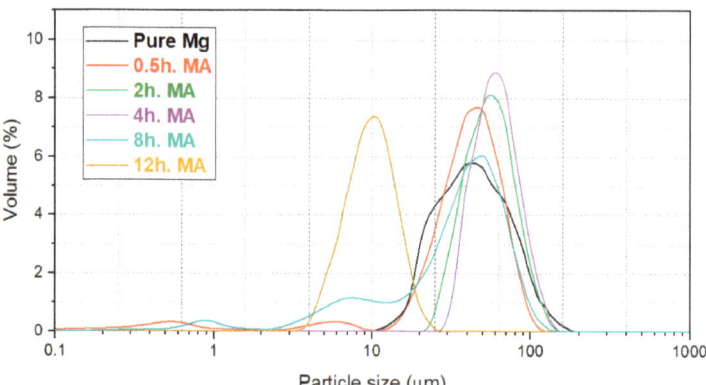

Figure 4. Particle size distribution of each sample, ball milled at different durations.

The crystallite size is also decreased as the particles break apart, as observed in SEM analysis. So, SEM micrograph analysis supports the XRD data analysis. The changes in crystallite size for the (002) reflection are inconsistent with the particle size variation as a function of the milling time. As represented in Figure 6, the crystallite size improved from 43 nm to 63 nm with the increasing milling duration of up to 4 h, according to the calculation with the (002) peak reflection parameters. On the contrary, the dislocation density and localized lattice strain are reduced as expected. However, a prolonged milling procedure of up to 12 h has a negative effect on crystallite size variation, and the crystallite size of the final powder is measured at 40 nm, which is lower than at the initial stage. Lattice strain and dislocation density values are increased up to %0.29 and 0.62×10^{15} line/m², respectively.

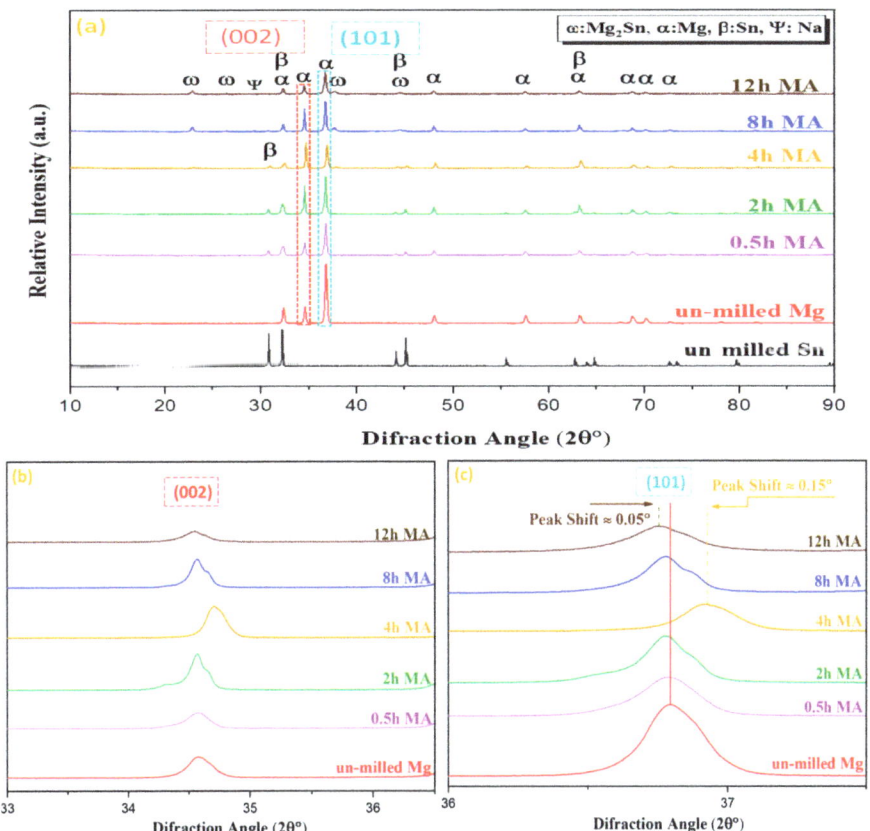

Figure 5. XRD analysis for all samples (**a**) and peak profile evolution on (002) (**b**) and (101) (**c**) reflection for different milling durations.

3.2. Bulk Material Characterization

The SEM micrograph examination, along with the elemental line scan analysis, is illustrated in Figure 7 with different milling durations. Secondary phases in the Mg matrix appear in a lighter color in SEM illustration due to the higher atomic number of Sn element. Agglomerated secondary phase existence on the Mg matrix is observed for all conditions except 12 milling durations. Secondary phases also precipitate vicinity of grain boundaries. Individual secondary phases create their island structure on the Mg matrix with a range of 40–60 microns. This behavior of the secondary phase suggests that the ball-milling process of up to 8 h does not provide a proper mixture in the Mg-Sn-Na material system. However, the microstructure of the 12-h milled conditions conveys a homogenously dispersed secondary phase with a smaller size in the Mg matrix. It is important to have homogenously distributed secondary phases to have improved physical and mechanical properties for multiphase material systems. According to our previous study, the homogenous distribution of the secondary phase in a binary Mg-Sn composition reached an 8-h milling condition, which advised that the existence of the Na phase also delayed the homogenous mixing. Line scan for each condition is collected via both the Mg matrix and the secondary phase. The lighter area in the SEM illustration has a higher concentration of the Sn phase with the Mg phase's existence, which suggests that the island structure secondary phases are mainly the Mg_2Sn intermetallic phases, as detected via the XRD analysis as well. The line scan also reveals surprising results regarding the existence of Na in this material system. Na elements mainly precipitate on grain boundary instead of dissolving into an Mg matrix to create the solid solution. This situation of Na elements could have an important role in this ternary material system to improve mechanical properties by providing a precipitation hardening mechanism. EDS mapping analysis for the 12-h condition shows the homogenous distribution of each phase with no impurities, as represented in Figure 8.

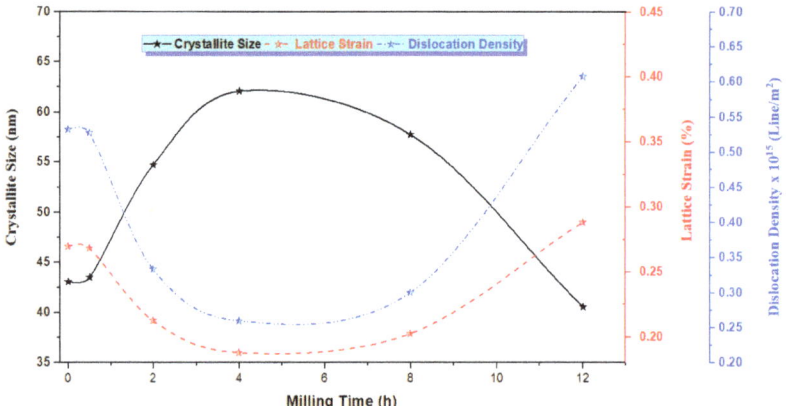

Figure 6. Illustration of changes for crystallite size, lattice strain, and dislocation density at different milling times.

Figure 9 represents the hardness value variation of sintered milled powders at different milling durations. The hardness values at first tend to increase with increasing milling process. However, an 8-h ball-milling duration has a negative effect on this trend, and the hardness value decreases in an 8-h milling duration. Then, the hardness value reaches the highest value of 90 HV for the 12-h milling condition. The increment in hardness value has a correlation with the milling duration because the higher milling duration generally provides homogenous mixing and dispersion of the additive elements in the Mg matrix. As explained before, the SEM micrograph examination on bulk samples suggests that the island-like precipitation of the secondary phase is observed instead of the homogenous dispersion in the matrix [63]. So, the incompatibility at the interface between the bigger secondary phase and matrix is the weak point, which reduces the mechanical properties [63–65]. The bigger size of the secondary phase creates a transition between small angles to high-angle interface formation [25,66] through the grain boundaries. Another explanation for variation in the mechanical properties is the Orowan mechanism, which is also involved and eventually affects the mechanical properties of the end product [67]. The slight reduction in the 8-h milling conditions can be explained by both the initial powder structure and the microstructure formation of bulk or sintered samples. The powder form of the 8-h condition suggests the bimodal particle size distribution. Repetitive collision in the ball-milling process results in a high-stress concentration in particles, which eventually hinders the diffusion or mass transport during the sintering procedure [68]. So, the density, elastic modulus, and hardness values for the 8-h condition are decreased. The second reason causing this disparity is the generation of a secondary phase in the Mg matrix after the sintering process. As explained, the island-like formation of a secondary phase instead of homogenous distribution causes the diminishment in physical and mechanical properties. According to the previous study, the hardness value for the binary system of Mg-Sn was found to be 71.2 HV [48], but in this ternary system, the hardness value is improved to the 90 HV range (Figure 9). So, the presence of Na element in grain boundaries causes such an impact on hardness by triggering a precipitation hardening mechanism [69,70]. The elasticity modulus and density values for this material system show the same behavior as the hardness values. The density value for each sample is the sign of the sinterability of the milled powder, which is determined by the ball-milling duration in this study. So, the particle morphology and the nature of mixing of secondary phases in the matrix are the parameters determining the sinterability of the powder. Overall, the highest density and modulus of elasticity are acquired as 96% and 36 GPa for the 12-h milling condition (Figure 10). In this material system, the Na element has a huge effect on powder morphology, which eventually determines the physical and mechanical properties of bulk samples.

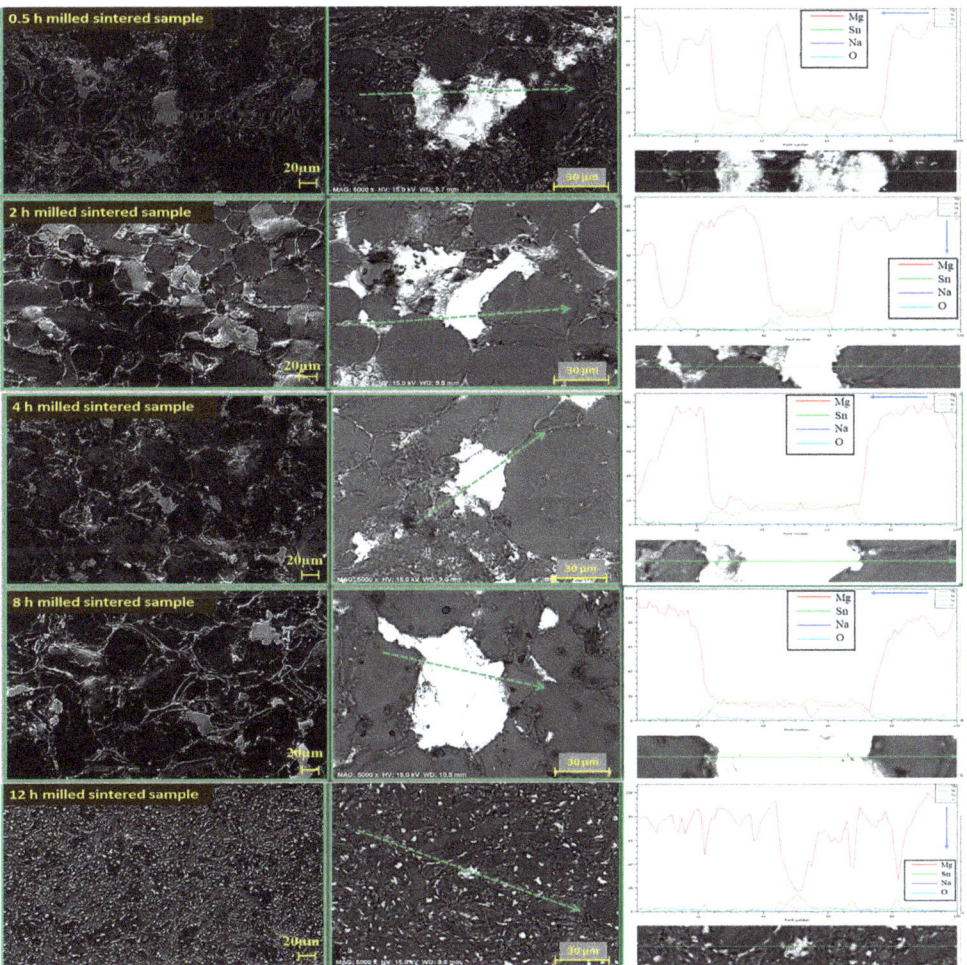

Figure 7. SEM micrograph representation on bulk ternary material system (Mg-Na-Sn) as a function of milling time.

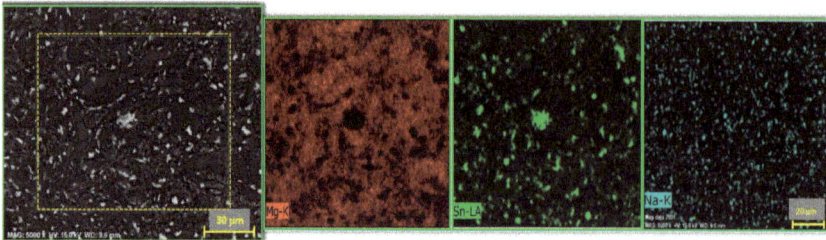

Figure 8. Mapping analysis of 12-h ball-milled 6 wt% Sn-Na-Mg sintered composite alloy.

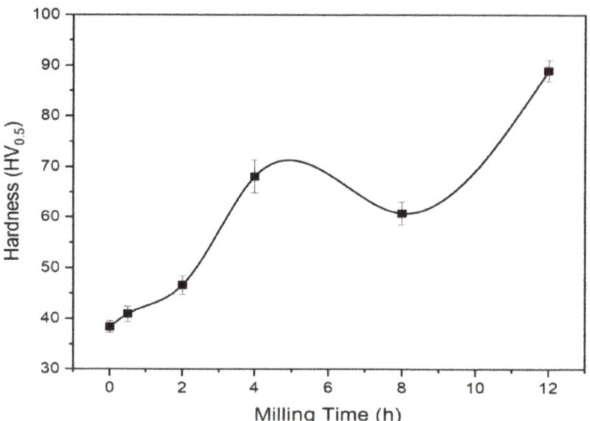

Figure 9. Representation of hardness values with different milling duration.

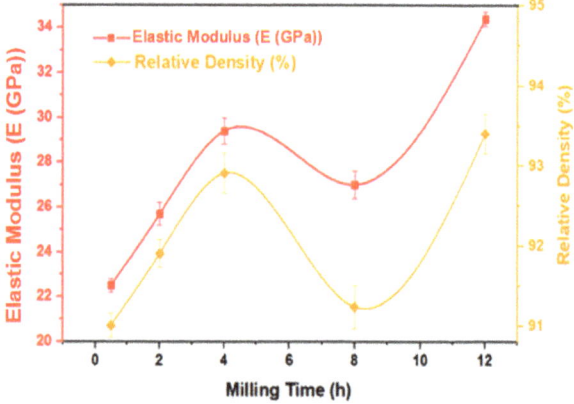

Figure 10. Variation of relative density and elastic modulus as a function of milling duration.

4. Conclusions

In this study, we aim to have a proper solid solution and mechanical alloying in the ternary system of (Mg-Sn-Na) by changing high-energy ball-milling duration. The following outcomes are acquired according to our experimental results:

- The milling process first causes the formation of platelet structure in this material's composition. Then, a further milling process of up to 12 h leads to breaking the flake form of powders into small spherical particles with monomodal distribution;

- XRD data reveal that the Sn phase starts to both dissolve and involve a chemical reaction for intermetallic phase (Mg$_2$Sn) formation with an 8-h milling condition;
- X-ray diffractometry study on (002) reflection of the Mg phase conveys the texture formation in the Mg matrix, which helps us to understand the nature of the platelet structure, as observed in SEM micrograph examination;
- The island structure of the secondary phase affects the physical and mechanical properties of bulk or sintered samples for samples milled up to 8 h;
- In 12 milling conditions, the homogenous secondary phase distribution is achieved, which eventually supplies the highest relative density (95%), modulus of elasticity (34.5 GPa), and hardness (89 HV) values in this ternary material system.

Author Contributions: H.S., investigation, methodology, data curation, writing—original draft; İ.Ş., investigation, conceptualization, supervision, writing—review and editing. All authors have read and agreed to the published version of the manuscript.

Funding: The financial support provided to this study by the Scientific Research Projects Coordination Unit (SRPCU) of Selçuk University through contract# 18401032 and the Scientific Research Projects at Konya Technical University through contract# 191019035.

Institutional Review Board Statement: Not applicable.

Informed Consent Statement: Not applicable.

Data Availability Statement: Not applicable.

Conflicts of Interest: The authors declare no conflict of interest.

References

1. Aalipour, Z.; Zarei-Hanzaki, A.; Moshiri, A.; Abedi, H.; Waryoba, D.; Kisko, A.; Karjalainen, L. Strain dependency of dynamic recrystallization during thermomechanical processing of Mg–Gd–Y–Zn–Zr alloy. *J. Mater. Res. Technol.* **2022**, *18*, 591–598. [CrossRef]
2. Zhong, X.; Wong, W.; Gupta, M. Enhancing strength and ductility of magnesium by integrating it with aluminum nanoparticles. *Acta Mater.* **2007**, *55*, 6338–6344. [CrossRef]
3. Koike, J.; Kobayashi, T.; Mukai, T.; Watanabe, H.; Suzuki, M.; Maruyama, K.; Higashi, K. The activity of non-basal slip systems and dynamic recovery at room temperature in fine-grained AZ31B magnesium alloys. *Acta Mater.* **2003**, *51*, 2055–2065. [CrossRef]
4. Atrens, A.; Song, G.-L.; Cao, F.; Shi, Z.; Bowen, P.K. Advances in Mg corrosion and research suggestions. *J. Magnes. Alloys* **2013**, *1*, 177–200. [CrossRef]
5. Kwak, T.; Kim, W. Effect of refinement of grains and icosahedral phase on hot compressive deformation and processing maps of Mg–Zn–Y magnesium alloys with different volume fractions of icosahedral phase. *J. Mater. Sci. Technol.* **2019**, *35*, 181–191. [CrossRef]
6. Zhao, C.; Pan, F.; Zhao, S.; Pan, H.; Song, K.; Tang, A. Preparation and characterization of as-extruded Mg–Sn alloys for orthopedic applications. *Mater. Des.* **2015**, *70*, 60–67. [CrossRef]
7. Jiang, W.; Wang, J.; Zhao, W.; Liu, Q.; Jiang, D.; Guo, S. Effect of Sn addition on the mechanical properties and bio-corrosion behavior of cytocompatible Mg–4Zn based alloys. *J. Magnes. Alloys* **2019**, *7*, 15–26. [CrossRef]
8. Tejeda-Ochoa, A.; Kametani, N.; Carreño-Gallardo, C.; Ledezma-Sillas, J.; Adachi, N.; Todaka, Y.; Herrera-Ramirez, J. Formation of a metastable fcc phase and high Mg solubility in the Ti–Mg system by mechanical alloying. *Powder Technol.* **2020**, *374*, 348–352. [CrossRef]
9. Cheng, J.; Cai, Q.; Zhao, B.; Yang, S.; Chen, F.; Li, B. Microstructure and Mechanical Properties of Nanocrystalline Al–Zn–Mg–Cu Alloy Prepared by Mechanical Alloying and Spark Plasma Sintering. *Materials* **2019**, *12*, 1255. [CrossRef]
10. Singh, D.; Suryanarayana, C.; Mertus, L.; Chen, R.-H. Extended homogeneity range of intermetallic phases in mechanically alloyed Mg–Al alloys. *Intermetallics* **2003**, *11*, 373–376. [CrossRef]
11. Fadonougbo, J.O.; Kim, H.-J.; Suh, B.-C.; Yim, C.D.; Na, T.-W.; Park, H.-K.; Suh, J.-Y. On the long-term cyclic stability of near-eutectic Mg–Mg$_2$Ni alloys. *Int. J. Hydrogen Energy* **2022**, *47*, 3939–3947. [CrossRef]
12. Zhong, H.; Xu, J. Tuning the de/hydriding thermodynamics and kinetics of Mg by mechanical alloying with Sn and Zn. *Int. J. Hydrogen Energy* **2019**, *44*, 2926–2933. [CrossRef]
13. Lesz, S.; Hrapkowicz, B.; Karolus, M.; Gołombek, K. Characteristics of the Mg–Zn–Ca–Gd Alloy after Mechanical Alloying. *Materials* **2021**, *14*, 226. [CrossRef] [PubMed]
14. Raducanu, D.; Cojocaru, V.D.; Nocivin, A.; Hendea, R.; Ivanescu, S.; Stanciu, D.; Trisca-Rusu, C.; Drob, S.I.; Cojocaru, E.M. Mechanical Alloying Process Applied for Obtaining a New Biodegradable Mg–xZn–Zr–Ca Alloy. *Metals* **2022**, *12*, 132. [CrossRef]

15. Al-Aqeeli, N.; Mendoza-Suarez, G.; Suryanarayana, C.; Drew, R. Development of new Al-based nanocomposites by mechanical alloying. *Mater. Sci. Eng. A* **2008**, *480*, 392–396. [CrossRef]
16. Lala, S.; Maity, T.; Singha, M.; Biswas, K.; Pradhan, S. Effect of doping (Mg, Mn, Zn) on the microstructure and mechanical properties of spark plasma sintered hydroxyapatites synthesized by mechanical alloying. *Ceram. Int.* **2017**, *43*, 2389–2397. [CrossRef]
17. Wang, X.; Tu, J.; Wang, C.; Zhang, X.; Chen, C.; Zhao, X. Hydrogen storage properties of nanocrystalline Mg–Ce/Ni composite. *J. Power Sources* **2006**, *159*, 163–166. [CrossRef]
18. Wang, C.; Dai, J.; Liu, W.; Zhang, L.; Wu, G. Effect of Al additions on grain refinement and mechanical properties of Mg–Sm alloys. *J. Alloys Compd.* **2015**, *620*, 172–179. [CrossRef]
19. Yang, L.; Huang, Y.; Peng, Q.; Feyerabend, F.; Kainer, K.U.; Willumeit, R.; Hort, N. Mechanical and corrosion properties of binary Mg–Dy alloys for medical applications. *Mater. Sci. Eng. B* **2011**, *176*, 1827–1834. [CrossRef]
20. Hao, H.; Ni, D.; Huang, H.; Wang, D.; Xiao, B.; Nie, Z.; Ma, Z. Effect of welding parameters on microstructure and mechanical properties of friction stir welded Al–Mg–Er alloy. *Mater. Sci. Eng. A* **2013**, *559*, 889–896. [CrossRef]
21. Lee, P.-Y.; Kao, M.C.; Lin, C.K.; Huang, J.C. Mg–Y–Cu bulk metallic glass prepared by mechanical alloying and vacuum hot-pressing. *Intermetallics* **2006**, *14*, 994–999. [CrossRef]
22. Peng, Q.; Wang, L.; Wu, Y.; Wang, L. Structure stability and strengthening mechanism of die-cast Mg–Gd–Dy based alloy. *J. Alloys Compd.* **2009**, *469*, 587–592. [CrossRef]
23. Liu, Y.; Li, K.; Luo, T.; Song, M.; Wu, H.; Xiao, J.; Tan, Y.; Cheng, M.; Chen, B.; Niu, X.; et al. Powder metallurgical low-modulus Ti–Mg alloys for biomedical applications. *Mater. Sci. Eng. C* **2015**, *56*, 241–250. [CrossRef] [PubMed]
24. Le Caër, G.; Delcroix, P.; Bégin-Colin, S.; Ziller, T. High-Energy Ball-Milling of Alloys and Compounds. *Hyperfine Interactions* **2002**, *141*, 63–72. [CrossRef]
25. Suryanarayana, C. Mechanical alloying and milling. *Prog. Mater. Sci.* **2001**, *46*, 1–184. [CrossRef]
26. Kishimura, H.; Matsumoto, H. Fabrication of Ti–Cu–Ni–Al amorphous alloys by mechanical alloying and mechanical milling. *J. Alloys Compd.* **2011**, *509*, 4386–4389. [CrossRef]
27. Qiu, W.; Pang, Y.; Xiao, Z.; Li, Z. Preparation of W-Cu alloy with high density and ultrafine grains by mechanical alloying and high pressure sintering. *Int. J. Refract. Met. Hard Mater.* **2016**, *61*, 91–97. [CrossRef]
28. Kristaly, F.; Sveda, M.; Sycheva, A.; Miko, T.; Racz, A.; Karacs, G.; Janovszky, D. Effects of milling temperature and time on phase evolution of Ti-based alloy. *J. Min. Met. Sect. B Met.* **2022**, *58*, 141–156. [CrossRef]
29. Liang, G.; Schulz, R. Synthesis of Mg–Ti alloy by mechanical alloying. *J. Mater. Sci.* **2003**, *38*, 1179–1184. [CrossRef]
30. Wilkes, D.; Goodwin, P.; Ward-Close, C.; Bagnall, K.; Steeds, J. Solid solution of Mg in Ti by mechanical alloying. *Mater. Lett.* **1996**, *27*, 47–52. [CrossRef]
31. Fecht, H.J.; Hellstern, E.; Fu, Z.; Johnson, W.L. Nanocrystalline metals prepared by high-energy ball milling. *Met. Trans. A* **1990**, *21*, 2333–2337. [CrossRef]
32. Salur, E.; Acarer, M.; Şavkliyildiz, I. Improving mechanical properties of nano-sized TiC particle reinforced AA7075 Al alloy composites produced by ball milling and hot pressing. *Mater. Today Commun.* **2021**, *27*, 102202. [CrossRef]
33. Révész, Á.; Gajdics, M. Improved H-Storage Performance of Novel Mg-Based Nanocomposites Prepared by High-Energy Ball Milling: A Review. *Energies* **2021**, *14*, 6400. [CrossRef]
34. Salleh, E.M.; Ramakrishnan, S.; Hussain, Z. Synthesis of Biodegradable Mg–Zn Alloy by Mechanical Alloying: Effect of Milling Time. *Procedia Chem.* **2016**, *19*, 525–530. [CrossRef]
35. Razzaghi, M.; Kasiri-Asgarani, M.; Bakhsheshi-Rad, H.R.; Ghayour, H. Microstructure, mechanical properties, and in-vitro biocompatibility of nano- NiTi reinforced Mg–3Zn–0.5Ag alloy: Prepared by mechanical alloying for implant applications. *Compos. Part B Eng.* **2020**, *190*, 107947. [CrossRef]
36. Salur, E.; Nazik, C.; Acarer, M.; Şavklıyıldız, I.; Akdoğan, E.K. Ultrahigh hardness in Y_2O_3 dispersed ferrous multicomponent nanocomposites. *Mater. Today Commun.* **2021**, *28*, 102637. [CrossRef]
37. Salur, E.; Aslan, A.; Kuntoğlu, M.; Acarer, M. Effect of ball milling time on the structural characteristics and mechanical properties of nano-sized Y_2O_3 particle reinforced aluminum matrix composites produced by powder metallurgy route. *Adv. Powder Technol.* **2021**, *32*, 3826–3844. [CrossRef]
38. Al, S.; Iyigor, A. Structural, electronic, elastic and thermodynamic properties of hydrogen storage magnesium-based ternary hydrides. *Chem. Phys. Lett.* **2020**, *743*, 137184. [CrossRef]
39. Wang, R.; Fang, C.; Xu, Z.; Wang, Y. Correlation of milling time with phase evolution and thermal stability of Mg-25 wt%Sn alloy. *J. Alloys Compd.* **2022**, *891*, 162014. [CrossRef]
40. Son, H.-T.; Lee, J.-B.; Jeong, H.-G.; Konno, T.J. Effects of Al and Zn additions on mechanical properties and precipitation behaviors of Mg–Sn alloy system. *Mater. Lett.* **2011**, *65*, 1966–1969. [CrossRef]
41. Poddar, P.; Sahoo, K.L.; Mukherjee, S.; Ray, A.K. Creep behaviour of Mg–8% Sn and Mg–8% Sn–3% Al–1% Si alloys. *Mater. Sci. Eng. A* **2012**, *545*, 103–110. [CrossRef]
42. Celikyürek, I.; Baksan, B.; Torun, O.; Arıcı, G.; Özcan, A. The Microstructure and Mechanical Properties of Friction Welded Cast Ni_3Al Intermetallic Alloy. *Trans. Indian Inst. Met.* **2018**, *71*, 775–779. [CrossRef]
43. Mendis, C.L.; Bettles, C.J.; Gibson, M.A.; Gorsse, S.; Hutchinson, C.R. Refinement of precipitate distributions in an age-hardenable Mg–Sn alloy through microalloying. *Philos. Mag. Lett.* **2006**, *86*, 443–456. [CrossRef]

44. Pradeep, N.; Hegde, M.R.; Rajendrachari, S.; Surendranathan, A. Investigation of microstructure and mechanical properties of microwave consolidated TiMgSr alloy prepared by high energy ball milling. *Powder Technol.* **2022**, *408*, 117715. [CrossRef]
45. Demirel, A.; Çetin, E.C.; Karakuş, A.; Ataş, M.; Yildirim, M. Microstructural Evolution and Oxidation Behavior of Fe–4Cr–6Ti Ferritic Alloy with Fe$_2$Ti Laves Phase Precipitates. *Arch. Met. Mater.* **2022**, *67*, 827–836. [CrossRef]
46. Yogamalar, R.; Srinivasan, R.; Vinu, A.; Ariga, K.; Bose, A.C. X-ray peak broadening analysis in ZnO nanoparticles. *Solid State Commun.* **2009**, *149*, 1919–1923. [CrossRef]
47. Miranda, M.A.R.; Sasaki, J.M.; Sombra, A.S.B.; Silva, C.C.; Remédios, C.M.R. Characterization by X ray diffraction of mechanically alloyed tripotassium sodium sulfate. *Mater. Res.* **2006**, *9*, 243–246. [CrossRef]
48. Sübütay, H.; Şavklıyıldız, I. The relationship between structural evolution and high energy ball milling duration in tin reinforced Mg alloys. *Mater. Today Commun.* **2023**, *35*, 105868. [CrossRef]
49. Shahmoradi, Y.; Souri, D.; Khorshidi, M. Glass-ceramic nanoparticles in the Ag$_2$O–TeO$_2$–V$_2$O$_5$ system: Antibacterial and bactericidal potential, their structural and extended XRD analysis by using Williamson–Smallman approach. *Ceram. Int.* **2019**, *45*, 6459–6466. [CrossRef]
50. Alshammari, Y.; Yang, F.; Bolzoni, L. Mechanical properties and microstructure of Ti–Mn alloys produced via powder metallurgy for biomedical applications. *J. Mech. Behav. Biomed. Mater.* **2019**, *91*, 391–397. [CrossRef]
51. Chaudhri, M. Subsurface strain distribution around Vickers hardness indentations in annealed polycrystalline copper. *Acta Mater.* **1998**, *46*, 3047–3056. [CrossRef]
52. Oral, I.; Kocaman, S.; Ahmetli, G. Characterization of unmodified and modified apricot kernel shell/epoxy resin biocomposites by ultrasonic wave velocities. *Polym. Bull.* **2023**, *80*, 5529–5552. [CrossRef]
53. Oral, I.; Ekrem, M. Measurement of the elastic properties of epoxy resin/polyvinyl alcohol nanocomposites by ultrasonic wave velocities. *Express Polym. Lett.* **2022**, *16*, 591–606. [CrossRef]
54. Altintas Yildirim, O.; Atas, M.S. Synthesis and characterization of spherical FeNi$_3$ metallic nanoparticles based on sodium dodecyl sulfate. *J. Mater. Manuf.* **2022**, *1*, 33–40.
55. Frost, M.; McBride, E.E.; Schörner, M.; Redmer, R.; Glenzer, S.H. Sodium-potassium system at high pressure. *Phys. Rev. B* **2020**, *101*, 224108. [CrossRef]
56. Şavklıyıldız, İ.; Akdoğan, E.K.; Zhong, Z.; Wang, L.; Weidner, D.; Vaughan, M.; Croft, M.C.; Tsakalakos, T. Phase transformations in hypereutectic MgO–Y$_2$O$_3$ nanocomposites at 5.5 GPa. *J. Appl. Phys.* **2013**, *113*, 203520. [CrossRef]
57. Şavkliyildiz, İ. In-Situ Strain Measurement on Al7075 Plate by Using High Energy Synchrotron Light Source. *Avrupa Bilim Ve Teknol. Derg.* **2021**, *23*, 435–439.
58. Akdoğan, E.K.; Şavkliyildiz, İ.; Berke, B.; Zhong, Z.; Wang, L.; Vaughan, M.; Tsakalakos, T. High-pressure phase transformations in MgO–Y$_2$O$_3$ nanocomposites. *Appl. Phys. Lett.* **2011**, *99*, 141915. [CrossRef]
59. Akdoğan, E.; Şavkliyildiz, İ.; Berke, B.; Zhong, Z.; Weidner, D.; Croft, M.C.; Tsakalakos, T. Pressure effects on phase equilibria and solid solubility in MgO–Y$_2$O$_3$ nanocomposites. *J. Appl. Phys.* **2012**, *111*, 053506. [CrossRef]
60. Gehrmann, R.; Frommert, M.M.; Gottstein, G. Texture effects on plastic deformation of magnesium. *Mater. Sci. Eng. A* **2005**, *395*, 338–349. [CrossRef]
61. Nayyeri, M.J.; Ganjkhanlou, Y.; Kolahi, A.; Jamili, A.M. Effect of Ca and Rare Earth Additions on the Texture, Microhardness, Microstructure and Structural Properties of As-Cast Mg$_{-4}$Al$_{-2}$Sn Alloys. *Trans. Indian Inst. Met.* **2014**, *67*, 469–475. [CrossRef]
62. Huot, J.; Skryabina, N.Y.; Fruchart, D. Application of Severe Plastic Deformation Techniques to Magnesium for Enhanced Hydrogen Sorption Properties. *Metals* **2012**, *2*, 329–343. [CrossRef]
63. Atas, M.S.; Yildirim, M. Morphological development, coarsening, and oxidation behavior of Ni–Al–Nb superalloys. *J. Mater. Eng. Perform.* **2020**, *29*, 4421–4434. [CrossRef]
64. Ye, H.Z.; Liu, X.Y. Review of recent studies in magnesium matrix composites. *J. Mater. Sci.* **2004**, *39*, 6153–6171. [CrossRef]
65. Chen, J.; Wei, J.; Yan, H.; Su, B.; Pan, X. Effects of cooling rate and pressure on microstructure and mechanical properties of sub-rapidly solidified Mg–Zn–Sn–Al–Ca alloy. *Mater. Des.* **2013**, *45*, 300–307. [CrossRef]
66. Huang, K.; Marthinsen, K.; Zhao, Q.; Logé, R.E. The double-edge effect of second-phase particles on the recrystallization behaviour and associated mechanical properties of metallic materials. *Prog. Mater. Sci.* **2018**, *92*, 284–359. [CrossRef]
67. Sun, S.; Deng, N.; Zhang, H.; He, L.; Zhou, H.; Han, B.; Gao, K.; Wang, X. Microstructure and mechanical properties of AZ31 magnesium alloy reinforced with novel sub-micron vanadium particles by powder metallurgy. *J. Mater. Res. Technol.* **2021**, *15*, 1789–1800. [CrossRef]
68. Li, Y.; Chen, C.; Deng, R.; Feng, X.; Shen, Y. Microstructure evolution of Cr coatings on Cu substrates prepared by mechanical alloying method. *Powder Technol.* **2014**, *268*, 165–172. [CrossRef]
69. Bamberger, M.; Dehm, G. Trends in the Development of New Mg Alloys. *Annu. Rev. Mater. Res.* **2008**, *38*, 505–533. [CrossRef]
70. Arici, G.; Acarer, M.; Uyaner, M. Effect of Co addition on microstructure and mechanical properties of new generation 3Cr-3W and 5Cr-3W steels. *Eng. Sci. Technol. Int. J.* **2021**, *24*, 974–989. [CrossRef]

Disclaimer/Publisher's Note: The statements, opinions and data contained in all publications are solely those of the individual author(s) and contributor(s) and not of MDPI and/or the editor(s). MDPI and/or the editor(s) disclaim responsibility for any injury to people or property resulting from any ideas, methods, instructions or products referred to in the content.

MDPI

St. Alban-Anlage 66

4052 Basel

Switzerland

www.mdpi.com

Crystals Editorial Office

E-mail: crystals@mdpi.com

www.mdpi.com/journal/crystals

Disclaimer/Publisher's Note: The statements, opinions and data contained in all publications are solely those of the individual author(s) and contributor(s) and not of MDPI and/or the editor(s). MDPI and/or the editor(s) disclaim responsibility for any injury to people or property resulting from any ideas, methods, instructions or products referred to in the content.